ELECTRIC POWER SYSTEMS

Advanced Forecasting Techniques and Optimal Generation Scheduling

T0239940

ELECTRIC POWER SYSTEMS

Advanced Forecasting Techniques and Optimal Generation Scheduling

Edited by
João P. S. Catalão

CRC Press
Taylor & Francis Group
Boca Raton   London   New York

CRC Press is an imprint of the
Taylor & Francis Group, an **informa** business

CRC Press
Taylor & Francis Group
6000 Broken Sound Parkway NW, Suite 300
Boca Raton, FL 33487-2742

First issued in paperback 2017

Version Date: 20111201

ISBN 13: 978-1-138-07398-2 (pbk)
ISBN 13: 978-1-4398-9394-4 (hbk)

Visit the Taylor & Francis Web site at
http://www.taylorandfrancis.com

and the CRC Press Web site at
http://www.crcpress.com

I would like to dedicate this book to my wife, Carla Cruz,
who truly inspires me with her love and care
to perform better and better every day, aiming to be
as exceptional as she already is.

Contents

Preface

A wide-ranging impression about the subjects discussed in this book is that the topics are pivotal for understanding and solving some of the problems flourishing in the second decade of the twenty-first century in the field of management of electric power generation systems. Noticeably, the chapters begin with some knowledge from the last decade to uncover lines of research on some of the present knowledge, and, in due course, anticipate some of the admissible lines for future research in management of electric power generation systems.

The scope of the book is well defined and of significant interest. Indeed, the development of new methodologies carrying away an improved forecasting and scheduling of electric power generation systems is crucial under the new competitive and environmentally constrained energy policy. The capability to cope with uncertainty and risk will benefit significantly generating companies. It is a fact that to avoid losing advantages of participating in the electricity market or negotiating bilateral contracts, a power producer should self-schedule its power system in anticipation. In recognition of this fact, hydro and thermal scheduling are relevant topics today. Already, wind power generation is playing an important role in some countries and will be even more important in the nearby future of energy supply in many countries. Thus, optimal coordination between hydro, thermal, and wind power is of utmost importance. Deterministic and stochastic modeling frameworks are allowing the development of the next generation of computational tools to help successful management of electric power generation systems. Research is underway to conquer the capability to cope with the present and the future of electric power generation systems as shown in this book.

The book fills a need in its field by having adequate strong points to fulfill not only the graduate learning task, written by qualified university professors in a pedagogical and systematic way with excellent quality, but also expands on many of the latest results that are adequate for engineers and researchers working in this field today. Many parts of the book are based on the author's and other's current research and some parts have never appeared elsewhere in text books; most of these discussions have been proposed as PhD theses, postdoctoral research, and industrial development carried out by the authors, but have been selected with opportunity, written with accuracy, and are well balanced in theory and practice. Each chapter is organized with adequate emphasis, coherently and effectively, presenting large-value, up-to-date research showing novelty, launching into the mind of interested readers' future lines for cutting-edge research on forecasting and scheduling issues regarding electric power generation systems.

The text has three main parts. The first part, Chapter 1, constitutes indispensable knowledge and embarks on the report of real-world problems, concerning the present technology of electric power generation systems, regarding both the structure and repercussion of these problems. The second part, Chapters 2 through 5, conveys the items on uncertainty, risk, and short-term forecast to systematize the development of information management systems, helping power system decision-makers to rise above the unknown and capricious behavior of the present and nearby future of power generation systems. Both, the importance of accessing those items and the formulation of the problems are discussed. No doubt that this part is of major importance and a crucial input for the scheduling task, aiming at the most favorable use of the energy sources available, which is the scope of the next part of the text. The third part, Chapters 6 through 10, is devoted to the rationality studies for developing information management systems to help take decisions to avoid losing advantages offered by playing in energy markets or negotiating bilateral contracts. Hydro and thermal scheduling are discussed both by identifying the main variables and parameters and by the formulation of the corresponding mathematical programming problems to achieve optimal decision. Wind power generation is addressed in coordination with thermal and hydropower generations. The thrust point, concluding this last part, is operation of multigeneration systems. Both the formulation and solution methodologies embodying a component supported by the theory of multi-objective programming and planning are discussed.

A final observation about this book is that no claim is made that it is a compendium of all known results about management of electric power generation systems. However, the editor and the authors have attempted to include a rich and varied selection of subjects that not only are of current interest but also, I believe, will be main research lines for the following two decades to achieve an enhanced management performance in the field of electric power generation systems.

MATLAB® is a registered trademark of The MathWorks, Inc. For product information, please contact:

The MathWorks, Inc.
3 Apple Hill Drive
Natick, MA 01760-2098 USA
Tel: 508 647 7000
Fax: 508-647-7001
E-mail: info@mathworks.com
Web: www.mathworks.com

Acknowledgments

This book would not be possible without the extraordinary contributions of several worldwide recognized experts in the power systems field. I specially thank all the contributors of each chapter for their acceptance to contribute to the book, in a timely and thorough manner, and for their enthusiastic and dynamic participation. It was an honor and a privilege for me to be able to work in collaboration.

I would like to thank all the professors who encouraged me to perform this challenging editorial task, and for their inspirational leadership and guidance. Particularly, I would like to thank Professors Carlos Cabrita, Luís Marcelino Ferreira, João Santana, and Antonio Conejo. Special thanks goes to Professor Victor Mendes, my mentor and friend for several years now, to whom I am indebted for sharing his wisdom and for advice and incentives.

I would like to thank all my graduate students for their keen interest and insightful ideas, constantly unveiling new research paths. Particularly, I would like to thank Rui Melício, Rafael Rodrigues, Hugo Pousinho, and Nelson Batista.

I would like to thank the Portuguese Science and Technology Foundation (FCT), and FEDER through COMPETE, for financial support.

I would also like to thank the CRC Press staff, particularly Nora Konopka for her kindness and professionalism.

I hope that the book will be interesting and useful for the power systems community worldwide, not only for graduate students but also for professionals in the field taking decisive steps toward a sustainable and smart grid of the future.

Chapter Abstracts

Chapter 1: Overview of electric power generation systems.

Chapter 2: The problem of generation self-scheduling in electricity markets can be approached via the tools of modern portfolio optimization theory. This chapter presents scheduling frameworks that are based on the Value-at-Risk (VaR) and Conditional Value-at-Risk (CVaR) measures. It also discusses worst-case risk formulations that account for data uncertainty within the decision-making process. In all formulations, the problem is cast as a semidefinite program that can be solved efficiently. The optimization programs are clearly illustrated on a small-scale test system.

Chapter 3: The focus on load forecasting application has moved from large aggregated loads toward bus demands, because of economic and security motivations. Therefore, with hundreds of load points to be simultaneously predicted in an online environment, nonautomatic procedures for model estimation are no longer convenient. This chapter presents a data-driven approach for the design of neural network-based load forecasters. The automatic procedure starts with data multiresolution decomposition via wavelets, followed by input and model structure selections. Input variables are chosen by a combination of techniques suitable for nonlinear models. Neural network structures are determined by Bayesian inference. A nonparametric technique to estimate prediction intervals completes the framework, providing the required information to feed operational aid tools.

Chapter 4: With the restructuring of electric power industry, electricity is traded as a commodity in the new environment of open electricity markets. As in many other financial markets, electricity market participants have paid attention to price forecast of this commodity. However, the great amount of traded volume and high volatility of electricity price usually discriminate electricity market from other financial markets. These two factors also motivate many research works on the area of electricity price forecasting in recent years.

Based on the forecast horizon, electricity price prediction can be classified into short term (few hours/days), mid-term (few weeks/months), and long term (few years). The most common form is short-term electricity price forecast, which is the focus of this chapter. This kind of forecast is used, for instance, for the preparation of sell/buy bids by the market participants. Accurate short-term electricity price forecast in the spot market helps power producers to adjust their bidding strategies to achieve the maximum benefit with minimum financial risk. Similarly, customers can derive a plan to maximize their purchased electricity from the pool, or use self-production capability to protect themselves against high prices. This chapter also focuses on the price forecast of the market participants (and not independent system operator), which means predicting the price before submitting the bids.

Most of electricity price forecast methods have been designed for normal price prediction. However, price spikes are an important aspect of electricity price impacting its forecast accuracy. Electricity price spikes can also have serious economical effects on the market participants. Recently, some researchers have paid attention to electricity price spike forecast.

In this chapter, the most accurate and robust electricity price forecast methods are described. These methods consist of pre/post processors, feature selection techniques and forecast engines. In the last part, some strategies for price spike forecasting, including prediction of price spike occurrence and value, are discussed.

Chapter 5:   Wind Power Forecasting has come a long way since the first papers in the 1980s and the first operational approaches in the early 1990s. Typically, the next-day forecast carries the largest economic weight, and is derived from online SCADA data are state of the art, and can be used in decision-making tools directly to fully utilize the full predicted distribution. Additional to the wind power forecast, ramps or variability on different timescales can be predicted as an additional help for the dispatcher or trader.

Chapter 6:   Optimal economic operation of power systems has always been one very important subject in the planning and operation of power systems. While the minimization of the overall cost considering both the investment cost as well as the operating cost has been the foundation of most planning approaches, the minimization of operation cost by operating the power system at the minimum marginal cost was the most common basis of optimal power system operation. These principles have played invaluable role in the optimal operation of the traditional power systems thus far. This basic tenet of these principles has been challenged by two fundamental changes in the power industry around the world: the development of competitive power markets, and the emergence of the renewable energy sources.

Incorporation of renewable energy posed some difficulty in the purely cost-based approaches, as these sources of energy incurred only capital costs and exhibited very little operating cost so that the marginal cost of energy production is virtually zero.

Deregulation of power markets leading to independent generating companies (Gencos) participating in the grid operated by the independent system operator (ISO) has brought the price as the basis for market operation. The system-wide operation is carried out by ISO through market clearing decided on the basis of the bids submitted by the Gencos. The individual Gencos are responsible for their own unit commitment and generation scheduling according to the market clearing by the ISO.

This chapter describes some studies carried out to investigate the price-based unit commitment and the price-based scheduling for Gencos utilizing a variety of approaches.

Chapter 7: We study the most appropriate modeling approaches for the analysis and design of the optimal operation strategies for a hydroelectric generating company that owns several plants along a cascaded reservoir system and operates in short-term electricity markets. The analysis is based upon a detailed representation of the hydroelectric generation facilities in the context of a day-ahead market. This approach is particularly attractive to strengthen competition and to increase the technical and economic efficiency of hydro power plants.

An appropriate Mixed Integer Nonlinear Programming (MINLP) algorithm that maximizes profit is developed considering the technical efficiency of a hydroelectric generating unit as a function of the discharge for a given net head of the associated reservoir. The developed methodology allows for an appropriate representation of the technical efficiency of a hydroelectric generating unit, where the power output of the unit considers the variable head effects. We apply the methodology described to a hydraulic chain of the Spanish electric system along the Duero river basin.

Chapter 8: This chapter presents the short-term hydrothermal producer self-scheduling problem. The problem objective is the maximization of the producer profits from his participation in the day-ahead energy and reserves markets. An integrated 0/1 mixed-integer linear programming (MILP) formulation is provided, which combines both thermal and hydro subsystems in a single portfolio for a hydrothermal producer who acts either as a price-taker or a price-maker in the day-ahead market. A detailed modeling of the operating constraints of thermal and hydro generating units is presented. Thermal unit constraints, such as unit operating limits, minimum up and down times, ramp rate limits, start-up and shutdown sequences, fuel limitations and so on, are discussed. Hydro constraints ranging from simple energy limit constraints to complex hydraulically coupled reservoir constraints with time lags, head dependent conversion efficiencies, hydro unit prohibited operating zones and discrete pumping are presented. Residual demand curves for energy and reserves are used to model the effect of the price maker producer's interactions with its competitors. Uncertainty of market conditions is also modeled within a two-stage stochastic programming framework, while a specific risk measure is also incorporated. Postprocessing techniques are applied for the construction of the generating units optimal offer curves. Numerical results from the application of the MILP-based solution to the short-term self-scheduling problem of a hydrothermal producer participating in the day-ahead market of a medium-scale real power system (the Greek interconnected power system) are presented and discussed.

Chapter 9: This chapter describes how wind power can be incorporated into traditional unit commitment and economic dispatch (UC-ED) algorithms used by operators of power plants and power systems market operators. Application of such an algorithm gives greater certainty about cost, technical limitations and emissions, and gives some insight into potential challenges from wind power. The information in this chapter is useful for power market participants who wish to optimize the operation of units in their portfolio. The knowledge is also key for transmission system operators or policy makers who are conducting studies to assess the feasibility of future wind power development scenarios. The interaction with international trade must also be taken into account, in addition to the usual constraints on system reserves and the operation of conventional units. A multiarea model with constraints on the interarea import or export capacity of transmission corridors is recommended. Special attention is given in this chapter to the detailed modeling of estimated wind power time series, starting with scenarios for locations and installed capacity of current and future wind power plants. As far as thermal units are concerned, special attention is given to the modeling of combined heat and power plants. Results are presented for a case study involving North-Western Europe, with a focus on a 12 GW wind power scenario for the Netherlands and 32 GW for Germany. The simulation results focus on typical outcome indicators such as operating cost savings, emissions reduction, and curtailed wind power. The results highlight the importance of well-functioning international markets, preferably with one-hour ahead gate-closure time, for the optimal integration of wind power in large power systems.

Chapter 10: Multigeneration (MG) of different energy vectors, such as electricity, heat, cooling, and others, represents a viable alternative to improve energy generation efficiency and decrease the environmental burden of energy systems. In particular, trigeneration plants can be efficiently deployed to supply complex energy services in urban areas, with typically high heat demand of heat in winter and different levels of cooling demand throughout the year, depending on the specific application. MG could be applied through a number of solutions exploiting for instance generators for small-scale distributed CHP (combined heat and power, or cogeneration), heat-fired absorption chillers, electrical heat pumps, and so forth. Managing MG systems is a challenging task due to the energy flow interactions among the manifold pieces of equipment within the plant and with external energy networks. In addition, different objectives could be pursued, for instance of economic, technical, or environmental nature, or a combination of the above. Therefore, robust methodologies for MG optimisation are needed, to cope with most general cases. In this context, this chapter presents a comprehensive introduction to modeling, analysis, and assessment of MG systems in the operational time frame, with special focus to cogeneration and trigeneration. It is shown how to formulate, in a compact and systematic form, suitable operational optimisation problems of different kinds. In particular, also relying upon relevant literature recently published in the field, the main variables involved in the analysis and the complexity of the operational optimization problem formulations and solutions are highlighted, including how to handle possible conflicting objectives within multiobjective optimization and relevant solution approaches.

Editor

João P. S. Catalão received an MSc degree from the Instituto Superior Técnico (IST), Lisbon, Portugal, in 2003 and a PhD from the University of Beira Interior (UBI), Covilha, Portugal, in 2007. He is currently a professor at UBI and a director of the Master Program in Electromechanical Engineering. His scientific area of interest at the Center for Innovation in Electrical and Energy Engineering (CIEEE), IST, is related to power and energy systems. He coordinates a research group in renewable energies and sustainability at UBI. He is an IEEE and an IET member, a senior member of the Portuguese Council of Engineers, and a member of the IASTED Technical Committee on Energy and Power Systems. He is the author/coauthor of more than 160 scientific papers, including 50 papers in international journals and 100 papers in conference proceedings. He is an editor of *IEEE Transactions on Sustainable Energy*, and an editorial board member of *Electric Power Components and Systems*.

Contributors

Alexandre P. Alves da Silva received a PhD in electrical engineering from the University of Waterloo, Waterloo, Ontario, Canada, in 1992. Currently, he is a full professor in electrical engineering at the Federal University of Rio de Janeiro (COPPE/UFRJ), Brazil. His research interests include intelligent systems application to power systems. He is a senior member of the IEEE.

Nima Amjady received a PhD in electrical engineering from Sharif University of Technology, Tehran, Iran, in 1997. Currently, he is a full professor with the Electrical Engineering Department, Semnan University, Semnan, Iran. His research interests include security assessment of power systems, reliability of power networks, load and price forecasting, and artificial intelligence and its applications to the problems of power systems. He is a senior member of IEEE.

Anastasios G. Bakirtzis received a PhD from Georgia Institute of Technology, Atlanta, in 1984. Currently, he is a full professor and director of the Power Systems Laboratory at the Aristotle University of Thessaloniki, Thessaloniki, Greece. His research interests include power system operation, planning, and economics. He is a senior member of the IEEE.

Pandelis N. Biskas received a PhD from the Aristotle University, Thessaloniki, Greece, in 2003. Currently, he is a lecturer at the same university. His research interests include power system operation and control, electricity market operational and regulatory issues, and transmission pricing. He is a member of the IEEE.

Gianfranco Chicco received his PhD in electrotechnics engineering in Italy, in 1992. He is a professor of electrical energy systems at Politecnico di Torino, a senior member of the IEEE Power and Energy Society, and an associate editor of the *Energy International Journal*. His research interests include power systems and distribution systems analysis and optimization, distributed resources and multigeneration systems planning, energy and environmental efficiency characterization, electrical load management and control, reliability and power quality applications. He is the author of a book, four book chapters, and has over 40 papers published in international journals, and over 120 papers published in the proceedings of international conferences.

Javier Contreras received a PhD from the University of California, Berkeley, in 1997. Currently, he is a full professor at the Universidad de Castilla–La Mancha, Ciudad Real, Spain. His research interests include power systems planning and economics and electricity markets. He is a senior member of the IEEE.

Michael Denhard received a PhD in meteorology from the J.-W.-Goethe University Frankfurt, Germany, for work on nonlinear time-series analysis in climatology. Currently, he is a senior scientist at Deutscher Wetterdienst in Germany. His research interests are ensemble prediction and communication of probabilistic weather forecasts.

F. Javier Díaz received a PhD in systems engineering from the Universidad Nacional de Colombia, Medellín, in 2008. Currently, he is an associate professor at the Universidad Nacional de Colombia, Medellín. His research interests include power systems planning, operations and economics, and optimization and simulation.

Vitor H. Ferreira received a PhD in electrical engineering from the Federal University of Rio de Janeiro (COPPE/UFRJ), Brazil, in 2008. Currently, he is an assistant professor at the Fluminense Federal University, Brazil. His research interests include time-series forecasting and neural networks.

Madeleine Gibescu received a PhD from the University of Washington, in 2003. She has worked as a research engineer for ClearSight Systems, and as a power systems engineer for the AREVA T&D Corporation of Bellevue, Washington. Currently, she is an assistant professor with the Electrical Power Systems Group at Delft University of Technology, the Netherlands. She is a member of the IEEE.

Gregor Giebel received a PhD in physics from Oldenburg University, Germany, for work done at Risø National Laboratory, Roskilde, Denmark, in 2000. Currently, he is a senior scientist at Risø DTU (Technical University of Denmark), National Laboratory for Sustainable Energy, Wind Energy Division. His research interests include distributed generation of wind energy and software development for a short-term prediction tool for utilities.

Rabih A. Jabr received a PhD in electrical engineering from Imperial College London, London, United Kingdom, in 2000. Currently, he is an associate professor in the Department of Electrical and Computer Engineering at the American University of Beirut, Beirut, Lebanon. His research interests include mathematical optimization techniques, design optimization, optimal power flow, and state estimation. He is a senior member of the IEEE.

Wil L. Kling received an MSc degree in electrical engineering from the Eindhoven University of Technology, the Netherlands, in 1978. Currently, he is a full professor and chair of the Electrical Power Systems Group at the Eindhoven University of Technology. He is leading research programs on distributed generation, integration of wind power, network concepts, and reliability issues. He is involved in CIGRE and IEEE.

Pierluigi Mancarella received his PhD in electrical engineering (power systems) from Politecnico di Torino, Italy, in 2006. After working as a research associate at Imperial College London, UK, he joined the University of Manchester in 2011 where he currently is a lecturer in sustainable energy systems. His research interests include modeling and analysis of multienergy systems, sustainable development of energy systems, energy markets, business models for distributed energy systems, and risk analysis of generation and network investment. He is the author of a book, several book chapters, and over 50 papers in peer-reviewed international conferences and journals.

Cláudio Monteiro received a PhD from FEUP, Porto, Portugal in 2003. Currently, he is an assistant professor in the Department of Electrical and Computer Engineering at FEUP. His research interests include forecasting applied to energy systems and dispatch optimization integrating forecast in market paradigm.

Engbert Pelgrum received a BSc in electrical engineering from the Technical College of Zwolle, the Netherlands, in 1982. Currently, he is with the Ownership, Business Development and Regulation Department of the Dutch Transmission System Operator, TenneT, Arnhem, the Netherlands. He has more than 20 years of experience in the area of chronological system modeling and power system operation and planning for the Dutch and foreign generation and transmission systems.

Songbo Qiao received an MS in power engineering in 2008 and is now working toward a PhD in power engineering at Nanyang Technological University, Singapore.

Barry G. Rawn received a PhD in electrical engineering from the University of Toronto, Toronto, Ontario, Canada, where he also received the BASc and MASc degrees in engineering science and electrical engineering in 2002 and 2004, respectively. His research interests include nonlinear dynamics and sustainable energy infrastructure. He is currently a postdoctoral researcher in the Electrical Power Systems Group at the Delft University of Technology, the Netherlands.

Govinda B. Shrestha received a PhD in electrical engineering from Virginia Tech, USA, in 1990. Currently, he is an associate professor at Nanyang Technological University, Singapore. His main areas of interest are power system operation and planning and power markets. He is a senior member of the IEEE.

Christos K. Simoglou received the diploma in electrical engineering from the Aristotle University of Thessaloniki, Greece, in 2005. Currently, he is a PhD candidate with the same university. His research interests include power system operation, planning, and energy bidding strategies.

Bart C. Ummels received a PhD in electrical engineering in 2009 from Delft University of Technology, the Netherlands. Currently, he is with BMO-Offshore BV. He is a member of the IEEE.

1

Overview of Electric Power Generation Systems

Cláudio Monteiro

1.1 Introduction

Power systems are one of the largest and most complex engineering systems created by mankind. The importance of these systems is unquestionable, giving us a product that was, is, and will be the support for the development of modern society.

Historically, all power systems have developed in a similar way although with some technical variants [1]. They started their history as small isolated systems, powered by small production units with autonomous control and local distribution networks with small extension. These small systems have evolved in size and extension clustering in interconnected systems, raising the size of generation units to increasingly larger and more efficient units, using various energy sources. Large generation systems feed larger service areas, and consequently need technological solutions to transmit power through long distances, requiring transmission and subtransmission systems with multiple stages of voltage level. Over the years, power systems have continued to grow and expand to the remotest locations where electricity is needed. This evolution resulted in a very efficient generation system, interconnected by transmission systems capable of delivering energy for hundreds of miles with little energy loss, and versatile distribution systems, with high reliability and quality, safe for end users.

Thus, the traditional power system is composed by a large centralized generation system, a very high-voltage transmission, a high-voltage subtransmission system, and a distribution system. Because this power system structure is an optimized solution, this configuration has successfully persisted through decades with high efficiency and quality of service. In recent decades, however, the threat of the unsustainability of the system was evident, from the perspective of the environment and the security of the supply. The conventional energy sources threaten not to be enough to cover the increasing consumption and are not sufficiently clean to mitigate the present environmental problems of the planet. This is how a new paradigm of dispersed generation based on renewable energies emerged [2]. The purpose of this new paradigm is to collect an endogenous and clean energy resource. Because of the dispersed nature of this resource, the change of the power system at a conceptual, technological, and organizational level is inevitable. This new paradigm develops dispersed and renewable generation on a larger scale, which is intermittent and difficult to control, requiring new technical approaches to control the systems.

It is amazing how actual power systems, with their dimension and complexity, are controlled. As there is a minimum storage component in power systems, it is necessary to ensure a perfect balance between production and consumption for every millisecond. We can imagine the difficulty and complexity of control in a system where most variables are not directly controllable. Power systems have the particular characteristic of being controlled mainly from the power flow itself rather than through a separated and dedicated information system. This type of control is possible when the generating units in a system are controllable; however, in recent decades, many new components that are not controllable have emerged, especially dispersed generation. The challenge of the power systems of the future is to maintain the same quality control with less controllable variables and less direct control over the components of the system.

Not only the changes in technological paradigm but also organizational changes have brought uncertainty to power systems. The change to deregulated organizational systems, unbounded organizational structures, and market-oriented approaches originate the loss of uniqueness and centralization in the management and control system. The number of independent agents in the system (e.g., independent power producers, market agents) with the capability influencing the control is increasing. Moreover, these agents work in a competitive environment which means that information, relevant for strategic decisions, is kept secret among agents. All this additional uncertainty in the information can only result in a lower capacity to control and optimize the management of power systems.

Throughout this chapter we will provide a brief overview of electrical power systems and give a perspective of its characteristics and control variables in order to lay the foundations for understanding the issues addressed in this book. Here, we will also introduce the challenges and possible solutions for this environment in order to change the power systems.

1.2 Power Generation Technologies

1.2.1 Generation, the Heart and Brain of the Power System

Generation is certainly the heart of the power system—it is from this unit that the power flows through the whole system to reach the consumers. In contrast with other network systems, the flow is not controlled at the transmission level but is mainly controlled in the generation units. Therefore, we may say that generation is the heart and brain of energy systems.

The generation system is very diversified. It has integrated different types of energy conversion technologies: thermal, wind, solar, and hydropower. It has various types of primary energy; for instance, in the case of thermal plants, the energy source can be gas, coal, fuel oil, geothermal, biomass, or solar. The generation system is also varied with regard to the size of the generation units: the system can integrate GW units, like nuclear power plants, or small generation, like residential photovoltaic microgeneration. The generation system is also diversified in terms of the geographic distribution; big conventional units have the advantage of being centralized, but renewable resources must be collected from dispersed areas with small-sized units. The different sizes of the systems respond to the optimal economical solution but are restricted by technical, geographical, and environmental constraints. This diversification of the generation technologies is a major advantage for the power system, mixing alternative energy resources and technologies, allowing more independence from fossil energy sources and energy markets providing security of supply. On the other hand, the management of such technological diversity implies greater complexity in managing and controlling the system [3]. We will address this problem and provide solutions throughout this book.

Some of the technologies can be scheduled. For fuel-based technologies, we can manage the storage and usage of the resource, as is the case for most thermal and some hydropower generation technologies. But for other technologies, we are unable to control the resource, as is the case for wind and solar photovoltaic technologies. In these cases there are independent variables that cannot be controlled. There are two possible ways of controlling them: one is to waste the energy, not using it when it is available, or forecasting the resource availability [4]; the second is the approach followed in this book, which is obviously the most intelligent approach for optimizing the internal resources.

The seasonality of the resource is an important issue for the scope of scheduling solutions. This seasonality differs from region to region and its impact depends on the proportion of different technologies. Different energy sources have different availability throughout the year. For instance, in Europe, during the winter months, the availability of water resources is very high and sometimes excessive. On the other hand, during the summer months, the availability of solar resource is high. This complementarity is very valuable for optimizing the system, but in some cases there is seasonal coincidence of the resources. For instance, wind resource in Europe is generally higher during the winter months, overlapping with the period of high penetration of hydropower generation. This can be a problem for power systems that use reservoirs to store excess wind energy, because this seasonal period reservoir no longer has the capacity to store wind power by pumping water to high reservoirs [5].

The variability and intermittency of the resource are important challenges to be overcome in generation scheduling. Some technologies, such as solar and wind, have very fast variability. In an area of 500×500 km^2 the wind resource can vary about 10% of the maximum capacity in 1 h [6]. Generation based on solar resources can be even faster, with variations of 30%. The aggregation of a wide geographical extent of dispersed generation softens this variability, but in a different way for each technology; for instance, it softens the variability, but this effect is more important for photovoltaic power than for wind or hydropower. This is due to the differences in the correlation of production at different distances.

However, the intermittency and variability do not depend only on meteorological aspects. Sometimes the mechanisms of tariffs and market signals cause undesirable variability for generation. For instance, step tariff periods in small hydropower generation, with lower price for off-peak and higher price for non-off-peak times, cause an artificial step variation of 40% generation in less than 1 h. Signals from market prices may also cause artificial variations in scheduling generation. This is more evident in technologies with more ability to store the primary resource, such as hydropower, biomass, and cogeneration. However, usually, market price signals are a positive contribution to the system control and a mitigation of the generation variability.

The regulatory aspects and the ownership of the generation also have influence on the management of the production mix. For instance, there are some generation units that are available to be changed according to the needs of the system. There are other generation units that just follow the rules of the market and the power system operators are only able to impose certain restrictions. There are also independent power producers that, in most cases, are totally uncontrollable and system operators can only impose a few restrictions in extreme cases of system noncontrollability.

1.2.2 Thermal Power Generation

As mentioned earlier, there are various characteristics for thermal power stations. In thermal generation, we include power plants based on fossil fuels and nuclear power plants. Thermal plants based on fossil fuels are classified according to the type of fuel, which can be coal, fuel oil, or gas [7]. The principle of operation follows a sequence of energy transformation. Initially, the fuel is burned in a boiler that produces water vapor. In the second stage, steam—at different pressure levels—is transformed into mechanical energy through a steam turbine. Finally, the mechanical energy is converted into electrical energy. The efficiency of the plant depends on the calorific value of fuel, but in general is less than 45% for steam-cycle power plants. For environmental or economic reasons, in many cases thermal generation units are converted to use other types of fuel. Some plants that were originally designed for coal were later converted to oil, converted back to coal, and then converted to gas. Because of thermal inertia of steam boilers, which is usually more than 6 h, but can reach 10 h for a completely cold start, thermal power stations are restricted to the temporal switch-on and switch-off, and consequently are slow and inflexible, conditioning the strategy of scheduling (unit commitment). For this reason, usually thermal power stations operate frequently on standby, without production, keeping warm for quick starts; of course, this has an associated cost to be considered in scheduling strategy.

Due to the limits of combustion stability of the boilers, steam power plants have a technical minimum in the order of 30–40% of nominal power; it is not recommended to operate the plant below this value, as it could cause rapid decrease in efficiency. Some plants are more flexible and can operate under a stop–start daily cycle. Other plants with higher thermal inertia are slower and can only operate on a weekly cycle, stopping and starting once a week. The thermal constants of the boiler impose the speed with which the power plant can vary the generation level. Thus, each plant has its own characteristic of ramp-up or ramp-down.

In addition to steam thermal power plants, there are two other types of plants based on fossil fuels. One of these is the gas turbine power plants, in which turbines gas is burned with air under pressure and the turbine converts the high temperature and pressure into mechanical energy converted into electricity by the power generator coupled to the same axe. The other type of thermal power plant is the combined-cycle type. This type combines a closed-loop steam cycle turbine with an open-cycle gas turbine. The main cycle is the gas turbine cycle, in which a compressor, coupled to the turbine axe, absorbs, compresses, and injects air into the combustion chamber. The hot gas expands in the turbine, making the first extraction of mechanical energy. From this first stage, the resulting gas that remains at a relatively high temperature is used to produce steam and operate the steam turbine, taking full advantage of the calorific value of fuel. The combined-cycle power plant has a high efficiency of about 60%. The plant also has the advantage of flexibility—it can operate with a fast start or fast ramping similar to typical gas turbines. In terms of operation, a combined-cycle power plant can have a cold start in just 1–2 h, but if needed, this time can be just a few minutes starting as a simple gas turbine. For this reason, and for economic and environmental advantages associated with the use of natural gas as fuel, the use of the combined-cycle plant is growing. The good competition of investment and operating costs makes a combined-cycle power plant a very interesting solution for countries that have natural gas available with some security of supply.

Nuclear power plants are large-sized generation units, with about 1000 MW, in contrast with other thermal power plants with a typical size of 500 MW. There are power plants that produce at a constant pattern, because of the dangers of the variation in the operation conditions of the refrigeration system. Basically, a nuclear power plant consists of a nuclear reactor based on a fission process producing a lot of heat. This heat is extracted through a heat transfer fluid and is transferred through a heat exchanger to a steam circuit. Steam thermal energy is converted first into mechanical energy and then into electricity as in the conventional steam power plants. For reasons associated with the risk of failure of the cooling system, nuclear plants have very little flexibility: they can never stop and production can vary with very slow rates. For this reason, they have very important restrictions for scheduling optimization; it is not an easily controllable variable despite being a thermal generation. However, in market environments, scheduling is very influenced by nuclear generation, through its influence on market prices. This influence may be direct, when markets integrate nuclear generation, or indirect, when the price signals appear in the electricity import transactions with neighborhood markets and networks. During off-peak periods, the price signal influenced by nuclear generation can be very low. In fact, these prices are not the cost of generation with nuclear but the cost associated with the risk of nongeneration.

For the scheduling problem, specific characteristics for each thermal unit are required:

- Maximum generation limits (MW) correspond to the maximum overload operation of the power plant; generally, the specific fuel consumption is high for overload operation and optimization avoids these extreme operation points.
- Minimum generation limits (MW) correspond to the minimum value required to guarantee the thermal stability of the boilers.
- Start-up time (h) is the time to start the power plant until it reaches the minimum generation. It can be a nonlinear curve as a function of the initial boiler temperature. Lower boiler temperature corresponds to a higher start-up time.
- Ramping-up limits (MW/h) correspond to the maximum power that can be taken up per hour. It can be a nonlinear curve, with lower ramping values when the boiler is colder.
- Ramping-down limits (MW/h) correspond to the maximum power that can be decreased per hour. It depends on the inertia of the boiler and the capacity of the refrigeration system. Larger boilers generally have more restrictive ramping-down limits.
- Specific fuel consumption curve (m^3/h) is generally a quadratic polynomial curve as a function of the power operation point. It reaches the maximum efficiency for the nominal power.
- Fuel consumption (m^3/h) in stand-by mode is the fuel consumption without power production to keep the power plant ready for a hot start.
- Start-up fuel consumption curve (m^3) is a nonlinear curve as a function of the temperature of the boiler. A lower boiler temperature corresponds to higher start-up fuel consumption.

This scheduling optimization depends on thermal generation, because this generation states the main operation cost of the power system in relation to fuel consumption. The cost of fuel consumption depends on the price of the fuel used by the generation unit and on the efficiency characteristics of the generation unit. Generation costs of the thermal power plant can be computed based on the fuel consumption and the cost of fuel. The resulting fuel cost is often referred in the literature approximated by a quadratic curve; however, due to varying conditions at certain levels of generation, like open or close step level valves, the real relationship between power operation point and fuel cost may be more complex than a quadratic equation [8,9].

1.2.3 Hydropower Generation

Hydropower generation involves the potential and kinetic energy on the water that flows in the rivers. The conversion of hydraulic energy into mechanical energy is done by a hydraulic turbine, and the conversion of mechanical energy into electrical energy occurs in an electric generator [10].

Hydropower is an important source of renewable energy. The most mature of renewable technologies, it provides significant added value in the control of the power system, it can be used as a base load and peak, and it is the most important energy storage technology in big power systems. The initial capital costs are high, but operation and maintenance costs are very low. Hydropower generation is a very simple and robust technology with a long service life and high reliability. Hydropower turbines have a very fast ramp-response technology with the ability to do a black start and, for reversible groups, to change from generation to pumping mode. All these characteristics make hydropower one of the most interesting technologies and one of the most important for power system control and scheduling optimization.

With regard to the size of generation units, hydropower plants have a range of capacities from very small hydropower with several kW to more than 10 GW. In this book, we are mainly interested in big hydropower plants ranging from 10 MW up to 10 GW. But we are also interested in small hydropower plants, ranging between 100 kW and 10 MW, generally operated by independent power producers.

With regard to the regularization of hydraulic flow, there are three main types of hydropower plants. The first type is the run-of-the-river plant that can store small quantities of water and shows little regularization of the flow. For this kind of hydropower plant generation is considered, in the scheduling problem, as an almost independent variable that depends on the water flow and meteorology. The second type is the storage plant that can store large quantities of water and thus regularizes the flow through the plant on a daily or seasonal basis. Large storage dams in relatively small-flow rivers have a higher capacity of regularization, but the annual energy production is low and, consequently, the leveled electricity production cost is high. The third type is the pumped storage plant that is based on a reversible turbine, pumping during off-peak from a lower reservoir to an upper reservoir, storing energy for later generation during peak hours. This third type of plant is in fact a particular case of the second type. From the perspective of operation, pumping energy is an interesting solution if the differences of prices between peak and off-peak are high enough to compensate for the losses in the pumping-generation cycle.

The generating capacity of a hydroelectric plant is a function of the head (m) and flow rate of water (m^3/s) discharged through the turbines. The efficiency of the plant is another important aspect that depends on the length and size of the ducts, but it depends essentially on the type of hydraulic turbine. The typical efficiency of a hydraulic turbine is very high, higher than 95%, and this high efficiency can be obtained between 20% and 125% of the nominal turbine flow.

There are several types of hydraulic turbines, and these are based on the head and flow rate. There are two classifications of hydraulic turbines: impulse turbines and reaction turbines. Impulse turbines use, at atmospheric pressure, the kinetic energy of high-velocity jets of water striking spoon-shaped buckets on the runner. The most common impulse turbines are the Pelton turbines that are struck by perpendicular jets. There are also the Turgo or cross-flow turbines that are struck by diagonal jets. Impulse turbines are used for high heads, ranging between 50 and 500 m, but with low flow rates of up to 2 m^3/s. In a reaction turbine, the water passes from a spiral casing through stationary radial guide vanes, through control gates, and onto the runner blades at above atmospheric pressures. There are two categories of reaction turbines: the Francis and the propeller. Both turbines may be arranged in bulb, tubular, slant, and rim generator configurations. In the Francis type, the water impacts on the runner blades tangentially and exits axially. The propeller type can be based on a fixed blade known as Kaplan type or on variable pitch blades known as double-regulated type. The Francis turbine is used for medium-range heads, between 10 and 300 m, and for medium-range flow rates, between 1 and 15 m³/s. The propeller turbine is used for low-range heads, between 3 and 30 m, and for high-range flow rates, up to 50 m³/s.

The hydropower plants have an important role in the control of the power system. They can produce power with the same high efficiency, having the advantage of flexibility without a decrease in the efficiency. Another important advantage is the ability of a quick start and a fast variation in generation justifying the role of hydropower plants in the control of primary and secondary reserves. Moreover, hydropower plants combine the advantage of fast flexibility with high inertia, supporting the stability of power systems.

Different operating strategies are used in hydropower plants, depending on the hydrological seasonal period. During periods of heavy rain, when reservoirs are at their maximum, the maximum possible power can be generated in order to avoid wasting water. During these periods, thermal power generation is replaced with hydropower as much as possible to avoid fuel costs even in off-peak hours. During seasonal shortage of water, hydropower generation is minimized, being used only in cases where the production cost is very high. In the intermediate seasonal period, production is controlled in order to produce power in the maximum head. It is in this intermediate period that schedule optimization can make the most of hydropower. Optimization of the value of hydropower is achieved by replacing the most expensive generation technique during the hours with high energy price [11].

To optimize scheduling, information about the characteristics of hydropower plants is required. These characteristics are as follows:

- Predicted water inflow in each period (m³/s) is not a characteristic of the hydropower plant but is an essential independent variable associated with each plant.
- Downstream minimum flow rate (m³/s) is the minimum ecological or reserved flow restriction that must be guaranteed downstream for different periods. We admit that this flow can pass through the turbine.
- Derived flow (m³/s) is an ecological, irrigation, or reserved minimum flow that cannot pass through the turbine. It can be specified for different periods.
- Maximum generation limits (MW) correspond to the maximum of the hydraulic valves, up to 125% of the nominal capacity.
- Minimum generation limits (MW) correspond to the minimum of the acceptable efficiency of the turbine, approximately 20% of the nominal capacity.
- Generation efficiency curve (%) is the efficiency as a function of the flow rate (m³/s); it aggregates efficiency in the turbine and loss in the penstock and valves as a function of the flow rate. The curve is nonlinear and can have some irregularity due to multiple water injectors.
- With or without a head regularization scheme, not all power plants are barrage type with the capacity of direct regularization at dam intake. There are hydropower schemes with a canal and forebay tank upstream the penstock; in these cases the head is always the same, independent of the level of water in the dam. The power production is a direct function of the head value for each instant.
- Head/volume curve shows the volume reservoir characteristics. With this curve it is possible to estimate the head in each instant as a function of the inflow and outflow.

- Minimum gross head (m) or minimum volume (m³) is used as a minimum-level restriction for the generation of power, as the hydropower plant should not produce less than minimum power.
- Maximum gross head (m) or maximum volume (m³) is used as a maximum-level restriction for the generation of power and to estimate water losses through the spillway.
- Minimum and maximum pumping flow limits (m³/s) are used as restrictions to the pumping decision variables.
- Curve of pumping efficiency (%) is the efficiency of the generator turbine, operating in reverse mode; it is a curve of efficiency as a function of the reverse flow rate (m³/s). This curve is used to compute the energy consumed for some head and flow operation points.

The optimal scheduling of hydropower production differs from the optimization of thermal power production. The optimization of an individual hydropower plant has as its first objective the minimization of wasted water and as its second objective the maximization of generation efficiency. The minimization of wasted water is achieved simply by generating the maximum possible power, restricted to constraints of the hydropower plant. Maximizing the efficiency can be reached by generating power at the maximum possible head and with flow rates that maximize turbine efficiency, but far more important than individual optimization is the optimization of the overall hydrothermal system. The value of hydropower in the overall system depends on the cost of replacing the most expensive thermal power generation for the periods with higher energy prices, in a market context. This is discussed in more detail in Section 1.3.

1.2.4 Wind Power Generation

Wind power generation is a new renewable and dispersed generation technique that has reached maturity and is becoming one of the most important types of power generation in countries such as Denmark, Spain, Germany, and Portugal [12]. In some of these countries, for windy days, energy generation reaches a percentage of the total load higher than 80%. This is obviously a challenge for the control of power systems because of the variable nature of wind power. High wind power penetration is in fact the great motivation for the discussions covered in this book. It is also the reason for new paradigms of power system operation.

For the purpose of this book, it is important to understand the technical aspects of technology that condition the impact, integration, and control of power systems. The geographical variability and dispersion of a resource are important characteristics that attenuate the impact on power systems. The greatest problem is the intermittence, when the wind farm is instantaneously disconnected from the grid. Reconnection is not a problem because it is a gradual increase of generation from zero to the maximum possible. The other problem is the variability. Wind cannot be stored; we can generate wind power when it is available or we can control the generation only by wasting, generating less than possible, which is not an intelligent strategy. To understand the interaction of wind power with the electric grid we will describe the main technical characteristics of wind power technology [13], dividing the discussion in terms of the concepts of wind turbine and wind farm.

Because wind is a dispersed resource, it needs to be collected by dispersed wind energy conversion units. These units have been suffering a convergent evolution to the present aspect of the propeller wind generator, with three blades, collecting wind power from a typical swept area of 6000 m^2, which is then converted into electricity by electric generation located in the nacelle mounted in a tower at 100 m. This is the typical aspect of a big wind generator, with an approximate rate power of 2 MW. In fact, generator capacity can range up to 6 MW, but the most versatile size to be installed in mountain areas is 2 MW. For offshore wind farms, there are other installation problems and the optimum size can in fact reach 5 MW. Wind turbines are electrically clustered in wind farms, with sizes that range from a few megawatts to a few hundred megawatts. A wind farm concentrates the power in a wind farm substation connected to the grid at 30–150 kV. At a regional level, wind farms can be clustered interconnecting several substations. As it is natural, the geographical concentration of wind farms is patterned by the geographical availability of wind resources, usually more intense on mountain regions [14].

From the electrical perspective, wind turbine technology has been developing since the last few decades from simple constant-speed turbines to fully variable-speed systems, enabling active output control. For the constant-speed category, the rotational speed of the electrical generator is imposed by the electrical grid frequency and, at most wind speeds, the turbine operates below its peak efficiency. The variable-speed category use of power electronic converters decouples the grid frequency from the rotational frequency of the generator imposed by the wind speed; this allows more flexibility in the wind turbine control system for frequency, voltage, active and reactive power, and optimization of performance and efficiency. Decoupling of the electrical and rotor frequencies absorbs wind speed fluctuations, allowing the management of kinetic energy and using the wind generator as a flywheel, thus smoothing power and voltage. Another evolution in wind generators is the active pitch control system on blades—an alternative to stall control, which is a fixed blade aerodynamic control. The pitch control allows full control of the aerodynamic power of the turbine.

Wind farms are a cluster of wind turbines that are connected to the grid in a single power plant. This is the concept of a virtual power plant that responds in the same way as a thermal or hydropower plant. This concept aims at the wind power plant providing ancillary services in addition to power production. The idea is to control actively the ramping-down and ramping-up; however, this can only be done by sacrificing the performance of the wind farm.

There is effective development in regulating the control capabilities at wind farms. This is implemented based on grid codes, stating the specific rules and limitations in the operation variables for wind power plants. The grid codes vary from country to country but in general there are rules related to: active power control, frequency control, voltage control, frequency range, voltage range, tap-changing transformers, and wind farm protection. The active power control and wind farm protection are especially important for scheduling.

The active power control is basically ramp rate limitations, positive ramps, and even negative ramps of active power output. The objective is to mitigate frequency fluctuations caused by extreme wind variations or by shutdown and startup of wind farms. To maintain the power balance, ramps in wind power generation are limited by symmetrical ramps in dispatchable power generation available in power systems. Positive ramps in wind power generation require a negative ramp response in thermal and hydropower generation; the typical value is 10% of rated powers per minute. Negative ramps in wind power generation require a positive ramp in thermal power generation, which is generally slower with typical values near 5% per minute. Systems with higher integration levels of wind power require more severe and lower ramp limits. In normal operation, wind variability is lower than 2% per minute for a typical 20 MW wind farm. However, with extreme winds, wind generators are disconnected for safety and negative ramps can be higher than 5% per minute. Fortunately, the disconnection time of wind generators can be controlled and the problem can be mitigated with adequate wind-farm control procedures. In the same way, forcing the generation to a maximum is also possible and useful when the system has high wind penetration, but it must respect the negative ramp limits. The startup of the wind farm is not a serious issue because it is slow and easy to control in the generator and in the sequencing of wind turbine startup.

The forced shutdown of the wind farm could be a serious problem. A frequency or voltage drop in the network, caused by a severe fault in the electric network, can result in an outage of the wind farms in a vast area, thus inducing a cascading outage of the system and a complete blackout. The problem is not in the wind farm itself but in the protection of the wind farm that disconnects the wind farm from the grid. The protection is parameterized to actuate in those conditions but in the case of faults occurring far from the wind farm, the outage contributes to exacerbating, and not mitigating, the problem. To solve this problem, actual wind farms are prepared to remain stable and stay connected to the network when faults occur on the transmission network. This is known as the fault ride-through capability. The detailed requirements of duration of the fault and voltage level often differ between countries. The existence of a fault ride-through capability allows the reduction of the reserves necessary to maintain the system securely in case of a fault and/or high levels of wind generation.

The variability of wind resource is clearly the most important issue for the scheduling of wind power generation, with high penetration of renewable energy. But not all variability of wind resource is important. For very short time intervals, up to 30 min, the synchronous variations in wind farms are uncorrelated, and they cancel out. For wind resource, variations of up to 30 min could have an impact on the secondary reserve. This impact can be mitigated with ramp control measures in wind power generation, with some but not significant losses in generation performance. Another solution is to increase the secondary reserve in about 5% of the wind power capacity. This is a better solution for low levels of capacity penetration in the system.

The scheduling of thermal and hydropower plants is only affected when the variation is in the time scale of hours to days. The hourly variation depends on the geographical extent of the system: the correlation of wind resource decreases with distance, but in areas of less than 500 km^2 the simultaneous generation is still significant. For a region of 500 km^2 there are frequent variations of 10% of the wind power capacity per hour. This scale of wind power variation and impact requires efficient forecasting. However, even the best wind power forecasting can have frequently an absolute error of more than 30% of the installed capacity each day of the forecast, but the average error is about 10%. This error decreases significantly for forecast time horizons with less than 4 h, with an average error near 5%. Thus, scheduling must be revised on the basis of forecast and real information whenever possible. The uncertainty in wind power forecast causes uncertainty in scheduling, and consequently has extra costs in secondary and tertiary reserves needed to mitigate the variability. Reducing the scheduling horizon from a day ahead to 6 h ahead can reduce the need for tertiary reserves in 50% of cases, but generally in a market context we need scheduling at least a day ahead.

Wind forecasting can be done for each wind power plant, for each region or for the entire system. For generation scheduling, only a forecast for the entire system is necessary. For scheduling a day ahead, the wind power forecast simply uses meteorological-based forecasts without integration of real-time measures of wind power generation. This forecast can be refreshed every 6 h with new meteorological information. For scheduling for the next 6 h, the integration of real-time wind generation measures is an important added value to reduce uncertainty. A forecast of extreme wind conditions is also very important in order to detect fast decreases in generation due to the disconnection of wind power plants.

1.2.5 Other Nonscheduled Power Generation

Wind power and run-of-the-river hydropower generations are technologies for which we have little capacity of scheduling. In fact, these generation technologies are independent variables for scheduling. To overcome the problem a good forecast is needed. But these are not the unique nonscheduled technologies. There are also the solar photovoltaic power [15], concentrated solar power (CSP) [16], wave energy [17], and other forms of generation that depend on the availability of the resource [18]. More generation types are nonscheduled because the power system is not controlled. This is the case of independent producers of cogeneration and biomass and geothermal power plants. These technologies are controllable thermal-based technologies, but because there are independent producers the system operator cannot schedule these generations and, from the perspective of the system operator, these generations are independent variables that must be forecast. There is also, in a market context, some loss of control: some power plants that are technically controllable result in independent variables that must be treated by forecasts where the main drivers are the market prices forecast.

Power from renewable energies, such as photovoltaic, wind, or run-of-the-river hydropower, is always described as an intermittent energy source. For wind and small hydropower generation, it is more correct to classify the technologies as variable output power sources instead of intermittent sources, because the power from these technologies does not start and stop on the basis of a minute time scale. However, for photovoltaic power plants, the term intermittent fits well because cloud shadowing can abruptly change the production. On the second to minute timescale, contrary to wind or hydropower, solar power can have a strong impact on reserves, even on primary reserves. This occurs because of the fast and deep effect of cloud shadowing, with a change in generation that can vary up to 80% in a few minutes for photovoltaic power and even a deep variation for concentrated photovoltaic power. For example, in the 46-MW capacity power plant in Moura, Portugal, a change of 40 MW in less than 1 min was registered, with a change in production from 45 to 5 MW, due to a simple and intermittent shadowing of clouds. Photovoltaic technologies instantaneously convert irradiance into electricity; this change in irradiance causes immediate changes in power generation. If the power production of the photovoltaic technology is more sensible to beam irradiance then the variation effect will be deeper, as is the case in concentrated photovoltaic power.

Photovoltaic power generation, although still relatively small, is growing fast. According to the International Energy Agency trends in 2020, 1% of the electricity consumed in the world will be produced by photovoltaic power, with an installed capacity of 200 GW worldwide. At present, more than 1900 large-scale photovoltaic power plants (with more than 200 kWp each) are installed worldwide. The cumulative power of all these photovoltaic power plants is more than 3600 MWp and the average power output of a plant is slightly more than 1800 MWp. More than 500 large-scale photovoltaic plants are located in Germany, more than 370 are in the United States, and more than 750 are in Spain. With information about very short-term forecast, the variations in photovoltaic power could be softened by anticipation actuating in the set points of the inverters. Inverters can easily control the ramping-up, but the ramping-down is only possible with information from a very short-term forecasting system.

A CSP plant uses the same principle as a thermal power plant, but using solar radiation as a source of heat. To achieve the high temperatures needed for an efficient energy conversion, optical concentration reflectors or lenses are used to concentrate the solar irradiance. Systems with high concentrating rates operate at high temperatures that can reach more than 1000°C in turbines. For technologies with high operation temperatures, heat is converted into mechanical energy by a steam turbine, followed by conversion into electricity by an electrical generator. For technologies with low temperatures and low concentrating rates, heat is converted into mechanical energy by a small-sized Stirling engine. There are several variants of CSP technology with different concentrating levels, designs, and sizes, namely, parabolic trough designs, power tower designs, dish designs, Fresnel reflectors, Fresnel lenses, and others. CSP technology is still in early development, but it uses medium-sized power plants with more than 50 MW. Because of the need for high solar resources it is expected that this technology will be installed in deserted or scarcely inhabited regions. This potential for installation causes high correlation of production, amplifying the problem of the strong variability of this kind of generation. For CSP, because it only captures beam irradiance, this intermittence is more abrupt, changing from the maximum production to zero in some seconds. In this type of solar power plant it is important to know the evolution of shading on the reflecting mirrors in order to avoid large variations in energy received at the solar receiver. Under normal temperature radiation, the receiver operates at a temperature of 800°C, but the temperature drops dramatically and very fast when there are shadows, and when the sun is beaming directly again, the temperature rises almost instantly. This can take the receiver to a solar thermal shock, permanently damaging the system. For CSP, the impact of fast variations can be solved by managing heat storage or the alternative energy source of the solar plant (e.g., natural gas). But the very-short-term forecast is essential for the management of CSP plants.

The trend of high penetration of solar power in the electric grid makes it necessary to integrate this generation type in the power system scheduling, and advanced and innovative forecasting tools are necessary to solve the intermittence and variability of solar power generation. Even on a clear day, without the effect cloud shadows, for sunrise and sunset, the solar-based generation varies 80% in 1h, simultaneously, for all solar power generation in the system. Clearly, with the growing popularity of this type of generation, this is a great challenge for scheduling. The possibilities and detailed strategies for managing intermittent-output solar power vary between national and regional power systems. Like any other form of power generation, solar power has an impact on power system reserves. It also contributes to a reduction in fuel usage and emissions. The impact of solar power depends mostly on the solar power penetration level, but it also depends on the solar power system size, generation capacity mix, and the degree of interconnection to neighboring systems and load variations. This impact is more critical for islands with a high penetration of solar power.

Methods, tools, and services for short-term forecasting of solar power on an hourly basis are available. These forecasting models are based on a mix of analytical forecast models, to model generation for clear days and to integrate meteorological information about solar radiation, cloud coverage, and temperature.

1.2.6 Storage Technologies

Storage is not a generation technology, but in fact sometimes storage technology generates electricity and can play an important role in the scheduling of power generation. The storage and load control are another form of matching the supply to the demand over a wide range of time periods. There are a wide variety of storage technologies [19]: pumped hydropower, heat storage, flywheels, batteries, fuel cells, regenerative redox systems, super capacitors, and compressed air. However, generally, all technologies are expensive and some inadequate and inefficient for long-term storage, from a few hours to a month. There are two kinds of storage applications in electrical systems: dedicated storage that is only used with this objective (e.g., pumped hydropower, flywheels, super capacitors, CSP heat storage, regenerative redox systems) and nondedicated storage that is dedicated to other kinds of applications but can be used to produce electricity when it is free from the main application (e.g., electric vehicles, cogeneration heat storage, backup storage systems). The use of dedicated storage systems normally results in an overall loss of energy, with efficiency commonly about 80%. Nondedicated storage has lower efficiencies because it is not optimized for this purpose, but it is the use of an available resource creating an important added value, for instance, for owners of electric vehicles, that offer the storage service, and for the electric grid, that uses this service.

Pumped hydropower is the most common and best-known technology for energy storage in power systems [20], and in fact it is the only large energy storage technique available in power systems. Pumped hydropower is usually composed of an upper reservoir, penstock or waterway, a pump, a turbine, a motor, a generator, and a lower reservoir.

The pump-turbine and generator motor can be composed of four, three, or two units. The three-unit set integrates a reversible generator-motor, the turbine, and the pump in the same shaft. The two-unit configuration integrates a reversible generator-motor and a reversible pump-turbine in the same shaft. The Francis turbine is the most common reversible pump-turbine. The complete hydropower storing cycle has an efficiency between 70% and 85%. Due to the low energy density of pumping storage, large reservoirs are needed to store significant energy; the size is reasonable for daily or weekly storage, but for monthly or seasonal storage the area of reservoirs is extremely large.

For scheduling optimization, we need to know the minimum and maximum flow limits, the head difference between upper and lower reservoirs, the curve of relation between head and volume of the upper reservoir, and the curve of efficiency of the reversible group as a function of the pumping flow.

The future role of electric storage technologies, such as battery and fuel cell systems, will depend on the development of the technologies. Electric storage is very expensive because of the material used; in fact, the complete life-cycle use of the energy stored is more expensive than the cost of generation. Thus, dedicated storage of electricity does not provide a solution in the short term. However, electric vehicles [21], working in a vehicle-to-grid mode could be an interesting storage solution, without additional cost for the power system and with advantages for the owners of the vehicles. An electric vehicle with typical autonomy has capacities of about 20 kWh and can charge and discharge this energy in a daily cycle. This capacity of 20 kWh is a typical daily consumption for one big house. This means that in a future scenario where most people have electric vehicles the owner will have a storage capacity for almost 20% of the daily power system generation. This could be very valuable for a daily power system balance, but this must be controlled indirectly by sending the right electricity price signal.

Another storage technology is thermal storage [22]. In a future scenario with high penetration of CSP technology, there will be large thermal storage based on several technologies such as pressurized steam, graphite heat storage, and phase change materials like molten salts or other inorganic or organic materials. Storing heat could be an interesting solution for all kinds of thermal power generation with high temperatures. If the temperatures are low the reversal of heat energy to electricity will be inefficient. Note that the average efficiency of the best thermal electricity generation is lower than 50% due to heat energy losses. For scheduling, the stored heat results in electricity generation, but this generation must be forecast as an independent variable that results from the operation strategy of the storage.

Another interesting form of nondedicated heat storage is available in the combined heat and power plant (CHP) [23]. A power system with a high level of penetration of CHP has the flexibility to manage, by sending the right price signals, the generation of electricity or heat. For hours of excess generation, CHPs receive signals to reduce generation and use thermal storage for heat; in this way, they can balance the electricity energy responding to the grid and balance the heat energy according to their internal needs. In this case, the heat does not need to be converted back to electricity because it is used in internal heat processes, having the advantage of a high efficiency for the overall process. For scheduling, the modeling of heat storage is done based on independent variables that result from the strategy of the operation of the CHP.

1.3 Operation of Power System Generation

1.3.1 Power Generation Control

The power system is controlled mainly at the point of generation, directly in the generation equipment or indirectly by defining the scheduling of generation and interconnections. The system is controlled by automatic control actions in the generation units, by control action signals sent by power plant operators or system operators, and suggestion or price signals sent by system operators or market operators [24]. The control signals can be either generated automatically by a knowledge system or sent after the case analysis by direct human decisions.

There are different types of power system organizations including systems with centralized control and those isolated from other systems. These power systems are a single control area and have only two levels of control: control in the generation units and control of the whole system. Most power systems control their own areas but have strong interaction and dependency of other control areas. These dependencies can be hierarchical, receiving exchanging control signals, or market dependent, with restriction of compliance at the level of market transactions. In these cases control exists at several levels: control in the generation units, control of the whole system, control in interconnections, and indirect control based on information from the market and from decisions of independent agents (e.g., independent power producers, market agents).

The primary priority driver for control is reliability and secondary priority decisions include economic criteria and restrictions. Economic optimization and restrictions are predefined conditions of control, but reliability must always be guaranteed with low levels of risk. For a monopoly oligopoly system, the economic optimization consists of optimization of the overall power system, minimizing the operation costs. In a liberalized and market context, the economic optimization of the system operation becomes more complex. In this context, there are multiple agents with different conflicts and economic interests. To solve this, the market and regulatory schemes must be arranged to guarantee that the optimization of the individual solutions for several agents converge on the global optimization of the power system. This can be reached with adequate exchange of information and economic signal between agents. This is also possible with the regulation restrictions applied to individual agents, but the solution deviates from its individual optimum. In a deregulated environment the real-time control must always be in an independent system operator (ISO) [25]. In real-time operation the system operator maintains a conventional control structure of the power system. The real-time control covers the period that corresponds to primary, secondary, and tertiary control, for a control horizon inferior to a few hours.

1.3.2 Scheduling

Scheduling generation is a planning process that specifies how the generation resources should be used to comply with security and economic criteria. As it is a planning process scheduling is done in advance and that is called the "scheduling horizon." Scheduling must provide information about

- The generation unit that will be spinning
- The amount of energy that will be generated when the unit is spinning
- The cost of this resource
- The uncertainty related with the availability of the resource
- The extra potential and services that the generation unit can provide if needed

The scheduling process is based on optimization algorithms, by optimizing the use of resources, based on economic criteria, subject to technical, security, environment, and regulatory restrictions. These algorithms integrate independent variables related to different kinds of predictions, including predictions of energy resources (wind, solar, and hydropower), consumption, market prices, and behaviors. Predictions are used when variable behavior cannot be explained by analytical models. For instance, in a market environment the generation behavior of other competing agents needs to be predicted when there is direct access to inside information about the rules that drive these behaviors. The scheduling algorithms, as any other optimization algorithm, should integrate uncertainty modeling. The uncertainty is intrinsic to any modeling that is based on forecasting and it is essential for the decision process. In the scheduling decision process, at the end, only one optimal solution is chosen; it is common for the agents to neglect the uncertainty. However, because in real scheduling there is uncertainty, in order to determine the optimum solution the agent needs to know the risk of not obtaining the optimal solution. Obviously, this kind of scheduling, with uncertainty modeling, needs an agent to decide between solutions or alternatively needs a decision system with decision rules.

Scheduling algorithms include several modules: unit commitment, economic dispatch, security constrained analysis, reserve assessment, reliability assessment, forecasting, market clearing, and risk analysis. Some of these modules will be introduced in this chapter, but they will be presented in detail in Chapters 2, 6 through 11 of this book.

1.3.3 Reserve Requirements

Because of the uncertainty and dynamic changes in real power systems, to have one scheduling solution is not enough. The power system needs to maintain a certain amount of operating reserves to run the system in a reliable and secure manner. There are several levels of reserves that respond to different time horizons. The operating reserve can be a regulating reserve, to guarantee the normal load and variable generation that follows. There can also be a contingency reserve, to guarantee the security of the system when a contingency occurs. The reserves can also include a spinning reserve, from generators that are spinning, and a complementary reserve, from generators that need to be started. The reserves are resources that can be controlled; usually they are generation based but they can also be consumption based, associated with interruptible loads, demand response, or consumer generation.

From the perspective of responsiveness of control, reserves are classified as primary, secondary, and tertiary. The first two kinds are used as contingency reserves and the third is used more for regulating control. The primary and secondary reserves use essentially spinning reserves and the tertiary reserve uses more complementary reserves. The consumption-based reserve controls are used mostly for tertiary reserves.

The primary reserve is the reserve needed for primary control that must respond in less than 30 s. This primary reserve is in kinetic energy, in the generation units, and in other forms of energy that are ready to be fed automatically by the power plant. The amount of primary reserve allocated in the power systems is determined by the reliability of the system, more specifically by the risk of fault in the generation system. For instance, it can be the capacity of the largest generation power plant in the area of control.

The secondary reserve should cover faster imbalances resulting from the primary control; it must start to react fast, between 30 and 60 s, providing the extra energy required to restore balance. The frequency deviation resultant from the primary control is used by some of the faster generators in the area for automatic generation control. The control signal uses the accumulated frequency deviation called area control error. The fast generation in the control area is used as a secondary reserve. The secondary reserve is available from hydropower plants, thermal gas turbines, flywheels, and storage batteries. Several generations that are available for the secondary reserve are selected by rank order, previously analyzed, because in 30 s it is not possible to run an economic dispatch. The quantification of the secondary reserve is done for the value of the primary reserve plus the generation with probability to be lost, tripping the frequency and voltage system protections. Note that a primary reserve is only temporary energy transference; the secondary reserve must restore that energy by injecting more energy into the system. For interconnected systems, part of the interconnection capacity can be used as secondary reserve.

The variable renewable energies, like wind and photovoltaic power, are not usual as a primary or secondary reserve. Also the consumption-based resources, like interruptibility, demand response, or consumer generation, are not usual as primary or secondary reserves. This is due to the fact that these resources are too expensive and complex to be used in a mode ready for reserve. However, with the larger integration of variable sources in the system, this is becoming an alternative that is being considered and also implemented in some systems. The variable renewable energies can contribute to the increase of secondary reserve. It is the case of cascading contingencies, like outages of wind farms without fault-ride-trough, disconnected from the grid by actuation of the protections. The case of fast cloud-shadow effect in photovoltaic power generation is an operational reserve, which is fast but very small compared with the reserve ready for contingencies.

For the tertiary reserve, the variable generation already has a significant impact. The main objective of the tertiary reserve is to guarantee the regulating control. The tertiary reserve regulates generation by adjusting scheduling. Deviations in scheduling can occur because of an error in forecast, a last-time unavailability of some scheduled generation, a deviation in generation, or a deviation resulting from contingencies. The tertiary reserve needed is not as important as the primary or secondary reserve. But because the contingency reserve must be maintained permanently, the regulating reserve (tertiary reserve) is an additional reserve with additional cost. The quantity of tertiary reserve is proportional to the percentage of variable generation and its rate of variability. Thus, for wind power generation there is 10% variation of the installed capacity. For photovoltaic power, there is an 80% variation of the clear-day maximum generation limit for that day per hour. For small hydropower plants with regularization tariffs, the variation can be 40% of the average daily generation limit. The most important is the variation in the shortest time, which is 15 min to 1 h for tertiary reserve. The typical high variation in consumption is less than 30% per hour. Sometimes several variations cancel each other, but the variation can also be summed up, requiring additional reserve. The additional tertiary reserve needed is the sum of the variation in load with the variation in variable generation. High penetration of wind can increase the tertiary reserve by 5%. The penetration of photovoltaic power is still small in power systems but, because of fast and high variability, for some hours, a tertiary reserve equivalent to 80% of the installed capacity is required.

Improving the forecast can be the cheapest solution to reducing the tertiary reserve. A good forecast could reduce the need of additional tertiary reserve by 50%. This represents a cost of tertiary reserve of about €5 per variable renewable energy megawatt-hour generation.

1.3.4 Unit Commitment

Unit commitment is one of the modules of the scheduling process; it defines which generation units will be "on" or "off" for each hour of the horizon. The algorithm of unit commitment optimizes the problem by using mixed integer programming, dynamic programming, or Lagrangian relaxation. The algorithm minimizes the operating costs, including the costs of startup, shutdown, and production. The economic optimization is based on a ranking of generation units that are in turn based on operation costs of generation, which in unit commitment are discrete for startup and shutdown and linear for generation costs. Thus, the optimization result is based on ranking and not on generation allocation, as occurs in economic dispatch algorithms.

The unit commitment algorithm also includes forecast information about load and independent generation; these are independent variables of the problem. It also includes restrictions like minimum-up and minimum-down time and minimum value for reserves. The unit commitment produces a generation schedule for a time horizon that is usually from 4 to 48 h, with a time interval between 5 min and 1 h. For long time horizons, wide time intervals are used, restricted by computational time requirements. The runs are refreshed every time new and more detailed information arrives. The computational effort depends on the number of simulation periods, the number of units, the number of restrictions, and the dependencies between periods.

When several types of constraints are included in the unit commitment algorithm the module is designated as "constrained unit commitment." A "reserve constrained unit" (RCU) includes constraints about the minimum value of reserves; a "security constrained unit commitment" (SCUC) includes minimum reserve and network constraints; and a "reliability constrained unit commitment" (RCUC) includes constraints about generation and network reliability.

Stochastic unit commitment approaches are very usual in literature, but not so usual in practical applications. This happens because these algorithms need to run very fast in real time, covering time horizons as long as possible with time intervals as detailed as possible; that is why stochastic approaches are very time-consuming. As commented previously, the integration of uncertainty in the algorithms is very important to evaluate risks and to help to decide between close solutions, but it is only useful if the agents are able to decide or if they have rules to help them decide. Thus, it is necessary to find a compromise between the detail and the robustness of the solution. Sometimes a model with high detail but no uncertainties is used and what happens is that the detail is lost in the noise. Or, on the other hand, we can have a robustness indicator by integrating uncertainty and risk analysis, but the model is not attractive for the operator because enough detail cannot be seen in the solution.

1.3.5 Economic Dispatch

The objective of economic dispatch is to allocate the generation among several generation units in such a manner as to minimize the costs of generation. The economic dispatch problem differs from the unit commitment problem because the characteristic of consumption are nonlinear and, consequently, the optimum is an allocation of generation among the units and not a simple ranking of priority. The optimal unit commitment solution results from the best combination of fixed costs, contrary to the economic dispatch for which the optimal solution depends on variable (or marginal) costs. One of the results of economic dispatch can be the locational marginal prices, which cannot be obtained from unit commitment. The optimization of economic dispatch uses as a basis the unit commitment solution, excluding from the optimization the generation units that are off. It is common to adopt approaches that merge the problem of unit commitment with the problem of economic dispatch using mixed integer programming optimization algorithms. This could have advantages because the nonlinear detail in economic dispatch could justify a change in the solution of unit commitment. On the other hand, economic dispatch could integrate unit commitment costs by combining the fixed costs of generation.

As for unit commitment, economic dispatch can include several types of constraints. The "security constrained economic dispatch" (SCED) includes reserve and network constraints. In the SCUC, the network constraints are usually simplified or omitted. There could also be approaches based on "reliability constrained economic dispatch" (RCED), where reliability is checked and, if it is out of the corresponding constraint limits, the reserve variables are adjusted. Thus, in this approach the reserves are modeled as dependent variables and not as restrictions, which means that we can optimize the reserves for each time interval. Other criteria can also be integrated in the objective function, like the cost of emissions for environmental criteria, or the cost of interruptibility or the value of demand response for demand-side criteria. The costs with network losses can be integrated in economic dispatch, but for that we need a simplified load flow analysis embedded in the optimization algorithms. With more detailed load flow it is possible to integrate the voltage and reactive flow constraints.

1.3.6 Hydrothermal Coordination

For power systems with large hydropower generation, the modules of unit commitment and economic dispatch are more complex. Hydropower can change on/off states very often and the unit commitment solution could be very dynamic. Consequently, the economic dispatch solution is more difficult to reach and has high sensibility to the unit commitment solution used as baseline. Therefore, a strong interaction between unit commitment and economic dispatch is needed in the algorithms.

Hydropower is characterized by the possibility of storage with higher time period dependencies to model water volume regulations. There are also dependencies between hydropower units because of the existence of cascading and water flow dependencies between them, with more constraints in the problem. All the dependencies lead to a very heavy optimization modeling that cannot be desegregated easily in suboptimization multiperiods. Additionally, hydropower systems have pumping units that add complexity with new kinds of variables and restrictions.

The operational cost characteristic of the hydropower plant is irrelevant in the scheduling algorithms. But, because one needs to comply with the power balance constraint, the optimization of hydropower generation results in the replacement of expensive thermal generation. The restrictions in the hydropower storage capability also restrict the advantages of the solution. Long-term storage capability allows better use of the water. Systems with long-term storage capability need longer horizons for the simulation and optimization or the use of time boundary conditions for water storage levels.

Pumping optimization is based on the difference between electricity prices in the pumping and generating instant. From a modeling perspective, pumping is a negative generation with a positive cost. This cost is the electricity price or, if unknown, the function of the generation cost of the system in that period. The use of pumping increases consumption in the cheapest electricity price hours. At least, one complete price-cycle period is needed for scheduling horizon optimization.

1.3.7 Scheduling with Nondispatchable Generation

Nondispatchable generation involves renewable sources with noncontrollable characteristics like wind power, photovoltaic power, CSP, small hydropower, and run-of-the-river hydropower. Independent power producers are also nondispatchable, notwithstanding the fact that they are controllable, like the cogeneration and biomass power plants.

Nondispatchable power is modeled in scheduling algorithms by independent variables. These independent variables are forecasts. Forecasts of renewable sources are based on meteorological forecasts with a 7-day time horizon, with a 15-min interval, refreshed four times a day. For very-short-term forecasts, less than 6 h, real measure information can be used as inputs of time series forecast. This nondispatchable modeling for scheduling, using forecast, is very similar to the traditional load modeling. When independent information is unknown it is necessary to rely on forecasts. This also happens for the independent power producers, for consumer response, and for generation from other agents with restricted information. For all these cases we need to forecast behavior, which generally is predictable because it responds to known and accessible independent variables.

1.3.8 Scheduling in a Market Context

In a market context, scheduling needs different modeling [4,26]. First of all, there are no unique agents with unique objective functions. There are multiple agents with multiple and noncommon goals. The market mechanisms and regulations resolve this issue with restrictions that force the convergence of individual goals and the common global goal. These restrictions are applied equally to all optimization perspectives for different agents. The different agents include the energy generation and distribution companies.

In a market context, based on spot market, generally there are the day-ahead markets and the intraday markets. Several intraday markets occur on the operating day and some even occur on the previous day. For each market, the corresponding bid is submitted some hours before the corresponding market clears.

To plan the biding, several market agents need to optimize their own scheduling. For generation side agents, the optimization can be from the perspective of a price-maker agent, a big-share agent with influence in the clearing price, or a price-taker agent.

Price-maker agents optimize scheduling in order to take advantage of their price influence capability. The objective is to maximize the accumulated difference between market price and generation cost. Market price is a function of the market share of different agents and the corresponding generation cost function for that agent. The generation cost function is obtained by forecast, inference, or a more complex model if the information is available. Price-taker agents cannot influence market price; individual scheduling optimization is based on the maximization of the integral of the difference between price forecast and generation cost.

For demand-side agents, optimization is based on the integral of the difference between the cost of electricity acquired in the market and the sum of the income from dispersed generation, with the income from the electricity sales to consumers. The market price of electricity, the consumption, and the dispersed generation must be forecast by appropriated models that will be discussed in Chapters 3, 4, and 5 of this book.

The market operator and the system operator need to do their scheduling, with different detail levels depending on the goal. The goal can be to check the constraints, to bring about economic optimization, to manage reserves, consumption, and independent power producers, and to guarantee reliability. Different types of unit commitment and economic dispatch algorithms are used: SCUC, RCU, RCUC, SCED, and RCED.

Scheduling and control actions are carried out at different time horizons. For an operation in a market context, more phases are needed. The following list describes these actions in a progressive close-up approach before and after the moment of operational control action.

- *Close-up 1*: 48 h before—market agents collect information about the availability of power plants and resources. They need information about independent variables for their own system and, if possible, information about the system managed by competing agents. They also need information about renewable forecast (wind, solar, and hydropower), system consumption forecast, availability, and technical constraints in power plants.
- *Close-up 2*: 42 h before—several agents prepare and submit the day-ahead market bids, using the scheduling obtained in close-up 1. To reduce uncertainty, the most recent information about forecasts and generation availability must be used. The ISO also receives information about scheduling constraints for each unit (e.g., ramping rates, startup costs/times, minimum down-time, efficiency curves, etc.).
- *Close-up 3*: 40 h before—the market operator clears transactions for the day-ahead market. The clearing process is done in coordination with the ISO in several steps: first, the requirements of reserves are defined based on bids and on generation variability; second, resources are committed for the day-ahead market based on SCUC; third, an SCED is carried out, finding the optimal scheduling and

complying with generation constraints, network restrictions, and multiperiod constraints.

- *Close-up 4*: 24 h before—before starting the operational day, an RCUC is done by the ISO, using updated load forecasts and renewable forecasts. In this phase, more detailed information about generation, only available for the ISO, could be integrated. After this analysis, the ISO can decide to change the scheduling. Mechanisms of compensation exist for deviation relatively to the clearing price.

- *Close-up 5*: Before each intraday market, 12 h before—the agents redo their internal scheduling optimization and use this information to bid in intraday markets. The information about errors in forecast and clearing prices in the previous market session will be very useful. Several intraday markets sessions are possible: six is a possible number on the operation day, but in some cases there are several intraday markets sessions on the day before the operation day.

- *Close-up 6*: After each intraday market, 8 h before—the same clearing transactions process in close-up 3 is repeated after receiving the bids for intraday market. For this phase, more detailed SCUC and SCED are done, providing the information about clearing prices and deviations; this information could be used by the agents in the next intraday market bids.

- *Close-up 7*: For each interval in the operation day, 1–4 h before—an SCED is permanently refreshed for each short period over the operating day. The scheduling is continuously refreshed (e.g., each 5 min) for the next hours (e.g., 4 h). Note that in this scheduling the optimization variables are highly constrained and most of the changes occur in the variables related with nondispatchable generation. This permanent and close-up scheduling adjustment has to guarantee the load and the adjustments to a very-short-term generation change prediction.

- *Close-up 8*: Operational control action—it is the instant of generation action for which all the scheduling has been planned before. After this instant, the primary, secondary, and tertiary controls act to adjust generation deviations.

- *Close-up 9*: Primary control, 0–30 s after—it is a dynamic control, carried out locally by individual synchronous generators, equilibrating the relation between rotation speed and grid frequency. This control responds to instantaneous local imbalances using kinetic energy in the generators. These imbalances are caused by generation or network faults. The energies used for energy balance in primary control include the primary reserve, the kinetic energy in machines, and the energy feed by a very fast control reaction (less than 30 s).

- *Close-up 10*: Secondary control, 30 s to 15 min after—it is a stationary automatic control response, for area or region, with very fast reaction usually in 30–60 s, done by automatic generation control. This control responds to frequency deviations resulting from the primary control. The control is done in a decentralized way for a specific area, and it is coordinated with specifications of the control role and reaction for each generator. Only some of the generators in a region are responsible for the secondary control, the other generators follow the frequency by using the primary control. Some fast-reaction generators like hydropower and simple gas turbines have an important role in this control. The control signal for the area is the area control error which is the accumulated frequency error measured at some specific points. There are economic mechanisms to adjust the deviations relatively to the last economic dispatch; these are part of the ancillary services. The energy used to balance the system is the secondary reserve, which is the energy controlled by fast automatic generation control, available in some of the power plants.

- *Close-up 11*: Tertiary control, 15 min to 1 h after—it is a centralized nonautomatic control with the objective to readjust the generation to the dispatched generations. The scheduling deviation results from the primary and secondary controls whose main goals are reliability and system security. The main goal of the tertiary control is to restore the economical optimum scheduling. When the restoration of the previous scheduling is not possible a close scheduling solution is adopted. The scheduling solution adopted is the one resulting from the SCED that is refreshed every 5 min.

1.4 Challenges of Future Power Generation

Power systems have been changing significantly during the last decades and will keep changing in the future. The changes occur due to several reasons: environmental obligations, security of supply, new generation technologies, technological development in communications and control, and the need for new market opportunities and deregulation.

Growing concerns about environmental issues create the need for power systems to change to clean technologies that use renewable resources that are dispersed; consequently, the system evolves for small and dispersed generation technologies that completely change the structure of the power system and the traditional control of generation. These changes have an additional motivation, the security of supply, to comply with consumption growth and decrease of fossil energy reserves. These main changes motivate the strong development of new generation technologies, mainly renewable sources that are or will be in short-term competition with conventional technologies. At the same time, new communication and control technologies have developed, and currently we are assisting a revolution in the integration and processing of information in power systems, leading to new paradigms of smart grids, smart equipment, and smart control. All these changes, with the tendency of decentralization, lead to new opportunities of market and decentralized investment, competitive markets, and a necessity for deregulation or regulation to frame this new economic environment.

In a dispersed generation paradigm, with small generation units, with a multitude of different agents, the system is less efficient and much more difficult to control. The great challenge of new power systems is to comply with the growing requirement of reliability and low price requirements of generation. Solutions to overcome this challenge include development of more intelligence in control, collection and use of more information, the use of more intelligent components in the power system, the increase in the storage in power systems, and the active participation of the demand-side agents.

The control will be done in a different way. Traditional control is mainly centralized and with direct actions. New control will be highly decentralized based on suggestive signals for action responses from independent agents. A wide network of information will be used to spread these signals. These information networks will be used to collect reaction responses and a huge amount of information that comes from every small component in the system.

Advanced forecasting techniques are the intelligence needed by the brain of the power system to manage these new challenges [27]. Forecast is always used when an important influence variable is unknown. In future power system generation, most of the influence variables are unknown, including wind generation, photovoltaic generation, hydropower generation, cogeneration behavior, consumption behavior, electricity price market, and market agent behavior. Research is being done in some of these forecasting techniques, and great effort is needed to improve these advanced forecasting techniques. This book is aimed at bringing a new impetus in this direction.

References

1. L. L. Grigsby, *Electric Power Generation, Transmission, and Distribution*. Boca Raton, FL: CRC Press, 2007.
2. F. A. Farret and M. G. Simões, *Integration of Alternative Sources of Energy*. Hoboken, NJ: John Wiley & Sons, 2006.
3. V. L. Nguyen, *Power Systems Modeling and Control Coordination: Steady-State Operating Mode with FACTS Devices*. Saarbrücken, Germany: LAP Lambert Academic Publishing, 2010.
4. M. Shahidehpour, H. Yamin, and Z. Li, *Market Operations in Electric Power Systems: Forecasting, Scheduling and Risk Management*. New York: John Wiley & Sons, 2002.
5. L. Freris and D. Infield, *Renewable Energy in Power Systems*. West Sussex, UK: John Wiley & Sons, 2008.
6. European Wind Energy Association (EWEA), *Large Scale Integration of Wind Energy in the European Power Supply: Analysis, Issues and Recommendations*, A report by EWEA, December 2005.
7. M. V. Deshpande, *Elements of Electrical Power Station Design*. New Delhi: PHI Learning Private Limited, 2010.
8. A. J. Wood and B. Wollenberg, *Power Generation, Operation, and Control*, 2nd ed., New Delhi: Wiley India Pvt. Ltd, 2006.
9. B. F. Hobbs, M. H. Rothkopf, R. P. O'Neill, and Hung-po Chao, eds., *The Next Generation of Electric Power Unit Commitment Models*. Norwell, MA: Kluwer Academic Publishers, 2001.
10. P. K. Nag, *Power Plant Engineering*, 3rd ed. New Delhi: Tata McGraw-Hill, 2008.
11. F. R. Førsund, *Hydropower Economics*. New York: Springer Science + Business Media LLC, 2007.
12. European Wind Energy Association (EWEA), *Wind Energy—The Facts: A Guide to the Technology, Economics and Future of Wind Power*. London, UK: Earthscan, 2009.
13. J. F. Manwell, J. G. McGowan, and A. L. Rogers, *Wind Energy Explained: Theory, Design and Application*, 2nd ed. West Sussex, UK: John Wiley & Sons, 2010.

14. T. Ackermann, *Wind Power in Power Systems*. West Sussex, UK: John Wiley & Sons, 2005.
15. Deutsche Gesellschaft für Sonnenenergie, *Planning and Installing Photovoltaic Systems: A Guide for Installers, Architects and Engineers*. London, UK: Earthscan, 2008.
16. K. Lovegrove and W. Stein, eds., *Concentrating Solar Power (CSP) Technology*. Cambridge, UK: Woodhead Publishing Limited, 2011 (forthcoming).
17. J. Cruz, ed., *Ocean Wave Energy: Current Status and Future Perspectives*. Berlin: Springer-Verlag, 2008.
18. International Energy Agency (IEA), *Harnessing Variable Renewables—A Guide to the Balancing Challenge*. Paris Cedex, France: IEA Studies, 2011.
19. R. A. Huggins, *Energy Storage*. New York: Springer Science + Business Media LLC, 2010.
20. Task Committee on Pumped Storage, American Society of Civil Engineers, *Hydroelectric Pumped Storage Technology: International Experience*. New York: ASCE Publications, 1996.
21. A. Emadi, M. Ehsani, and J. M. Miller, *Vehicular Electric Power Systems: Land, Sea, Air, and Space Vehicles*. Boca Raton, FL: CRC Press, 2003.
22. İ. Dinçer and M. Rosen, *Thermal Energy Storage: Systems and Applications*. London, UK: John Wiley & Sons, 2010.
23. M. P. Boyce, *Handbook for Cogeneration and Combined Cycle Power Plants*, New York: ASME Press, 2002.
24. N. V. Ramana, *Power System Operation and Control*. New Delhi: Dorling Kindersley (India) Pvt. Ltd, licencees of Pearson Education in South Asia, 2010.
25. X.-P. Zhang, *Restructured Electric Power Systems: Analysis of Electricity Markets with Equilibrium Models*. Hoboken, NJ: John Wiley & Sons, Inc., 2009.
26. G. S. Rothwell and T. Gómez, eds., *Electricity Economics: Regulation and Deregulation*. Hoboken, NJ: John Wiley & Sons, Inc., 2003.
27. R. Weron, *Modeling and Forecasting Electricity Loads and Prices: A Statistical Approach*, West Sussex, UK: John Wiley & Sons, 2006.

2

Uncertainty and Risk in Generation Scheduling

Rabih A. Jabr

2.1 Introduction

Power generation companies compete in a deregulated energy market to sell the amount of power that maximizes their profit. Within energy markets that use power pool trading, power producers and consumers submit production and consumption bids to the independent system operator so that these bids get cleared on the basis of an appropriate market-clearing procedure [1]. The clearing procedure is constrained by available transfer capacities and transmission congestion in the power market. For the self-schedule to be accepted by the independent system operator, an appropriate bidding strategy should be applied. Conejo et al. [2] describe a framework that price-taker generation companies can adopt to construct successful hourly bidding curves. The construction of these curves relies on a price forecast and a profit-maximizing self-schedule.

The concept of self-scheduling has recently received attention by the power systems community. Yamin et al. [3] argue that, in order to obtain successful generation bids, the generation companies have to self-schedule their units by taking into account the power flow constraints of both the intact and contingent networks. The formulation in the study by Yamin et al. [3] is deterministic in the sense that it uses the nominal forecasted locational marginal prices (LMPs) to maximize the company's profit. It is known that a deterministic solution taking the expected price as given will not in general produce the correct expectation as it ignores volatility [4]. Research by Jabr [5] builds on that by Yamin et al. [3] by incorporating price volatility in the self-scheduling formulation. This is achieved by explicitly quantifying the risk through a measure that maps the loss into a real number. Two measures of risk are common in the portfolio optimization theory: Value-at-Risk (VaR) and Conditional Value-at-Risk (CVaR). In VaR-based approaches, the volatility in LMPs is captured using a covariance matrix obtained from historical values of true and forecasted prices. The *RiskMetrics™—Technical Document* describes methods that are useful for estimating the covariance matrix from historical data [6]. In fact, the covariance matrix has been used by power system researchers investigating risk management within competitive energy markets [1,7,8]. The methodologies that use the covariance matrix are fundamentally based on Markowitz's seminal work in the area of portfolio selection [9]. CVaR-based approaches, on the other hand, model the volatility in LMPs through the explicit use of a probability density distribution [10]. Another technique for treating risk relies on modeling price uncertainty by using fuzzy numbers [11,12].

This chapter discusses recent advances in the modern portfolio theory in the context of generation self-scheduling. It discusses approaches based on VaR and CVaR together with their extensions to deal with the problem of data uncertainty. Data uncertainty is considered in the models that specify the volatility in LMPs, namely, the mean vector and covariance matrix in VaR-based scheduling and the probability density function in CVaR-based approaches. This leads to scheduling models based on worst-case VaR and worst-case CVaR. The schedule from each of the models is obtained by solving a semidefinite program, or, in some instances, a second-order cone program. Second-order cone programs are special forms of semidefinite programs that can be solved with efficiency close to that of linear programs [13]. Appendix A2.1 includes an introduction to semidefinite programming and second-order cone programming. The robustness of the schedules obtained from the risk-averse scheduling paradigms is illustrated with simple numerical examples.

2.2 Generation Self-Scheduling

The self-scheduling problem definition was first discussed by Yamin et al. [3]. This definition postulates several generation companies competing to sell their generated power in a pool market. In this setting, a specific generation company is therefore not obliged to supply the total forecasted system demand; this task is left to the independent system operator. The generation company aims to supply the portion of the demand that maximizes its profit. To simplify the presentation, only one scheduling period is considered and the behavior of the power company is modeled as a price-taker.

Each generating unit in the company is assumed to have a convex quadratic cost function:

$$C_i(P_{Gi}) = a_i + b_i P_{Gi} + c_i P_{Gi}^2, \quad i = 1, \ldots, N_G \tag{2.1}$$

where N_G is the number of units owned by the generation company. Let λ^n denote the $N_G \times 1$ vector of nominal (forecasted) LMPs; then, the expected (mean) profit is

$$(\lambda^n)^T P_G - \sum_{i=1}^{N_G} C_i(P_{Gi}) \tag{2.2}$$

where P_G is an $N_G \times 1$ vector containing the generation P_{Gi} and N_G is the number of units owned by the generation company. The deterministic self-scheduling problem is a specific instance of a security-constrained optimal power flow model:

$$\max_{P_G \in \Pi} \ (\lambda^n)^T P_G - \sum_{i=1}^{N_G} C_i(P_{Gi}) \tag{2.3}$$

The feasible region Π of the optimal power flow model is defined by the following set of constraints:

i. Power generation limits:

$$P_{Gi}^{\min} \leq P_{Gi} \leq P_{Gi}^{\max} \tag{2.4a}$$

ii. DC network model:

$$0 \leq P_i + \sum_{j \in k(i)} a_{ij}(\delta_i - \delta_j) \leq P_{Di} \tag{2.4b}$$

iii. Intact network line flow constraints:

$$-T_{ij}^{\max} \leq T_{ij} = -a_{ij}(\delta_i - \delta_j) \leq T_{ij}^{\max} \tag{2.4c}$$

iv. Security constraints following the outage of lines $m_1 k_1$ to $m_r k_r$ in terms of flows in the intact network:

$$-\hat{T}_{ij}^{\max} \leq \hat{T}_{ij} = T_{ij} + \sum_{l=1}^{r} \sigma_l T_{m_l k_l} \leq \hat{T}_{ij}^{\max} \tag{2.4d}$$

where $k(i)$ represents the set of nodes connected to node i, P_i the power injection at node i $(= 0$ or $P_{Gi})$, P_{Gi}^{max} the maximum limit of P_{Gi}, P_{Gi}^{min} the minimum limit of P_{Gi}, P_{Di} the forecasted power demand at node i, T_{ij} the intact power flow on line ij, $\hat{T}_{ij}$ the contingent power flow on line ij, T_{ij}^{max} the prefault rating of line ij, $\hat{T}_{ij}^{max}$ the postfault (emergency) rating of line ij, V_i the voltage magnitude at node i, y_{ij} the line admittance $(g_{ij} + \sqrt{-1}b_{ij}$, $a_{ij} = V_iV_jb_{ij})$, δ_i the voltage angle at node i $(\delta_1 = 0)$, and σ_l the lth element of the row vector of load transfer coefficients (see Reference [14] for details).

The above-mentioned deterministic self-scheduling problem can be formulated as a second-order cone program:

Maximize

$$(\lambda^n)^T P_G - \sum_{i=1}^{N_G}(a_i + b_iP_{Gi}) - p \qquad (2.5a)$$

subject to

$$w_i = \sqrt{c_i}P_{Gi} \quad \text{for } i = 1, \ldots, N_G \qquad (2.5b)$$

$$2pq \geq \sum_{i=1}^{N_G} w_i^2, \; q = \frac{1}{2}, \; p \geq 0 \qquad (2.5c)$$

$$P_G \in \Pi \qquad (2.5d)$$

The first relationship in Equation 2.5c is a rotated quadratic cone used to model $C_i(P_{Gi})$. The solution P_G^* to Equations 2.5a–d is referred to as the deterministic schedule.

It is obvious that the accuracy of forecasting the LMPs affects the accuracy of estimating the expected revenue as given by Equation 2.5a [15]. Because the accuracy of the forecast cannot be guaranteed, the actual LMPs can take random unexpected values, and the actual revenues may be well below the expected value in Equation 2.5a promised by the deterministic self-scheduling solution. Consequently, a method that balances risk and revenue is desired. Toward this end, Section 2.3 discusses how to quantify the risk by a monetary value.

2.2.1 Test System Model

The test system is the 5-bus network shown in Figure 2.1. It is small enough to allow documentation of the complete dataset, yet it is comprehensive to allow demonstration of the different aspects of the self-scheduling models. The generation, line, and bus data are given in Tables 2.1 through 2.3, respectively. They are based on the dataset originally appearing in Reference [16]. Both generators are assumed to be the property of the same generation company that is participating in a power pool market.

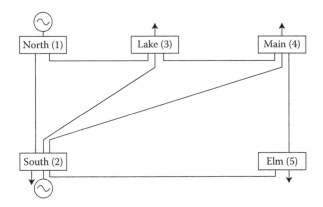

FIGURE 2.1   Benchmark 5-bus example.

TABLE 2.1   Generator Data of the 5-Bus Network

Bus i	P_{Gi}^{min} (MW)	P_{Gi}^{max} (MW)	a_i ($/h)	b_i ($/MWh)	c_i ($/MW²h)
1	10	200	0	3.4	0.004
2	10	200	0	2.4	0.003

TABLE 2.2   Line Data of the 5-Bus Network (100 MVA Base)

Bus i	Bus j	r (per unit)	x (per unit)	$b_{sh}/2$ (per unit)	Rating (per unit)
1	2	0.02	0.06	0.030	1.00
1	3	0.08	0.24	0.025	1.00
2	3	0.06	0.18	0.020	1.00
2	4	0.06	0.18	0.020	1.00
2	5	0.04	0.12	0.015	1.00
3	4	0.01	0.03	0.010	1.00
4	5	0.08	0.24	0.025	1.00
1	2	0.02	0.06	0.030	1.00

TABLE 2.3   Bus Data of the 5-Bus Network

Bus i	Type	P_{Di} (MW)
1	Slack	0
2	PV	20
3	PQ	45
4	PQ	40
5	PQ	60

The nominal (forecasted) LMPs at buses 1 and 2 are assumed to be 4.5 and 3.4 \$/MWh, respectively:

$$\lambda^n = \begin{bmatrix} 4.5 \\ 3.4 \end{bmatrix}$$

It is also assumed that the actual spot-market price deviation from the forecast market price follows the historical pattern of the price [17]. The corresponding price difference (actual LMP–forecasted LMP) distributions at buses 1 and 2 are given in Figure 2.2. In practice, the price difference distributions are obtained from historical values of the actual and forecasted LMPs. Each distribution has 168 data points. The corresponding discrete probability density distribution has a sample space given by the price difference values shifted by the nominal prices; that is, it would consist of the vectors $\lambda_1, \lambda_2, \ldots, \lambda_{168}$ with probabilities:

$$\Pr\{\lambda_k\} = \frac{1}{168} \quad \text{for } k = 1, \ldots, 168$$

Appendix A2.2 includes the sample space so as to allow testing of the methods and reproduction of the results of this chapter in future research work. The covariance matrix, obtained for consistency from the price difference distributions using the cov(.) function in MATLAB® [18], is

$$V_\lambda = \begin{bmatrix} 0.1299 & -0.0026 \\ -0.0026 & 0.0533 \end{bmatrix}$$

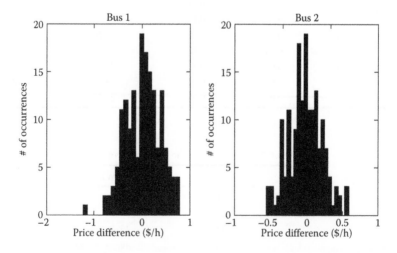

FIGURE 2.2 Price difference distributions.

2.3 Measures of Risk

In return-risk trade-off analysis, the risk is quantified by a risk measure that maps the loss to a monetary value. This approach is widely adopted in practical applications and theoretical studies because it facilitates the understanding of risk. The foundations of the modern portfolio theory have been set in 1952 by Markowitz through the framework of mean variance analysis [9]. Two risk measures, among others, have appeared since then: VaR and CVaR.

2.3.1 Value-at-Risk

VaR is a risk assessment tool that is used by financial institutions to measure the minimum occasional loss expected in a given portfolio within a stated time period [4,19]. In generation self-scheduling, VaR determines the monetary risk corresponding to a given generation schedule P_G. For a given probability of occurrence, VaR estimates how much the power company could lose due to fluctuations in LMPs. The probability level β represents the degree of certainty of the VaR estimate. In other words, the probability that the monetary loss exceeds VaR_β is $(1 - \beta)$ [5]:

$$\Pr\{(\lambda^n - \lambda)^T P_G > \text{VaR}_\beta\} = 1 - \beta \tag{2.6}$$

This is illustrated in Figure 2.3 which shows a general probability density distribution of the profit function

$$f(P_G, \lambda) = \lambda^T P_G - \sum_{i=1}^{N_G} C_i(P_{Gi}) \tag{2.7}$$

where λ is an $N_G \times 1$ vector that follows the probability density distribution $\rho(\lambda)$. The probability of the profit function $f(P_G, \lambda)$ not falling below a threshold t is therefore given by

$$\Psi(P_G, t) = \int_{f(P_G, \lambda) \geq t} \rho(\lambda) \, d\lambda \tag{2.8}$$

As a function of t and for a given value of P_G, $\Psi(P_G, t)$ is the cumulative distribution function of the profit associated with P_G. $\Psi(P_G, t)$ is in general nonincreasing with respect to t.

The robust profit (RP_β) is the lower $100(1 - \beta)$ percentile of the profit distribution. It is given by [10]:

$$\text{RP}_\beta(P_G) = \max\{t \in R : \Psi(P_G, t) \geq \beta\} \tag{2.9}$$

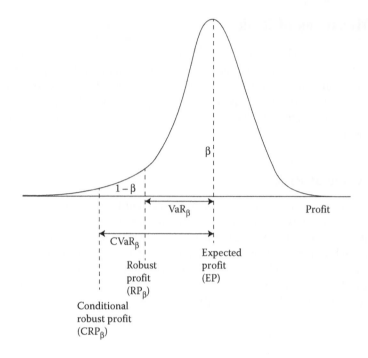

FIGURE 2.3 Graphical representation of Value-at-Risk (VaR$_\beta$) and Conditional Value-at-Risk (CVaR$_\beta$) measures.

Equation 2.9 indicates that RP$_\beta$(P_G) is the right endpoint of the nonempty interval consisting of the values of t such that $\Psi(P_G,t) = \beta$. VaR$_\beta$ is the difference between the expected profit (EP) and the RP$_\beta$:

$$\text{VaR}_\beta(P_G) = \text{EP}(P_G) - \text{RP}_\beta(P_G) \tag{2.10}$$

When the distribution $\rho(\lambda)$ is Gaussian with given mean λ^n and covariance matrix V_λ, the VaR$_\beta$ can be expressed as [5]:

$$\text{VaR}_\beta(P_G) = \kappa(\beta)\sqrt{P_G^T V_\lambda P_G} \tag{2.11}$$

where the safety parameter $\kappa(\beta)$ is given by the standard normal tables, for instance, $\kappa(0.95) = 1.65$. Even if the distribution of returns cannot be assumed to be Gaussian, Tchebycheff's inequality can be used to find an upper bound on the probability that the loss exceeds VaR$_\beta$:

$$\Pr\left\{(\lambda^n - \lambda)^T P_G > \kappa(\beta)\sqrt{P_G^T V_\lambda P_G}\right\} \leq \frac{1}{\kappa(\beta)^2} \tag{2.12}$$

By comparing with Equation 2.6, $\kappa(\beta) = 1/\sqrt{1 - \beta}$. However, the classical Tchebycheff bound is not exact, meaning that the upper bound is not achieved [20]. It can be replaced

TABLE 2.4 Safety Parameter $\kappa(\beta)$ for Selected Probability Values

β	Gaussian Distribution	Classical Tchebycheff Bound	Exact Tchebycheff Bound
0.90	1.2816	3.1623	3.0000
0.95	1.6449	4.4721	4.3589
0.99	3.3263	10.000	9.9499

by its exact version by simply setting $\kappa(\beta) = \sqrt{\beta/(1-\beta)}$. Table 2.4 shows the values of $\kappa(\beta)$ corresponding to different levels of the probability value.

2.3.2 Conditional Value-at-Risk

CVaR is an alternative risk assessment tool that gives a better indication of risk than VaR; it quantifies the losses associated with the tail of the profit distribution [21,22]. For the same confidence level used for VaR, CVaR is defined as the mean of the tail distribution exceeding VaR. Therefore, it provides an estimate of the average loss exceeding the VaR value. CVaR_β is defined graphically in Figure 2.3. The expected value of $100(1-\beta)$ percentile of the lowest profit values is the conditional robust profit (CRP_β) [10]:

$$\text{CRP}_\beta(P_G) = \frac{1}{1-\beta} \int_{f(P_G,\lambda)\leq RB_\beta(P_G)} f(P_G,\lambda)\rho(\lambda)\,d\lambda \qquad (2.13)$$

In other words, CRP_β is the conditional expectation of the profit associated with P_G given that the profit is RP_β or less. The corresponding CVaR_β is given by the EP minus the CRP_β in Equation 2.13:

$$\text{CVaR}_\beta(P_G) = \text{EP}(P_G) - \text{CRP}_\beta(P_G) \qquad (2.14)$$

The CRP_β can be obtained by maximizing an auxiliary function without predetermining the corresponding RP_β first, and, at the same time, RP_β is computed as a by-product. The auxiliary function is [21,23]:

$$F_\beta(P_G,t) = t + \frac{1}{1-\beta} \int_{\lambda\in R^{NG}} \left[f(P_G,\lambda) - t \right]^- \rho(\lambda)\,d\lambda \qquad (2.15)$$

where $z^- = \min(0,z)$. By assuming that $\Psi(P_G,t)$ is everywhere continuous with respect to t, the relationships between $\text{RP}_\beta(P_G)$, $\text{CRP}_\beta(P_G)$, and $F_\beta(P_G,t)$ are given as follows:

$$\text{RP}_\beta(P_G) \in A_\beta(P_G) = \arg\max_{t\in R} F_\beta(P_G,t) \qquad (2.16a)$$

$$\text{CRP}_\beta(P_G) = \max_{t\in R} F_\beta(P_G,t) = F_\beta(P_G,\text{RP}_\beta(P_G)) \qquad (2.16b)$$

In Equation 2.16a, if the interval $A_\beta(P_G)$ does not reduce to a single point, then $RP_\beta(P_G)$ is the right endpoint of the nonempty interval $A_\beta(P_G)$. The results in Equations 2.16a and 2.16b are based on the theorems originally proposed by Rockafellar and Uryasev [21,23].

2.4 Data Uncertainty

Self-scheduling formulations that attempt to balance risk and reward are based on maximizing the robust profit [5] or conditional robust profit [10]. These approaches assume perfect knowledge of the data, in particular, the mean vector and covariance matrix of prices for VaR-based scheduling and the probability density function of prices for CVaR-based scheduling. In reality, the data are often subject to error.

2.4.1 Data Uncertainty in VaR

VaR approaches are known to produce schedules that are extremely sensitive to errors in the mean vector and covariance matrix of prices. To deal with such errors within the scheduling framework, a model of the data uncertainty is required. One intuitive method to handle this uncertainty is to assume that λ^n and V_λ are only known within the interval bounds:

$$\lambda_-^n \leq \lambda^n \leq \lambda_+^n \tag{2.17a}$$

$$V_{\lambda-} \leq V_\lambda \leq V_{\lambda+} \tag{2.17b}$$

where λ_-^n and λ_+^n are given $N_G \times 1$ vectors, and $V_{\lambda-}$ and $V_{\lambda+}$ are given $N_G \times N_G$ matrices. In the interval uncertainty model of Equation 2.17, the inequalities are understood component-wise.

2.4.2 Data Uncertainty in CVaR

In CVaR approaches, an analytical expression for $\rho(\lambda)$ is not needed, rather a sampling procedure is used. The sample space of the discrete probability distribution consists of the vectors $\lambda_1, \lambda_2, ..., \lambda_{N_C}$ with probabilities

$$\Pr\{\lambda_k\} = \pi_k^0 \quad \text{for } k = 1, ..., N_C \tag{2.18a}$$

and

$$\sum_{k=1}^{N_C} \pi_k^0 = 1 \tag{2.18b}$$

In this case, the uncertainty is accounted for by requiring the vector of probability values $\pi = [\pi_1, ..., \pi_{N_C}]^T$ to lie in a given uncertainty set $\wp_\pi$. Two uncertainty structures are considered. The first is box uncertainty where it is assumed that π belongs to a box [24]:

$$\pi \in \wp_\pi^B = \left\{ \pi : \pi = \pi^0 + \varepsilon,\ e^T \varepsilon = 0,\ \breve{\varepsilon} \le \varepsilon \le \hat{\varepsilon} \right\} \tag{2.19}$$

In the above set definition, $\pi^0 = [\pi_1^0, \ldots, \pi_{N_C}^0]^T$ is the nominal (most likely) probability distribution, ε is an $N_C \times 1$ error vector, $\breve{\varepsilon}$ and $\hat{\varepsilon}$ are given ($N_C \times 1$) constant vectors, and e is an $N_C \times 1$ vector of ones. The condition $e^T \varepsilon = 0$ ensures that the sum of probabilities remains 1. Moreover, the nonnegativity constraint on π is captured by the lower limit $\breve{\varepsilon}$. The second uncertainty structure assumes that π belongs to an ellipsoid [24]:

$$\pi \in \wp_\pi^E = \left\{ \pi : \pi = \pi^0 + Av,\ e^T Av = 0, \pi^0 + Av \ge 0, \|v\| \le 1 \right\} \tag{2.20}$$

where π^0 is the nominal probability distribution at the center of the ellipsoid, A is the $N_C \times N_C$ scaling matrix of the ellipsoid, and $\|.\|$ represents the Euclidean norm. The constraints $e^T Av = 0$ and $\pi^0 + Av \ge 0$ ensure that π is a probability distribution.

Specifying the box and ellipsoidal structures requires knowledge of the nominal probability distribution together with a set of other possible distributions obtained from historical observations. In particular, it is assumed that the nominal distribution π^0 and a set of possible distributions π^i ($i = 1, \ldots, m$) are available. Under this assumption, the bounds of the box uncertainty set can be specified as

$$\breve{\varepsilon}_k = \min_{i \in \{1, \ldots, m\}} (\pi_k^i - \pi_k^0, 0) \quad \text{and} \quad \hat{\varepsilon}_k = \max_{i \in \{1, \ldots, m\}} (\pi_k^i - \pi_k^0, 0) \quad \text{for } k = 1, \ldots, N_C \tag{2.21}$$

The scaling matrix for ellipsoidal uncertainty can be more easily specified by considering the special case of a ball uncertainty set. For ball uncertainty, the scaling matrix is given by αI, where I is the $N_C \times N_C$ identity matrix and the radius of the corresponding uncertainty ball is

$$\alpha = \max_{i \in \{1, \ldots, m\}} \left\| \pi^i - \pi^0 \right\| \tag{2.22}$$

2.5 VaR Formulations

RP_β as given by Equation 2.10 is the EP minus the VaR_β. Therefore, the problem of maximizing RP_β is essentially a biobjective optimization problem in which equal weights are given to maximizing both EP and$-VaR_\beta$, or equivalently, maximizing EP and minimizing VaR_β. The solution to maximizing RP_β for a fixed β leads to a compromise between increasing EP and reducing VaR_β. The value of the parameter β, or alternatively the safety parameter $\kappa(\beta)$, dictates the trade-off between risk and reward.

2.5.1 VaR-Based Nominal Robust Schedule

To account for risk using VaR_β, the robust version of Equation 2.3 maximizes RP_β in place of the deterministic profit:

$$\max_{P_G \in \Pi} (\lambda^n)^T P_G - \sum_{i=1}^{N_G} C_i(P_{Gi}) - \kappa(\beta)\sqrt{P_G^T V_\lambda P_G} \qquad (2.23)$$

The above problem is equivalent to the following second-order cone program [5]: Maximize

$$(\lambda^n)^T P_G - \sum_{i=1}^{N_G} (a_i + b_i P_{Gi}) - p - \kappa(\beta)s \qquad (2.24a)$$

subject to

$$v = D_\lambda P_G \qquad (2.24b)$$

$$s \geq \sqrt{\sum_{i=1}^{N_G} v_i^2} \qquad (2.24c)$$

$$w_i = \sqrt{c_i} P_{Gi} \quad \text{for } i = 1,\ldots,N_G \qquad (2.24d)$$

$$2pq \geq \sum_{i=1}^{N_G} w_i^2, \quad q = \frac{1}{2}, \; p \geq 0 \qquad (2.24e)$$

$$P_G \in \Pi \qquad (2.24f)$$

The conic constraints are given in Equations 2.24c and 2.24e; Equation 2.24c is a quadratic cone whereas the first relationship in Equation 2.24e is a rotated quadratic cone. The covariance matrix V_λ is in general symmetric positive (semi-)definite. Therefore, the $N_G \times N_G$ square matrix D_λ required in defining the quadratic cone of Equation 2.24c can be obtained using eigenvalue or spectral decomposition [6]:

$$V_\lambda = Q\Delta Q^T \qquad (2.25)$$

where Q is a square orthogonal matrix of eigenvectors and Δ is a square diagonal matrix with eigenvalues of V_λ along its diagonal. It follows that

$$D_\lambda = \Delta^{1/2} Q^T \qquad (2.26)$$

The solution P_G^* to Equation 2.24 is referred to as the (VaR-based) nominal robust schedule. The term nominal refers here to the fact that data uncertainty is not accounted for.

2.5.2 VaR-Based Worst-Case Robust Schedule

By considering the uncertainty in the mean vector and covariance matrix as described in Equation 2.17, the worst-case robust counterpart of Equation 2.24 corresponding to the probability β can be derived by introducing additional variables [20]. Let Λ_- and Λ_+ denote unknown $N_G \times N_G$ matrices, P_{G-} and P_{G+} denote unknown $N_G \times 1$ vectors, and v an unknown real number. The worst-case robust problem involves optimization over linear constraints, conic quadratic, and semidefiniteness constraints [25]:

Maximize

$$\langle V_{\lambda-}, \Lambda_- \rangle - \langle V_{\lambda+}, \Lambda_+ \rangle + (\lambda_-^n)^T P_{G-} - (\lambda_+^n)^T P_{G+}$$
$$- \kappa(\beta)^2 v - \sum_{i=1}^{N_G}(a_i + b_i P_{Gi}) - p \tag{2.27a}$$

subject to

$$P_G = P_{G-} - P_{G+} \tag{2.27b}$$

$$P_{G-} \geq 0, \quad P_{G+} \geq 0, \quad \Lambda_- \geq 0, \quad \Lambda_+ \geq 0 \tag{2.27c}$$

$$\begin{bmatrix} \Lambda_+ - \Lambda_- & P_G/2 \\ P_G^T/2 & v \end{bmatrix} \succeq 0 \tag{2.27d}$$

$$w_i = \sqrt{c_i} P_{Gi} \quad \text{for } i = 1,\ldots,N_G \tag{2.27e}$$

$$2pq \geq \sum_{i=1}^{N_G} w_i^2, \quad q = \frac{1}{2}, \ p \geq 0 \tag{2.27f}$$

$$P_G \in \Pi \tag{2.27g}$$

In Equation 2.27a, $\langle A, B \rangle = \text{Tr}(AB)$ denotes the inner product in the space of symmetric matrices, and in Equation 2.27d, $X \succeq 0$ means that the matrix X is symmetric positive semidefinite. As in Equation 2.24, the quadratic terms of $C_i(P_{Gi})$ are modeled using a rotated quadratic cone in Equation 2.27f. The solution P_G^* to Equation 2.27 is the (VaR-based) worst-case robust schedule. In the case that λ^n and V_λ are exactly known, that is, $\lambda_-^n = \lambda_+^n = \lambda^n$ and $V_{\lambda-} = V_{\lambda+} = V_\lambda$, the above worst-case robust counterpart problem yields the same optimal solution as Equation 2.24. Problem 2.27 can be also used to compute the worst-case robust profit for a given schedule P_G^n.

2.5.3 Example of VaR-Based Robust Scheduling

Consider the 5-bus network whose data are specified in Section 2.1. The deterministic schedule obtained by solving Equation 2.5 is given (in MW) by

$$P_G = \begin{bmatrix} 77.86 \\ 87.14 \end{bmatrix}$$

This schedule does not consider possible fluctuations of the predicted LMPs; its corresponding EP is 125.76 \$/MWh. To account for this fluctuation, the covariance matrix of prices V_λ is used to give nominal robust schedules with different levels of risk aversion. This is achieved by solving the conic program in Equation 2.24 for three typical values of the probability level β: 0.90, 0.95, and 0.99. The corresponding value of $\kappa(\beta)$ that is used in Equation 2.24 is obtained from the second column of Table 2.4, that is, by assuming that the profit distribution is approximately Gaussian. The nominal robust schedules (P_G^n) are given in Table 2.5. Table 2.6 shows the values of EP, RP_β, and VaR_β for both the deterministic and nominal robust solutions. It is evident that the VaR-based schedule promises less profit but is however less risky. For the values of β considered in Tables 2.5 and 2.6, $\beta = 0.90$ corresponds to the most risky policy, while $\beta = 0.99$ corresponds to the least risky one. When $\beta = 0.95$ is chosen as a compromise, there is 5% chance that the nominal robust schedule loss will exceed 53.42 \$/MWh; this value is the 95% VaR. For the deterministic schedule, the 95% VaR increases to 55.94 \$/MWh. In all cases, the robust profit of the nominal robust schedule is higher than that of the deterministic schedule.

The schedules in Table 2.5 assume that the data are error free. For $\beta = 0.95$, the nominal robust schedule is contrasted with the worst-case robust schedule obtained from Equation 2.27, in which the forecasted LMPs (i.e., elements of the mean vector) and the elements of the covariance matrix are assumed to belong to uncertainty intervals as in Equation 2.17. Let r be a parameter that denotes the relative uncertainty on λ^n, understood in the sense of a component-wise, uniform variation. The uncertainty in the mean vector is described by the interval limits:

$$\lambda_-^n = \lambda^n - r|\lambda^n| \tag{2.28a}$$

$$\lambda_+^n = \lambda^n + r|\lambda^n| \tag{2.28b}$$

TABLE 2.5 Value-at-Risk-Based Nominal Robust Schedules

Power Generation (MW)	$\beta = 0.90$	$\beta = 0.95$	$\beta = 0.99$
P_{G1}	67.87	66.15	63.63
P_{G2}	97.13	98.85	101.37

TABLE 2.6 Performance Parameters of the Deterministic and Value-at-Risk-Based Nominal Robust Schedules

Schedule	Parameter ($/MWh)	$\beta = 0.90$	$\beta = 0.95$	$\beta = 0.99$
Deterministic	EP	125.76	125.76	125.76
	RP_β	82.18	69.82	46.65
	VaR_β	43.58	55.94	79.11
Nominal robust	EP	125.06	124.80	124.34
	RP_β	83.21	71.38	49.33
	VaR_β	41.85	53.42	75.01

Note: EP, expected profit; RP, robust profit; VaR, Value-at-Risk.

The mean vector is harder to estimate than the covariance matrix [20]. Therefore, the relative uncertainty on V_λ is assumed to be 10% that on λ^n:

$$V_{\lambda-} = V_\lambda - 0.1r\left|V_\lambda\right| \tag{2.29a}$$

$$V_{\lambda+} = V_\lambda + 0.1r\left|V_\lambda\right| \tag{2.29b}$$

The worst-case robust profit can now be computed for increasing values of r for both the nominal robust schedule and the worst-case robust schedule. The worst-case robust profit corresponding to the worst-case robust schedule is computed from Equation 2.27. For the nominal robust schedule, Equation 2.27g is replaced by $P_G = P_G^n$. The results are depicted in Figure 2.4 where the x-axis is the relative uncertainty r and the y-axis is the worst-case RP_β as a percentage of the nominal RP_β (corresponding to $r = 0$). The results suggest that the worst-case robust schedule is superior to the nominal robust schedule in

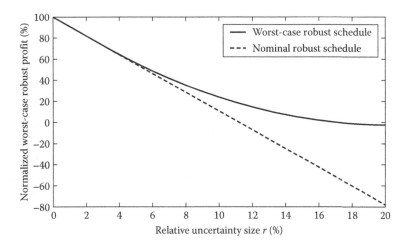

FIGURE 2.4 Normalized worst-case robust profit of the Value-at-Risk-based nominal and worst-case robust schedules ($\beta = 0.95$).

the sense that it has a higher worst-case robust profit. For example, at an uncertainty level of $r = 12\%$, the normalized worst-case robust profit of the nominal robust schedule drops to around -5%, whereas for the worst-case robust schedule it remains positive at around 10%.

2.6 CVaR Formulations

VaR approaches to risk management suffer from several shortcomings [10]. First, the VaR does not reveal the extent of losses that might be suffered beyond the amount quantified by this measure. Second, VaR is in general not coherent, as in Reference [26], that is, the VaR associated with a combination of two generation schedules can be deemed greater than the sum of risks of the individual schedules. Third, VaR is very difficult to work with when computations are based on historical observations or scenarios [21]. For the same confidence level used for VaR, CVaR provides an estimate of the average loss exceeding the VaR value. CVaR is also a coherent measure of risk, as in Reference [26], and can be easily integrated with models based on historical observations or scenarios. In addition, because CVaR is greater than VaR, schedules with low CVaR will also have a low VaR.

2.6.1 CVaR-Based Nominal Robust Schedule

Accounting for risk through CVaR_β rather than VaR_β translates into an optimization problem that maximizes CRP_β instead of RP_β:

$$\max_{P_G \in \Pi} \text{CRP}_\beta(P_G) \tag{2.30}$$

Using Equation 2.16b, it can be shown that [21,23]:

$$\max_{(P_G, t) \in \Pi \times R} F_\beta(P_G, t) = \max_{P_G \in \Pi} \max_{t \in R} F_\beta(P_G, t) = \max_{P_G \in \Pi} F_\beta(P_G, \text{RP}_\beta(P_G)) = \max_{P_G \in \Pi} \text{CRP}_\beta(P_G) \tag{2.31}$$

Equation 2.31 implies that the maximization of $F_\beta(P_G, t)$ over $(P_G, t) \in \Pi \times R$ produces a solution (P_G^*, t^*) such that P_G^* maximizes CRP_β. In the typical case where the interval $A_\beta(P_G^*)$ reduces to a single point, t^* gives the corresponding RP_β. Otherwise, RP_β can be obtained using a line search [23].

To simplify the optimization problem in Equation 2.31, $F_\beta(P_G, t)$ is commonly approximated by sampling the probability distribution of λ according to its density function $\rho(\lambda)$ [21,27]. Let the sampling generate the vectors $\lambda_1, \lambda_2, \ldots, \lambda_{N_C}$, then the corresponding approximation to $F_\beta(P_G, t)$ in Equation 2.15 is

$$\tilde{F}(P_G, t) = t + \frac{1}{N_C(1 - \beta)} \sum_{k=1}^{N_C} \left[\lambda_k^T P_G - \sum_{i=1}^{N_G} C_i(P_{Gi}) - t \right]^- \tag{2.32}$$

The maximization of $\tilde{F}_\beta(P_G,t)$ over $\Pi \times R$, which gives an approximate solution of the maximization of $F_\beta(P_G,t)$ over $\Pi \times R$, can be reduced to a second-order cone program [10]:

Maximize

$$t + \frac{1}{N_C(1-\beta)}\sum_{k=1}^{N_C} u_k \tag{2.33a}$$

subject to

$$u_k \leq \lambda_k^T P_G - \sum_{i=1}^{N_G}(a_i + b_i P_{Gi}) - t - p, \quad u_k \leq 0, \quad \text{for } k = 1,\ldots,N_C \tag{2.33b}$$

$$w_i = \sqrt{c_i} P_{Gi} \quad \text{for } i = 1,\ldots,N_G \tag{2.33c}$$

$$2pq \geq \sum_{i=1}^{N_G} w_i^2, \quad q = \frac{1}{2}, \quad p \geq 0 \tag{2.33d}$$

$$P_G \in \Pi \tag{2.33e}$$

The solution P_G^* to Equation 2.33 is the (CVaR-based) nominal robust schedule.

2.6.2 CVaR-Based Worst-Case Robust Schedule

The robust self-scheduling model based on CVaR, as in Equation 2.33, assumes exact knowledge of the density function $\rho(\lambda)$. This assumption may not be realistic, for instance, in cases where enough data samples are not available. To circumvent this problem, it is possible to relax this assumption by considering that the density function is only known to belong to a certain set $\wp$ of probability distributions:

$$\rho(.) \in \wp \tag{2.34}$$

The worst-case CRP_β (WCCRP_β) for a given $P_G \in \Pi$ with respect to $\wp$ can be defined as

$$\text{WCCRP}_\beta(P_G) = \inf_{\rho(.)\in\wp} \text{CRP}_\beta(P_G) \tag{2.35}$$

The WCCRP_β as defined above remains a coherent measure of risk [24]. The robust self-scheduling problem in Equation 2.31 in terms of WCCRP_β becomes

$$\inf_{\rho(.)\in\wp} \max_{(P_G,t)\in\Pi\times R} F_\beta(P_G,t) \tag{2.36}$$

Let $\wp_\pi$ denote $\wp$ in the case of discrete probability distributions. Moreover, assume that $\wp_\pi$ is a compact convex set. Then the discrete version of Equation 2.36 reduces to [24]:

$$\max_{(P_G,t)\in\Pi\times R} \min_{\pi\in\wp_\pi} \tilde{F}(P_G,t,\pi) \tag{2.37}$$

Problem 2.37 can be written in a form similar to Equation 2.33 [28]:
Maximize

$$\theta \tag{2.38a}$$

subject to

$$\theta \le \min_{\pi\in\wp_\pi} t + \frac{1}{1-\beta}\sum_{k=1}^{N_C}\pi_k u_k \tag{2.38b}$$

$$u_k \le \lambda_k^T P_G - \sum_{i=1}^{N_G}(a_i + b_i P_{Gi}) - t - p; \quad u_k \le 0, \quad \text{for } k = 1,\dots,N_C \tag{2.38c}$$

$$w_i = \sqrt{c_i}\,P_{Gi} \quad \text{for } i = 1,\dots,N_G \tag{2.38d}$$

$$2pq \ge \sum_{i=1}^{N_G}w_i^2, \quad q = \frac{1}{2}, \ p \ge 0 \tag{2.38e}$$

$$P_G \in \Pi \tag{2.38f}$$

However, the above problem cannot be yet optimized using conic programming due to the min operator in the constraint in Equation 2.38b. Conic program formulations can be obtained in cases where $\wp_\pi$ is assumed to be a box or ellipsoidal uncertainty set.

2.6.2.1 Box Uncertainty

Under the box uncertainty of Equation 2.19, it is possible to obtain the worst-case robust schedule from a quadratic cone program [24]. Let υ and ω denote additional unknown $N_C \times 1$ vectors and τ be an unknown real variable. The second-order cone program for worst-case robust scheduling is [28]:
Maximize

$$t + \frac{1}{1-\beta}(\pi^0)^T u - \frac{1}{1-\beta}(\bar{\varepsilon}^T\upsilon - \breve{\varepsilon}^T\omega) \tag{2.39a}$$

subject to

$$-e\tau - \upsilon + \omega = u, \quad \upsilon \geq 0, \ \omega \geq 0 \tag{2.39b}$$

$$u_k \leq \lambda_k^T P_G - \sum_{i=1}^{N_G}(a_i + b_i P_{Gi}) - t - p, \quad u_k \leq 0, \quad \text{for } k = 1,\ldots,N_C \tag{2.39c}$$

$$w_i = \sqrt{c_i} P_{Gi} \quad \text{for } i = 1,\ldots,N_G \tag{2.39d}$$

$$2pq \geq \sum_{i=1}^{N_G} w_i^2, \quad q = \frac{1}{2}, \ p \geq 0 \tag{2.39e}$$

$$P_G \in \Pi \tag{2.39f}$$

The solution P_G^* to the above problem is referred to as the (CVaR-based) worst-case robust schedule under box uncertainty. The formulation also allows computing the worst-case conditional robust profit for a prespecified generation schedule. In this case, Equation 2.39f is dropped from the constraint set.

2.6.2.2 Ellipsoidal Uncertainty

For the ellipsoidal uncertainty of Equation 2.20, the worst-case robust schedule can be also obtained from a second-order cone program [24,28]:

Maximize

$$t + \frac{1}{1-\beta}(\pi^0)^T u - \frac{1}{1-\beta}(\zeta + (\pi^0)^T \omega) \tag{2.40a}$$

subject to

$$\upsilon + A^T \omega - A^T e\tau = A^T u, \quad \omega \geq 0 \tag{2.40b}$$

$$\zeta \geq \sqrt{\sum_{i=1}^{N_c} \upsilon_i^2} \tag{2.40c}$$

$$u_k \leq \lambda_k^T P_G - \sum_{i=1}^{N_G}(a_i + b_i P_{Gi}) - t - p, \quad u_k \leq 0, \quad \text{for } k = 1,\ldots,N_C \tag{2.40d}$$

$$w_i = \sqrt{c_i} P_{Gi} \quad \text{for } i = 1,\dots,N_G \qquad\qquad (2.40e)$$

$$2pq \geq \sum_{i=1}^{N_G} w_i^2, \quad q = \frac{1}{2}, \ p \geq 0 \qquad\qquad (2.40f)$$

$$P_G \in \Pi \qquad\qquad (2.40g)$$

As in the case of box uncertainty, the program requires defining additional variables: υ and ω are unknown $N_C \times 1$ vectors, and τ and ζ are unknown real variables. The optimal generation schedule P_G^* to the above worst-case robust counterpart problem is referred to as the (CVaR-based) worst-case robust schedule under ellipsoidal uncertainty.

2.6.3 Example on CVaR-Based Robust Scheduling

The same system that was considered in the context of VaR-based scheduling is also studied with CVaR-based scheduling. The nominal robust schedules (P_G^n), obtained from solving Equation 2.33 for three levels of β, are shown in Table 2.7. In this computation, there is no requirement to assume a Gaussian distribution for the profit as was done in the VaR-based schedules. Table 2.8 shows the corresponding values of EP, RP_β, CRP_β, VaR_β, and $CVaR_\beta$ for both the deterministic and nominal robust solutions. The CVaR-based solution reveals more information than the VaR-based solution. For instance, when $\beta = 0.95$, there is 5% chance that the profit of the nominal robust schedule drops below 69.32 \$/MWh, and in such unfavourable cases the profit is on average 58.30 \$/MWh. The $CVaR_\beta$ indicates the average loss exceeding VaR_β for the given probability of occurrence; in all cases, $CVaR_\beta$ of the nominal robust schedule is less than that of the deterministic schedule.

The values in Tables 2.7 and 2.8 are valid provided that the actual probability distribution of prices is perfectly described by the discrete uniform distribution in Section 2.1 (also given in Table A2.2.1 Appendix A2.2). Because this assumption may not be entirely realistic, the worst-case robust schedule was calculated under box and ellipsoidal (ball) uncertainty structures and compared against the nominal robust schedule. The bounds

TABLE 2.7   Conditional Value-at-Risk-Based Nominal Robust Schedules

Power Generation (MW)	$\beta = 0.90$	$\beta = 0.95$	$\beta = 0.99$
P_{G1}	67.23	62.81	64.16
P_{G2}	97.77	102.19	96.23

TABLE 2.8 Performance Parameters of the Deterministic and Conditional Value-at-Risk-Based Nominal Robust Schedules

Schedule	Parameter ($/MWh)	$\beta = 0.90$	$\beta = 0.95$	$\beta = 0.99$
Deterministic	EP	125.78	125.78	125.78
	RP_β	80.84	67.52	48.85
	CRP_β	65.83	56.68	43.25
	VaR_β	44.94	58.26	76.93
	$CVaR_\beta$	59.95	69.10	82.53
Nominal robust	EP	124.64	123.70	122.20
	RP_β	82.52	69.32	46.35
	CRP_β	67.13	58.30	44.25
	VaR_β	42.12	54.38	75.85
	$CVaR_\beta$	57.51	65.40	77.95

Note: CRP, conditional robust profit; CVaR, Conditional Value-at-Risk; EP, expected profit; RP, robust profit; VaR, Value-at-Risk.

of the uncertainty structures were parameterized by a real number $r \in [0,1]$. For box uncertainty, the bounds can be described by the interval limits:

$$\hat{\varepsilon} = -r\pi^0 \quad \text{and} \quad \hat{\varepsilon} = r\pi^0 \tag{2.41}$$

The scaling matrix for ball uncertainty is

$$A = \frac{r}{N_C} I \tag{2.42}$$

Equations 2.41 and 2.42 show that the larger the value of r, the more uncertain the distribution. The parameter r can be chosen by the decision-maker to reflect the confidence in the nominal probability distribution.

For increasing values of r, the worst-case CRP_β was computed for both the nominal robust and worst-case robust schedules. Under box uncertainty, the worst-case CRP_β corresponding to the worst-case robust schedule is calculated from Equation 2.39. For the nominal robust schedule, which is prespecified, the constraint in Equation 2.39f is replaced by $P_G = P_G^n$. The numerical results are illustrated in the lower two curves in Figure 2.5. The x-axis represents the parameter r, whereas the y-axis is the worst-case CRP_β as a percentage of the nominal CRP_β (corresponding to $r = 0$). Figure 2.5 suggests that the worst-case robust schedule is superior to the nominal robust one because it has a higher worst-case CRP_β. Figure 2.5 also includes similar curves obtained under ellipsoidal uncertainty. In this case, the worst-case CRP_β is computed from Equation 2.40 with $P_G = P_G^n$ for the nominal robust schedule and $P_G \in \Pi$ for the worst-case robust schedule. The normalized worst-case CRP_β values for ellipsoidal uncertainty are higher than their counterparts for box uncertainty, though they follow the same trend. This

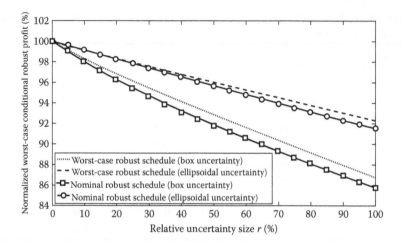

FIGURE 2.5 Normalized worst-case conditional robust profit of the Conditional Value-at-Risk-based nominal and worst-case robust schedules ($\beta = 0.95$).

result is expected because for the same value of r in Equations 2.41 and 2.42, the uncertainty ball would be contained in the uncertainty box.

2.7 Conclusions

This chapter presented specific optimization formulations for managing risk faced by power producers bidding in pool-based electricity markets. The methods are fundamentally based on the modern portfolio optimization theory and use two risk measures: VaR and CVaR. Uncertainty in the data is also addressed through optimizing the corresponding worst-case measures of risk. In all instances, convex problem formulations are presented. These problems can be solved using semidefinite programming or a special form of it known as second-order cone programming. Numerical results are presented on a small-scale system whose dataset is completely specified. This would help in the progress of future research based on the current methods.

There are several directions for future research. The current formulations consider only a single time period and future work can consider risk-averted unit commitment methodologies. For worst-case scheduling, a reasonable specification of the uncertainty set is of paramount importance for practical self-scheduling. This requires further investigation. Another direction of research could consider the effect of price-taker assumption on the optimization formulations.

Appendix A2.1

Semidefinite programming (SDP) is an extension of linear programming (LP). In LP, the variables are elements of a vector that is required to be component-wise nonnegative, whereas, in SDP, the variables form a symmetric matrix that is constrained to be positive semidefinite. Let A_0 and A_i, $i = 1,\ldots,m$, be given $n \times n$ symmetric matrices and

b denote a given $m \times 1$ vector. If X is an unknown $n \times n$ symmetric matrix, then the primal semidefinite programming problem is

Minimize

$$\langle A_0, X \rangle \tag{A2.1.1a}$$

subject to

$$\langle A_i, X \rangle = b_i, \quad i = 1,\ldots,m \tag{A2.1.1b}$$

$$X \succeq 0 \tag{A2.1.1c}$$

The inner product is defined by

$$\langle A_0, X \rangle = \sum_{i,j} A_{0ij} X_{ij} \tag{A2.1.2}$$

and the notation $X \succeq 0$ means that X is positive semidefinite. Polynomial time interior-point methods for LP have been shown to be extendable to SDP [29].

A special form of SDP is second-order cone programming (SOCP). This generalizes an LP problem by including a constraint of the form $x \in C$ in the problem definition where C is required to be a convex cone [13]. Let each element x_i of the vector x be a member of exactly one of the vectors $x^j, j = 1,\ldots,k$. Condition $x \in C$ is satisfied if each one of the vectors x^j belongs to one of the following cones:

i. The R set (set of real numbers).
ii. The quadratic cone:

$$C^q = \left\{ x \in R^m : x_1 \geq \sqrt{\sum_{j=2}^{m} x_j^2} \right\} \tag{A2.1.3a}$$

iii. The rotated quadratic cone:

$$C^r = \left\{ x \in R^m : 2x_1 x_2 \geq \sum_{j=3}^{m} x_j^2, x_1, x_2 \geq 0 \right\} \tag{A2.1.3b}$$

Conic quadratic problems can be solved by polynomial time interior-point methods at basically the same computational complexity as linear programming problems of similar size [13].

Because SOCP is a special case of SDP, it is possible to define convex optimization problems that include constraints of the type $x \in C$ and $X \succeq 0$. The solutions to all the optimization programs in this chapter were obtained using the SeDuMi solver [30] under the Matlab [18] environment.

Appendix A2.2

TABLE A2.2.1 Sample Space of the Discrete Probability Distribution of Locational Marginal Prices (5-Bus Example)

k	λ_1	λ_2	k	λ_1	λ_2	k	λ_1	λ_2	k	λ_1	λ_2
1	5.2336	3.4313	43	4.6246	3.2194	85	4.7948	3.6952	127	4.6362	3.3397
2	5.1754	3.3269	44	4.7232	3.5152	86	4.1516	2.8598	128	4.7056	3.4087
3	4.6084	3.4444	45	4.8130	3.3872	87	5.2833	3.3241	129	4.2909	3.3594
4	4.4587	3.2507	46	4.2922	3.4999	88	4.6366	3.3131	130	4.7495	3.4110
5	4.7912	3.3922	47	4.8643	3.3724	89	5.0421	3.5365	131	5.0289	3.5469
6	5.1339	3.3693	48	5.0301	3.0689	90	4.4212	3.2843	132	4.4769	3.4825
7	4.9245	3.3764	49	4.0500	3.7202	91	4.4812	3.1997	133	4.9105	3.3079
8	4.7471	3.1314	50	4.4570	3.0889	92	4.5133	3.6108	134	4.3376	3.6980
9	4.5193	3.4262	51	4.5896	3.5139	93	4.8968	3.3388	135	4.8958	3.9844
10	4.7337	3.3654	52	4.0924	3.5351	94	4.7557	3.4585	136	4.1553	3.2491
11	3.786	3.0474	53	5.0024	3.0512	95	5.0118	3.6060	137	4.2457	3.3992
12	4.2004	3.2557	54	4.8420	3.5037	96	4.6794	3.4563	138	3.8151	3.4960
13	3.966	3.7576	55	4.2243	3.5388	97	4.2132	2.9879	139	4.3596	3.6058
14	4.5323	3.3857	56	4.6617	3.1638	98	4.5779	3.2815	140	4.4783	3.2838
15	4.5904	3.5125	57	4.4414	2.9181	99	4.5781	3.6089	141	4.7717	3.4617
16	4.8808	3.5445	58	5.0383	3.4798	100	5.2484	3.2801	142	5.0498	3.1873
17	4.5895	3.4668	59	3.2712	3.6185	101	4.5838	3.2868	143	4.1604	3.4594
18	4.1233	3.5586	60	4.3299	3.4284	102	4.4951	3.5971	144	4.5623	3.5869
19	5.0925	3.4615	61	4.8296	3.7334	103	4.5331	3.3006	145	4.9073	3.8810
20	4.1818	3.1363	62	4.5530	3.6576	104	4.5288	3.0719	146	4.4972	3.2488
21	4.9071	3.5679	63	4.3851	3.5515	105	4.1076	3.5124	147	4.3152	3.6456
22	3.9075	3.2896	64	4.4790	3.6711	106	4.5856	3.6609	148	4.4805	3.5960
23	4.0906	3.8651	65	4.005	3.1675	107	4.6166	2.9259	149	4.7160	3.6293
24	4.6828	3.3800	66	4.5923	3.4953	108	4.7324	3.1751	150	3.7562	3.9414
25	4.1296	3.4175	67	4.8763	3.5199	109	5.0674	3.7204	151	4.8956	3.7964
26	4.5274	3.5845	68	4.9761	3.0838	110	4.6417	3.1722	152	4.5970	3.0445
27	3.985	3.6082	69	4.7134	3.4523	111	4.8715	3.3897	153	4.3552	3.1338
28	4.0836	3.5511	70	4.1606	3.0732	112	4.5855	3.3266	154	4.8622	3.3869
29	5.1871	3.3482	71	4.3060	3.1868	113	4.5047	3.8376	155	4.6976	3.4080
30	4.6149	2.9322	72	4.6128	3.0882	114	4.6353	3.2645	156	4.2014	3.2759
31	5.2518	3.2956	73	4.7809	3.4068	115	3.9153	3.8193	157	4.3597	3.6059
32	4.6424	3.1285	74	4.2658	3.3140	116	4.5178	3.3914	158	4.6345	3.6844
33	4.4999	3.3492	75	4.105	3.2375	117	4.0634	3.3120	159	4.4919	3.4106
34	4.1004	2.8745	76	3.7430	3.6413	118	4.2978	3.1357	160	4.1768	3.2398

TABLE A2.2.1 (**continued**) Sample Space of the Discrete Probability Distribution of Locational Marginal Prices (5-Bus Example)

k	λ_1	λ_2	k	λ_1	λ_2	k	λ_1	λ_2	k	λ_1	λ_2
35	4.6571	2.8429	77	4.3315	3.4316	119	4.1503	3.2512	161	4.9604	3.0451
36	4.0459	3.0641	78	4.2783	3.4687	120	4.1619	3.0224	162	4.4229	3.3805
37	4.0178	3.1390	79	4.1364	3.2818	121	4.7037	3.9519	163	4.3322	3.2570
38	4.6856	3.0574	80	4.4917	3.1123	122	4.7736	2.9756	164	4.1348	3.3023
39	4.5319	3.0518	81	4.9496	3.3989	123	4.2640	3.3355	165	3.8859	3.3396
40	4.2723	3.6589	82	4.4454	3.4604	124	4.1039	3.5643	166	5.1564	3.1470
41	4.0337	3.7290	83	4.5053	3.2974	125	4.8906	3.3596	167	4.9275	3.4700
42	4.3802	3.2996	84	3.9588	3.2792	126	4.2007	3.3977	168	4.4271	3.5445

References

1. A. J. Conejo, F. J. Nogales, J. M. Arroyo, and R. García-Bertrand, Risk-constrained self-scheduling of a thermal power producer, *IEEE Transactions on Power Systems*, 19(3), 1569–1574, 2004.
2. A. J. Conejo, F. J. Nogales, and J. M. Arroyo, Price-taker bidding strategy under price uncertainty, *IEEE Transactions on Power Systems*, 17(4), 1081–1088, 2002.
3. H. Yamin, S. Al-Agtash, and M. Shahidehpour, Security-constrained optimal generation scheduling for GENCOs, *IEEE Transactions on Power Systems*, 19(3), 1365–1372, 2004.
4. M. Denton, A. Palmer, R. Masiello, and P. Skantze, Managing market risk in energy, *IEEE Transactions on Power Systems*, 18(2), 494–502, 2003.
5. R. A. Jabr, Self-scheduling under ellipsoidal price uncertainty: Conic-optimisation approach, *IET Generation, Transmission & Distribution*, 1(1), 23–29, 2007.
6. J. P. Morgan: *RiskMetrics™—Technical Document*, 4th ed. New York: J. P. Morgan, 1996.
7. H. Y. Yamin and S. M. Shahidehpour, Risk and profit in self-scheduling for GenCos, *IEEE Transactions on Power Systems*, 19(4), 2104–2106, 2004.
8. R. Bjorgan, C.-C. Liu, and J. Lawarrée, Financial risk management in a competitive electricity market, *IEEE Transactions on Power Systems*, 14(4), 1285–1291, 1999.
9. H. Markowitz, Portfolio selection, *The Journal of Finance*, 7(1), 77–91, 1952.
10. R. A. Jabr, Robust self-scheduling under price uncertainty using conditional value-at-risk, *IEEE Transactions on Power Systems*, 20(4), 1852–1858, 2005.
11. P. Attaviriyanupap, H. Kita, E. Tanaka, and J. Hasegawa, A fuzzy-optimization approach to dynamic economic dispatch considering uncertainties, *IEEE Transactions on Power Systems*, 19(3), 1299–1307, 2004.
12. H. Y. Yamin, Fuzzy self-scheduling for GenCos, *IEEE Transactions on Power Systems*, 20(1), 503–505, 2005.
13. E. D. Andersen, C. Roos, and T. Terlaky, On implementing a primal-dual interior-point method for conic quadratic optimization, *Optimization Online*, 2000 [Online]. Available: http://www.optimization-online.org/DB_HTML/2000/12/245.html

14. R. A. Jabr, Homogeneous cutting-plane method to solve the security-constrained economic dispatching problem, *IEE Proceedings C—Generation, Transmission & Distribution*, 149(2), 139–144, 2002.
15. J. Bastian, J. Zhu, V. Banunarayanan, and R. Mukerji, Forecasting energy prices in a competitive market, *IEEE Computer Applications in Power*, 12(3), 40–45, 1999.
16. E. Acha, C. R. Fuerte-Esquivel, H. Ambriz-Pérez, and C. Angeles-Camacho, *FACTS: Modeling and Simulation in Power Networks*. Chichester: John Wiley & Sons, 2004.
17. M. Shahidehpour, H. Yamin, and Z. Li, *Market Operations in Electric Power Systems: Forecasting, Scheduling, and Risk Management*. New York: John Wiley & Sons, 2002.
18. MATLAB (Release 2006b), The Math Works, Inc., 3 Apple Hill Drive, Natick, MA 01760-2098 [Online]. Available: http://www.mathworks.com
19. K.-H. Ng and G. B. Sheblé, Exploring risk management tools, *Proceedings of the IEEE/IAFE/INFORMS 2000 Conference on Computational Intelligence for Financial Engineering*, pp. 65–68, 2000.
20. L. El Ghaoui, M. Oks, and F. Oustry, Worst-case value-at-risk and robust portfolio optimization: A conic programming approach, *Operations Research*, 51(4), 543–556, 2003.
21. R. T. Rockafellar and S. Uryasev, Optimization of conditional value-at-risk, *Journal of Risk*, 2, 21–41, 2000.
22. R. T. Rockafellar and S. Uryasev, Conditional value-at-risk for general loss distributions, Research Report No. 2001-5, ISE Department, University of Florida, 2001.
23. R. T. Rockafellar and S. Uryasev, Optimization of conditional value-at-risk, Research Report No. 99-4, ISE Department, University of Florida, 1999.
24. S. Zhu and M. Fukushima, Worst-case conditional value-at-risk with application to robust portfolio management, *Operations Research*, 57(5), 1155–1168, 2009.
25. R. A. Jabr, Worst-case robust profit in generation self-scheduling, *IEEE Transactions on Power Systems*, 24(1), 492–493, 2009.
26. P. Artzner, F. Delbaen, J.-M. Eber, and D. Heath, Coherent measures of risk, *Mathematical Finance*, 9(3), 203–228, 1999.
27. F. Andersson, H. Mausser, D. Rosen, and S. Uryasev, Credit risk optimization with conditional value-at-risk criterion, *Mathematical Programming, Series B*, 89, 273–291, 2001.
28. R. A. Jabr, Generation self-scheduling with partial information on the probability distribution of prices, *IET Generation, Transmission & Distribution*, 4(2), 138–149, 2009.
29. D. G. Luenberger and Y. Ye, *Linear and Nonlinear Programming*, 3rd ed. New York: Springer, 2008.
30. J. F. Sturm, Using SeDuMi 1.02, a Matlab toolbox for optimization over symmetric cones, *Optimization Methods & Software*, 11–12, 625–653, 1999.

3

Short-Term Load Forecasting

Alexandre P.
Alves da Silva

Vitor H. Ferreira

3.1 Introduction

Operational decisions in power systems, such as unit commitment, economic dispatch, automatic generation control, security assessment, maintenance scheduling, and energy commercialization, depend on the future behavior of loads. Therefore, several short-term load forecasting (STLF) methods, for which the load is sampled on an hourly (or half-hourly) basis, or even daily basis (peak load), have been proposed during the last four decades. Such a long experience in dealing with the load forecasting problem has revealed some useful models such as the ones based on multilinear regression [1], Box–Jenkins method [2], artificial neural networks (ANNs) [3], fuzzy systems [4], and hybrid models [5].

After the restructuring of the electric power industry, one of the major difficulties in applying nonautomatic methods is scalability. Aggregated load forecasts, well performed by parametric models, are used to provide sufficient information for operational planning purposes. However, deregulated energy markets have presented new challenges to decision-making, which requires more information dependent on accurate bus load forecasting. Therefore, the corresponding development and maintenance efforts for dealing with hundreds of irregular bus load series, which need to be simultaneously forecast for security and economical analyses, are beyond practical consideration for tailormade parametric models. Autonomous load forecasters are needed to avoid expert intervention and to extend the application to the bus load level.

Furthermore, the relationship between electric load and its exogenous factors is complex and nonlinear, making it quite difficult to be modeled through conventional techniques such as linear time-series and regression analyses. Classical methods are bias-prone, that is, they are based on theoretical guesses about the underlying laws governing the system under study. On the other hand, after some years of practical experience, it has been recognized that ANNs can provide superior forecasting performance when dealing with nonlinear and multivariate problems involving large datasets, such as short-term load prediction. ANNs have more flexible functional forms in which there are few *a priori* assumptions about the relationships between input and output variables.

Although usually more robust than traditional load forecasting models, ANNs have overcome several problems in order to become commercially successful [6]. Since the first proposals of ANN-based load forecasters, five major drawbacks have been tackled: heavy training burden, lack of prediction interval estimation, inference opacity, input space representation, and model complexity control. Fast training algorithms have been developed since the early 1990s [7], which have allowed the tracking of load nonstationarities. On the other hand, some time has passed until the recognition of the practical importance of prediction interval estimation [8]. Qualitative interpretations of the ANN's forecasts have been proposed by Iizaka et al. [9]. It seems that improvement on forecasting accuracy provided by ANNs cannot come without degrading model transparency. Therefore, with ANNs it is hard to achieve a level of interpretability comparable to the one extractable from linear models.

The last two drawbacks are critical for STLF. The ANN input representation and complexity control shall not be treated separately. The extent of nonlinearity required from an ANN is strongly dependent on the selected input variables. One of the advantages of neural network models is the universal approximation capability, that is, unlimited precision for continuous mapping. However, this theoretical advantage can backfire if data overfitting is not avoided. The main objective of model complexity control is to match the data regularity with the model structure, maximizing the generalization capacity.

A popular procedure for ANN complexity control is based on cross-validation with training early stopping, that is, the iterative updating of the connection weights until the error for the validation subset stops decreasing. This procedure is very heuristic because it is not easy to detect the right iteration for interrupting the training process. Besides, although cross-validation has been successfully applied to the design of neural classifiers, serial interdependence information can be lost when it is used in time-series forecasting.

This chapter aims to explain all phases involved in the development of an ANN-based STLF system. Section 3.2 deals with data preprocessing. A certain regularity of the data is an important precondition for the successful application of ANNs. In order to tackle the problem of nonstationarity, wavelets have been utilized because they can produce a useful local representation of the signal in both time and frequency domains [10]. Moreover, the wavelet decomposition can be used to unfold inner load characteristics, which are helpful for a more precise forecasting.

Section 3.3 describes the first stage of an input variable selection process. Input space representation is probably the most important subtask in load forecasting. It has been shown that input variable selection based on linear auto- and cross-correlation analyses is not appropriate for nonlinear models such as ANNs [11]. ANN-oriented input selection schemes are explained in Sections 3.3 and 3.4. They are able to capture important information about the linear and nonlinear interdependencies in the associated multivariate data. Section 3.4 presents the second stage of the recommended input selection process. It determines, through Bayesian training, useful explanatory input variables from the preselected set [12]. This ANN training technique uses Bayesian inference to minimize the out-of-sample prediction error without cross-validation. The training method includes complexity control terms in its objective function, which allow autonomous modeling and adaptation.

In Section 3.5, the Bayesian training algorithm is extended to include structure identification in the automatic ANN design. Nevertheless, one should not produce a forecast of any kind without an idea of its reliability. Point predictions are meaningless when the time series is noisy. However, there are many difficulties in computing those indices for nonlinear models. Prediction intervals should be as narrow as possible, while encompassing a number of true values according to its reliability. Section 3.6 shows a technique for the estimation of prediction intervals for ANN-based short-term load forecasters. Finally, Section 3.7 concludes the chapter with suggestions for future work.

3.2 Data Preprocessing

STLF can be enhanced by data preprocessing. This section is divided into two parts. The first is related to standardization and differencing. The second describes multiresolution representation through wavelet transform.

3.2.1 Standardization and Differencing

The basic motivation for normalizing input and output variables is to make them comparable for the training process. Standardization, that is, to transform a random variable such that its mean and variance become zero and one, respectively, usually helps to improve the training efficiency and the ANN mapping interpretability as well. Although ANNs have no strict hypothesis regarding the time-series stationarity, differencing can also help during ANN learning.

First-order differencing can be used to compute the differences between adjacent values of a time series. A linear trend in a load series can be removed by first-order differencing. Seasonal differencing also helps to improve stationarity. With seasonal differencing, the original series is subtracted by lagged values that correspond to the seasonal component to be attenuated (e.g., subtraction by the load value at the same time in the previous day). A load series usually presents daily, weekly, and yearly cyclic components, which can be mitigated with seasonal differencing. However, experience shows that ANNs prefer to model the seasonalities by themselves. Therefore, seasonal differencing has not been used.

3.2.2 Wavelet Filtering

The objective of wavelet filtering is to identify different sources of useful information embedded in a load time series. Electric load series are formed by the aggregation of individual consumers of different natures. A good piece of the information provided by a load series is useful for forecasting. The rest is related to a random component that cannot be predicted. Therefore, there are two main reasons for filtering a load time series. First, important regular behavior of the load series can be emphasized. Second, a partition into different components of the load series can be produced, decreasing the learning effort.

The multiresolution decomposition, through discrete wavelet transform and its inverse, splits up the load series into one low-frequency and some high-frequency components (see Appendix A3.1). Using this new representation of the original load signal, two different alternatives have been developed. The first one consists of creating a multilayer perceptron (MLP) model for STLF whose inputs are based on information from the original load sequence and from the subseries (components) produced by multiresolution decomposition, that is, the approximation (A) and detail (D) levels, as in Figure 3.1. The second alternative predicts the load's future behavior by independently forecasting each subseries through MLPs. The final forecast is obtained by combining the predictions for each subseries (Figure 3.2).

Before extracting the smoothed version of the load series (approximation level) and the detail levels (higher frequency bands), two choices must be made: selection of the mother wavelet and definition of the number of levels for decomposition. There are

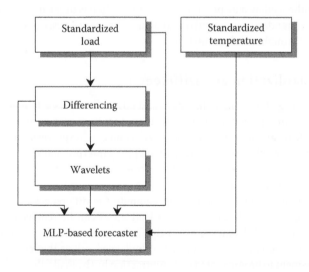

FIGURE 3.1   Combined inputs alternative.

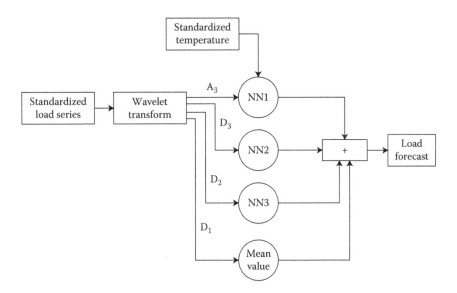

FIGURE 3.2 Combined models alternative.

many types of mother wavelets that can be used in practice. To choose the most suitable one, the attributes of the mother wavelet and the characteristics of the load series must be taken into account. Different wavelet families have been considered in the study by Reis and Alves da Silva [13]. This study shows that the Daubechies wavelet of order 2 is the most appropriate for treating load series.

It is also advisable to select a suitable number of decomposition levels based on the load series' dynamics. It has been concluded that the three-level decomposition is the most promising choice, because it has described the load dynamics in a more meaningful way than others. This conclusion is not based only on the approximation level, which is the most significant part of a load signal. Note that intraday seasonalities become evident from d2 and d3 in Figure 3.3. The three-level decomposition emphasizes the regular behavior of the load series. It reveals hidden patterns that are not clear in the original series. On the other hand, the highest frequency band concentrates the load's random component (d1).

In fact, the load approximation and details presented in Figure 3.3 have been produced by the so-called stationary wavelet transform, which is similar to the discrete wavelet transform, but preserves the original sampling rate for all subseries. This is achieved by never downsampling at each level of decomposition. The stationary wavelet transform is an inherently redundant scheme because a half-range of input frequencies needs only a half-sampling rate. However, it simplifies input selection, considering that each subseries contains the same number of points as the original load series. There are many useful software tools for multiresolution decomposition, as shown, for example, in the study by Misiti et al. [14].

There is an important issue for successful application of those procedures. Border distortions arise when one performs filtering of finite-length time series. Therefore,

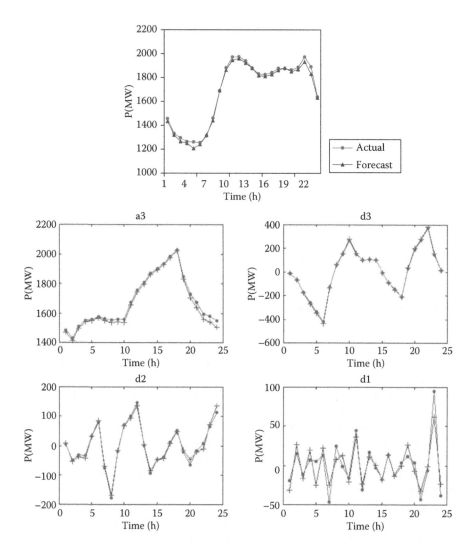

FIGURE 3.3 Forecasts for each load subseries (bottom graphics) and corresponding final predictions (top graphics).

multilevel load decomposition through wavelets corrupts the information at both sides of each subseries. Distortion on the left side (i.e., the *oldest* information is corrupted) degrades the forecast estimation. Corruption of the most recent load information on the right side affects both model estimation and forecasting. In order to deal with this problem, signal extension (known as padding) at the borders of the original load series is applied. The goal is to minimize the amount of distortion on both edges of the subseries.

Reis and Alves da Silva [13] propose a padding strategy that appends the previous load values at the beginning of the series to be used for model estimation and "naively" predicted values at the end of it. By using the proposed padding, one can expect no distortion at the beginning of the decomposed signals and a reduced amount of distortion at their ends. Another important question is related to the length of the attached information. Empirically, it has been found that the use of 72 padding values (at least) in each extremity of the load series is enough to reduce distortions.

3.3 Input Preselection

The first stage for input variable selection analyzes input relevance using statistical tests. The second stage evaluates the usefulness of the previously selected input set using the ANN training algorithm described in Section 3.4. Because of the feedforward structure and supervised learning of the neural networks of interest, input selection in this chapter is limited to determining significant delays in the time series and important dummy variables. The first stage of the input selection process (preselection) does not take into account the neural network structure and training algorithm. The method is based on a normalized mutual information measure.

Mutual information has been used for selecting inputs to neural networks [15]. The underlying motivation for its application is the capacity for detecting high-order statistical relationships among variables. In this chapter, instead of advocating the application of mutual information as in the study by Batitti [15], a normalized mutual information measure, called interdependence redundancy [16], estimated on subsets of interdependent outcomes, is used for preselecting input variables. Interdependence redundancy is capable of screening out statistically irrelevant information, making the detection of linear and nonlinear relationships more robust [17]. In previous works, interdependence redundancy has been used as the basis of statistical procedures for estimating missing values. Following the study by Alves da Silva et al. [18], this measure of information is recommended as a tool for preselecting inputs to ANNs. Interdependence redundancy provides two advantages for this particular purpose in comparison with the mutual information measure suggested by Batitti [15]. First, interdependence redundancy is bounded, with values between zero (independent) and one (strictly interdependent). Second, as interdependence redundancy is estimated considering different subspaces, it is a more reliable way of evaluating nonlinear interactions among variables.

The rest of this section describes the method for estimating the interdependence redundancy between pairs of random variables. Initially, the load series has its long-term trend removed by linearly detrending the entire extent of the series, to allow a better analysis of the relationship between endogenous and exogenous variables. Then, standardization has been applied to the detrended load series and to the exogenous variables.

Let $\underline{x}_1 = (x_1, x_2, \ldots, x_m), \underline{x}_2, \ldots, \underline{x}_M$ be an ensemble of M multiple observations of the initial set of input variables (e.g., load, temperature and price lags, temperature forecast, etc.) and corresponding output variables (i.e., load at the forecasting horizon) $X_1, X_2, \ldots, X_m$, where $x_1, x_2, \ldots, x_m$ represent the corresponding outcomes. Furthermore, $T_j = \{a_{jr} | r = 1, 2, \ldots, L_j\}$ represents the set of L_j discretized (with equal frequency intervals) outcomes of X_j ($j = 1, 2, \ldots, m$). Then, a multiple observation is described by simultaneous realizations x_j ($j = 1, 2, \ldots, m$), where x_j assumes a specific value from T_j.

Initially, the outcomes of the pairs of variables, that is, output against endogenous, output against exogenous, endogenous against endogenous, endogenous against exogenous, and exogenous against exogenous, are divided into two subsets. The first subset contains events that have interdependence for inference. The second subset contains events that cannot be used for inference. The following procedure is applied to separate the subsets.

Step 1: Estimate the expected frequency of all joint outcomes of X_k and X_j, (a_{ks}, a_{jr}), from the marginal frequency of each outcome in the ensemble, that is, obs(a_{ks}) and obs(a_{jr}), respectively. If independence of variables is assumed, then

$$\exp(a_{ks}, a_{jr}) = [\text{obs}(a_{ks})\text{obs}(a_{jr})]/M \tag{3.1}$$

Step 2: Obtain the observed frequencies of the joint outcomes (a_{ks}, a_{jr}). Calculate the degree to which an observed frequency deviates from the expected frequency under the independence hypothesis:

$$D_k^j = \sum_{r=1}^{L_j} [\text{obs}(a_{ks}, a_{jr}) - \exp(a_{ks}, a_{jr})]^2 / \exp(a_{ks}, a_{jr}) \tag{3.2}$$

Step 3: Test the statistical interdependency between a_{ks} and X_j, for $k \neq j$, at a presumed significance level. As D_k^j has an asymptotic χ^2 distribution with $L_j - 1$ degrees of freedom, statistical interdependency between a_{ks} and X_j can be determined by the following test:

$$h_k^j(a_{ks}, X_j) = \begin{cases} 1 & \text{if } D_k^j > \chi^2_{(\alpha, L_j - 1)} \\ 0, & \text{otherwise} \end{cases} \tag{3.3}$$

where $\chi^2_{(\alpha, L_j - 1)}$ is the tabulated χ^2 value with significance level α.

Step 4: Build, for all pairs of variables (X_k, X_j), two subsets of interdependent outcomes:

$$E_k^j = \left\{ a_{ks} \middle| h_k^j(a_{ks}, X_j) = 1 \right\} \tag{3.4}$$

and

$$E_j^k = \left\{ a_{jr} \middle| h_j^k(a_{jr}, X_k) = 1 \right\} \tag{3.5}$$

Next, the statistical interdependency between the restricted variables, X_j^k and X_k^j, with outcomes in the subspace spanned by $E_j^k \times E_k^j$, is estimated.

Step 5: Obtain the interdependence redundancy between X_j^k and X_k^j as follows:

$$R(X_j^k, X_k^j) = \frac{I(X_j^k, X_k^j)}{H(X_j^k, X_k^j)} \tag{3.6}$$

where $I(X_j^k, X_k^j)$ is the expected mutual information between X_j^k and X_k^j, and $H(X_j^k, X_k^j)$ is Shannon's entropy function, given by

$$I(X_j^k, X_k^j) = \sum_{E_j^k} \sum_{E_k^j} P(X_j^k, X_k^j) \log_2 [P(X_j^k, X_k^j)/(P(X_j^k) P(X_k^j))] \tag{3.7}$$

$$H(X_j^k, X_k^j) = -\sum_{E_j^k} \sum_{E_k^j} P(X_j^k, X_k^j) \log_2 P(X_j^k, X_k^j) \tag{3.8}$$

The statistical interdependencies between variables can be determined by a χ^2 test, with $(|E_j^k| - 1)(|E_k^j| - 1)$ degrees of freedom, on the scaled interdependence redundancy $2 \cdot N(X_j^k, X_k^j) \cdot H(X_j^k, X_k^j) \cdot R(X_j^k, X_k^j)$, where $|E_j^k|$ is the cardinality of E_j^k and $N(X_j^k, X_k^j)$ is the number of joint observations of the restricted variables (X_j^k, X_k^j). Interdependence redundancy is used not only for detecting a strong interaction between a possible input and the output variable. It is also used to eliminate redundant input variables. Experience has shown that $L_j = 20$ is a reasonable choice for discretizing the m standardized variables.

3.4 Refining Input Selection for ANNs

Neural network models commonly used in load forecasting are of the MLP type, with one hidden layer only. To introduce the adopted nomenclature, this section describes the general structure of an MLP with one hidden layer and one output neuron, under supervised learning.

Let $\underline{x} \in \mathfrak{R}^n$ be a vector representing input signals and $\underline{w} \in \mathfrak{R}^M$ the vector with the ANN connection weights, where $M = mn + 2m + 1$ and m is the number of neurons in the hidden layer. The biases of the sigmoidal activation functions of the hidden neurons are represented by b_k, $k = 1, 2, \ldots, m$, where b stands for the bias of the output neuron linear activation function. The final mapping is

$$y = f(\underline{x}, \underline{w}) = \sum_{k=1}^{m} (w_k c_k) + b \tag{3.9}$$

where $c_k = \varphi\left(\sum_{i=1}^{n} (w_{ik} x_i) + b_k\right)$.

Given a dataset U with N input/output pairs, $U = \{X, D\}$, for $X = (\underline{x}_1, \underline{x}_2, \ldots, \underline{x}_N)$ and $D = (d_1, d_2, \ldots, d_N)$, where $d_j \in \mathfrak{R}$ represents the desired outputs, the MLP training objective usually is the estimation of the weight vector $\underline{w}$ such that the empirical risk (training error) is minimized:

$$\min_{\underline{w}} \left\{ E_s(\underline{w}, U) = \frac{1}{2} \sum_{j=1}^{N} \left[d_j - f(\underline{x}_j, \underline{w}) \right]^2 \right\} \tag{3.10}$$

There are several algorithms for minimizing Equation 3.10. Independent of using the classical error back-propagation, or any other training method, the main drawback of this unconstrained training criterion is the absence of any concern regarding model complexity. A robust approach for controlling the MLP complexity is based on regularization theory, in which analytical methods adjust its extent of nonlinearity without necessarily changing the model structure.

3.4.1 Regularization Methods

A balance between training error and generalization capacity can be obtained through the minimization of the total risk:

$$\min_{\underline{w}} \left\{ R(\underline{w}) = E_s(\underline{w}, U) + \lambda E_c(\underline{w}) \right\} \tag{3.11}$$

In Equation 3.11, $E_s(\underline{w}, U)$ denotes the empirical risk, given by Equation 3.10, whereas $E_c(\underline{w})$ estimates the model complexity. The factor λ is known as the regularization parameter, which weighs the bias-variance trade-off, that is, training error versus model complexity. The regularization parameter λ can be estimated by the MLP training procedure described in the next section.

3.4.2 Bayesian MLP Training

One way to define the functional form of $\lambda E_c(\underline{w})$, in Equation 3.11, is through the application of Bayesian inference [19]. Using Bayes' rule, the conditional probability density function (PDF) of $\underline{w}$, given a dataset U, $p(\underline{w}|D, X)$, is estimated by

$$p(\underline{w}|D, X) = \frac{p(D|\underline{w}, X) p(\underline{w}|X)}{p(D|X)} \tag{3.12}$$

As X is conditioning all probabilities in Equation 3.12, it will be omitted from this point on. Therefore, in Equation 3.12, $p(D|\underline{w})$ is the likelihood of D given $\underline{w}$, $p(\underline{w})$ is $\underline{w}$'s *a priori* PDF, and $p(D) = \int p(D|\underline{w}) p(\underline{w}) \, d\underline{w}$ is enforcing $\int p(\underline{w}|D) \, d\underline{w} = 1$.

It is initially assumed that $\underline{w}$ presents a Gaussian distribution with zero mean and diagonal covariance matrix equal to $\alpha^{-1}\underline{\underline{I}}$, where $\underline{\underline{I}}$ is the $M \times M$ identity matrix:

$$p(\underline{w}) = \frac{1}{Z_{\underline{w}}(\alpha)} e^{-\left(\frac{\alpha}{2}\|\underline{w}\|^2\right)}, \quad \text{where } Z_{\underline{w}}(\alpha) = \left(\frac{2\pi}{\alpha}\right)^{\frac{M}{2}} \tag{3.13}$$

The desired outputs can be represented by $d_j = f(\underline{x}_j, \underline{w}) + \zeta_j$, where ζ is Gaussian white noise with zero mean and variance equal to β^{-1}. The regularization factors α and β (learning parameters, also called hyperparameters), in contrast to other regularization techniques, are estimated along with the model parameter $\underline{w}$. Considering the previous hypotheses and assuming that the dataset patterns are independent, then

$$p(D|\underline{w}) = \frac{e^{\left\{-\frac{\beta}{2}\sum_{j=1}^{N}\left[d_j - f(x_j,\underline{w})\right]^2\right\}}}{Z_Y(\beta)}, \quad \text{where } Z_Y(\beta) = \left(\frac{2\pi}{\beta}\right)^{\frac{N}{2}} \tag{3.14}$$

Consequently, based on Equation 3.12,

$$p(\underline{w}|D) = \frac{e^{[-S(\underline{w})]}}{\int e^{-S(\underline{w})}d\underline{w}} \tag{3.15}$$

where

$$S(\underline{w}) = \frac{\beta}{2}\sum_{j=1}^{N}\left[d_j - f(x_j,\underline{w})\right]^2 + \frac{\alpha}{2}\sum_{l=1}^{M}w_l^2 \tag{3.16}$$

Therefore, the maximization of the *a posteriori* distribution of $\underline{w}$, $p(\underline{w}|D)$, is equivalent to the minimization of $S(\underline{w})$. Dividing $S(\underline{w})$ by β and making $\lambda = \alpha/\beta$ in Equation 3.11, the equivalence between $S(\underline{w})$ and $R(\underline{w})$ can be verified if

$$E_c(\underline{w}) = \frac{1}{2}\|\underline{w}\|^2 \tag{3.17}$$

The regularization term in Equation 3.17, known as *weight decay*, favors neural models with small magnitudes for the connection weights. Small values for the connection weights tend to propagate the input signals through the almost linear segment of the sigmoidal activation functions. Note that the requirement of prior information in Bayesian training is the primary instrument for controlling the MLP complexity.

One of the advantages of Bayesian training of an MLP is the embedded iterative mechanism for estimating λ, that is, α and β, which avoids cross-validation. For multivariate problems such as load forecasting, the use of one single hyperparameter α for dealing with all connection weights is not recommended. Load- and weather-related input variables, such as temperature, require different priors. Even among the same type of variables, different levels of interdependence are involved [e.g., $P(k)$ against $P(k+1)$ and $P(k-23)$ against $P(k+1)$, for an hourly basis load].

In the following section, each group of connection weights directly related to an input variable receives a different α_i. The same idea is applied to the groups of weights associated with the biases (one α_i for the connections with the hidden neurons and another for the output neuron connection). One last α_i is associated with all connection weights between the hidden and output layers. Therefore, for n-dimensional input vectors $\underline{x}$, the total number of α_is is $n + 3$.

3.4.3 Input Selection by Bayesian Training

For a given model structure, the magnitudes of the α_is can be compared to determine the relevance of the corresponding input variables. As $p(\underline{w}_i)$ is supposed to be normally distributed with zero mean and $\alpha_i^{-1}\underline{\underline{I}}$ covariance, then, the largest α_is lead to the smallest $\underline{w}_i$s. For estimating the *a posteriori* PDF of $\underline{w}$, Bayesian training combines the *a priori* PDF with the information provided by the training set (Equation 3.12). If an α_i is large, the prior information about $\underline{w}_i$ is almost certain, and the effect of the training data on the estimation of $\underline{w}_i$ is negligible. Another way to see the influence of α_i on $\underline{w}_i$ is through Equation 3.16.

The impact on the output caused by input variables with very small $\underline{w}_i$s, that is, very large α_is, is not significant. However, a reference level for defining a very large α_i has to be established. For STLF, two different references of irrelevance are needed: one for continuous variables, such as loads and temperatures, and another for dummy variables, such as hours of the day and days of the week. Uniformly distributed input variables can be used to define the references of irrelevance. For continuous input variables, a uniform random variable with lower and upper limits equal to $-\sqrt{3}$ and $\sqrt{3}$, respectively, is used as reference of irrelevance, as continuous variables have been standardized (zero mean and unit variance). For dummy variables, the reference is a binary random variable with uniform distribution. These two reference variables are added to the preselected set of inputs.

After training the model with the preselected set of input variables, continuous and dummy variables are separately ranked. For each rank, the variables with corresponding α_is larger than α_{ref} (irrelevance level) are disregarded. After input selection, the MLP is retrained with the selected variables.

3.4.4 Illustrative Results

For a dataset, with hourly load and temperature values, containing 6 years of information, the task is to forecast up to 168 h (steps) ahead. With training data from the month to be forecast and from two months earlier, along with the data corresponding to the same "window" in the previous year, seven models are estimated, one for each day of the week.

As the initial set of inputs, the following variables are tested: 24 dummy variables codifying the hour of the day; lags $S(k-1)$, $S(k-2)$, ..., $S(k-6)$, $S(k-24)$, $S(k-25)$, ..., $S(k-29)$, $S(k-168)$, $S(k-169)$, ..., $S(k-173)$ for load, temperature, and temperature square series; the temperature forecast for hour k and its square value, that is, $T(k)$ and $T^2(k)$, respectively; the daily maximum temperature forecast and its square value, $T_{max}(d)$ and $T_{max}^2(d)$; and the daily maximum temperature for the previous day and its square value, $T_{max}(d-1)$ and $T_{max}^2(d-1)$. Therefore, a total of 84 initial inputs (including dummies) have been chosen. The output is the predicted hourly load $L(k)$. As weather services can provide quite precise forecasts for the horizon of interest, true temperatures have been used as "perfect" predictions. The forecasts are provided by recursion, that is, load forecast feed inputs.

Figure 3.4 presents an illustrative result for this database. It shows predictions based on inputs selected by interdependence redundancy analysis and predictions provided by the set of inputs obtained from Bayesian selection. To establish a fair comparison, in both cases the initial set of inputs is the same, that is, the 84 variables previously

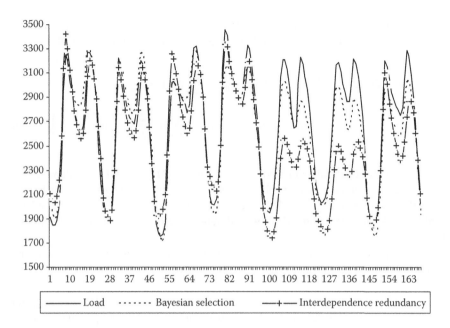

FIGURE 3.4 One-week-ahead hourly forecasts.

described. The improvement produced by the Bayesian selection is clear. Although interdependence redundancy analysis has the advantage of being model-independent, selection criteria based on relevance of input variables do not guarantee good-quality forecasts, as particular predisposition of ANN models and corresponding training algorithms are not taken into account. In practice, as already mentioned, the two methods should be used in a sequential cooperative way, that is, interdependence redundancy preselects and the Bayesian training refines the input set.

3.5 MLP Structure Identification in Bayesian Training

Bayesian inference can also be used to determine the best MLP structure among a predefined set of possibilities; for example, $H = \{H_1, H_2, \ldots, H_K\}$, for which the corresponding inputs have been previously selected:

$$P\left(H_h|D\right) = \frac{p\left(D|H_h\right)P\left(H_h\right)}{p\left(D\right)} \qquad (3.18)$$

In Equation 3.18, $P(H_h)$ represents the *a priori* probability of model H_h and $p(D|H_h)$ is given by

$$p\left(D|H_h\right) = \iint p\left(D|\alpha,\beta,H_h\right)p\left(\alpha,\beta|H_h\right)d\alpha\,d\beta \qquad (3.19)$$

Using Gaussian approximation around the estimated hyperparameters (from training), analytic integration of Equation 3.19 is possible, leading to Equation 3.20:

$$\ln p\left(D|H_h\right) = -S(\underline{w}) - \frac{1}{2}\ln\left|\nabla\nabla S(\underline{w})\right| + \frac{1}{2}\sum_{i=1}^{n+3} M_i \ln\alpha_i$$

$$+ \frac{N}{2}\ln\beta + \ln(m!) + 2\ln m + \frac{1}{2}\sum_{i=1}^{n+3}\ln\left(\frac{2}{\gamma_i}\right) + \frac{1}{2}\ln\left(\frac{2}{N-\gamma}\right) \quad (3.20)$$

where m denotes the number of hidden neurons in the MLP model H_h. As all models, *a priori*, are assumed equally probable, H_h is selected by maximizing $P(D|H_h)$, which is equivalent to maximizing $\ln p(D|H_h)$. Consequently, Equation 3.20 can be used for ranking and selecting among MLPs with different numbers of neurons in the hidden layer.

The complete Bayesian training algorithm for selecting inputs and identifying the MLP structure can be summarized by the following steps:

1. Set the minimum and maximum number of neurons in the hidden layer N_{min} and N_{max}, respectively. Experience has shown that $N_{min} = 1$ and $N_{max} = 10$ can model the dynamics of any load series.
2. Make the number of neurons in the hidden layer $m = N_{min}$.
3. Add the reference variables to the preselected set of inputs. Therefore, if dummy variables are used, the extended set will contain $n = n + 2$ inputs. Otherwise, if only continuous inputs are allowed, $n = n + 1$.
4. Set $l = 0$ and initialize $\underline{w}(l) = [\underline{w}_1(l), ..., \underline{w}_{n+3}(l)]^t$, $\underline{\alpha}(l) = [\alpha_1(l), ..., \alpha_{n+3}(l)]^t$ and $\beta(l)$.
5. Minimize $S(\underline{w})$ on $\underline{w}(l)$ to obtain $\underline{w}(l + 1)$.
6. Calculate $\alpha_i(l + 1)$, $\beta(l + 1)$ and $\gamma_i(l + 1)$ using the following equations:

$$\nabla\nabla S(\underline{w})\Big|_{\underline{w}=\underline{w}(l+1)} = \beta(l)\nabla\nabla E_s(\underline{w},U)\Big|_{\underline{w}=\underline{w}(l+1)} + \alpha(l)\underline{\underline{I}}$$

$$\underline{\underline{B}}_i(l + 1) = \left[\nabla\nabla S(\underline{w})\Big|_{\underline{w}=\underline{w}(l+1)}\right]^{-1}\underline{\underline{I}}_i$$

$$\gamma_i(l + 1) = M_i - \mathrm{trace}\left\{\underline{\underline{B}}_i(l + 1)\right\}$$

$$\alpha_i(l + 1) = \frac{\gamma_i(l + 1)}{\left\|\underline{w}_i(l + 1)\right\|^2} \quad (3.21)$$

$$\beta(l + 1) = \frac{N - \sum_{j=1}^{n+3}\gamma_i(l + 1)}{\sum_{j=1}^{N}\left[d_j - f\left(\underline{x}_j,\underline{w}(l + 1)\right)\right]^2}$$

7. Make $l = l + 1$ and return to step 5 until convergence is achieved.
8. Put into two different vectors the hyperparameters α_i associated with continuous and discrete inputs and sort them in descending order.

9. For each sorted vector, select the inputs situated above the corresponding reference, that is, $\alpha_i < \alpha_{ref}$, with α_{ref} been the hyperparameter associated with the irrelevant added signal.
10. Repeat steps 4–7 using only the inputs selected in step 9, with n equal to the number of variables selected, to obtain the trained model H_m.
11. Evaluate the log evidence of the hypothesis H_m using Equation 3.20.
12. If $m = N_{max}$, go to step 13. Else, $m = m + 1$ and return to step 3.
13. Select hypothesis H_k with the largest log evidence and make predictions.

In Equation 3.21, $\underline{I}_i$ is an $M \times M$ diagonal matrix with ones at the positions corresponding to the ith group of weights and with zeros otherwise. M_i is the number of connection weights in each group. Details on how to calculate the Hessian $\nabla\nabla E_s(\underline{w}, U)$ can be found in the study by Bishop [20]. In fact, a straightforward adaptation of backpropagation can be used to evaluate the second-order partial derivatives of the error with respect to the connection weights. Nabney [21] provides useful computational routines to implement the complete Bayesian training algorithm.

3.6 Estimation of Prediction Intervals

Although prediction intervals can be derived from Bayesian training, experience shows that the Gaussian distribution hypothesis is not reliable for this particular purpose. Alves da Silva and Moulin [8] have presented a prediction interval estimation method that is not dependent on the data distribution. In fact, it is not necessary to estimate density functions for computing prediction intervals. Cumulative distribution functions provide enough information for this purpose. The recommended technique is based on resampling.

3.6.1 Resampling Method

The sampling of prediction errors for each forecasting lead-time can be conducted in the following way. Consider that the resampling set is representative of the loads to be found in the future. In addition, assume that the error samples are independent with the same, although unknown, probability distributions.

Let us suppose that Figure 3.5 represents the available (known) test data. A recursive forecasting process (i.e., forecasts feeding the ANN), using three lagged inputs for predicting one to four steps ahead, is considered. The available load values for instants 1, 2, and 3 are used to predict the load for instant 4. As the true load value for instant 4 is known, the prediction error for this one-step-ahead forecast can be computed. Afterward, using the known values for instants 2 and 3, and the previous prediction for instant 4, a two-steps-ahead forecast is achieved, allowing the calculation of the corresponding prediction error. The known value for instant 3 and the forecasts for instants 4 and 5 are used to predict the load for instant 6, and so on. One prediction error measurement for each lead-time has been gathered once the maximum desired prediction horizon, instant 7, is reached.

The same procedure is repeated for collecting one more sample for each lead-time using the known values for instants 2–8 (upper dotted line). This process is repeated

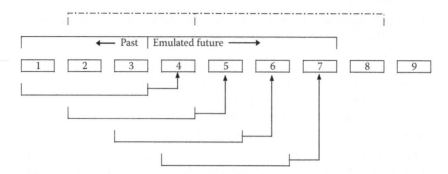

FIGURE 3.5 Example of the resampling process for prediction interval estimation.

until, for a certain window, the maximum desired prediction horizon reaches the end of the available load series.

Afterward, by sorting the n errors in ascending order (considering the signs) and representing them by $z_{(1)}, z_{(2)}, \ldots, z_{(n)}$, the cumulative sample distribution function of the prediction errors can be estimated as follows:

$$
S_n(z) = \begin{cases} 0, & z < z_{(1)} \\ r/n, & z_{(r)} \leq z < z_{(r+1)} \\ 1, & z_{(n)} \leq z \end{cases} \tag{3.22}
$$

where $S_n(z)$ is the fraction of the collection of errors less than or equal to z. When n is large enough, $S_n(z)$ is a good approximation of $F(z)$, the true cumulative probability distribution. Therefore, prediction intervals can be estimated by keeping the intermediate $z_{(r)}$s and discarding the extreme ones, according to the desired confidence degree. The intervals are computed in order to be symmetrical in probability (not symmetric in z). The number of cases to discard in each tail of the prediction error distribution is np, where p is the probability in each tail. As np is generally a fractional number, it is conservatively truncated, and $(np - 1)$ is taken as the number of cases to discard in each tail.

Denoting Z_p such that $F(Z_p)$ is equal to p, then there is a probability p that an error is less than or equal to Z_p. Therefore, Z_p is the lower confidence limit for future forecast errors. Consequently, Z_{1-p} is the upper limit and there is a $(1 - 2p)$ confidence interval for future errors. The value $nS_n(Z_p)$ represents the estimate of how many elements in the collection of errors are less than or equal to Z_p. As the prediction errors are assumed to be independent of each other, then $m = nS_n(Z_p)$ follows a binomial distribution

$$
B(m, n, p) = \frac{n!}{m!(n-m)!} p^m (1-p)^{n-m} \tag{3.23}
$$

independent of the distribution F.

In Equation 3.23, $B(m,n,p)$ represents the probability that exactly m, among n randomly sampled cases, are less than or equal to Z_p. In fact, if $B(m,n,p)$ is computed for $m = 0, 1, \ldots, n$, it can be shown that the largest probability is obtained when $m = np$.

3.6.2 Illustrative Results

Figures 3.6 and 3.7 show a week-ahead daily load peak forecasts for two 138 kV buses from the Brazilian system. Two years of data have been provided. To avoid recursion because of the jumpy dynamics of the peak load series, seven forecasting models have been estimated, one for each step ahead, using all data before the first day to be predicted. For the jth model, the initial inputs are related to the 7 most recent daily peak load values, plus $j + 7$ lagged temperature variables, and 19 dummy variables, 7 for the days of the week and 12 for the months. Therefore, a total of $33 + j$ initial inputs have been considered for each model. The lags for the load and temperature variables are $L(d - j)$, $L(d - j - 1)$, ..., $L(d - (6 + j))$ and $T(d)$, $T(d - 1)$, ..., $T(d - (6 + j))$, respectively, where d denotes the day to be predicted (i.e., $d = 1, 2, ..., 7$). As before, true temperatures for the forecasting horizon are used as "predictions." Each model output is the daily load peak $L(d)$ of interest. The available temperature-related variables have been disregarded because of lack of significant interdependence.

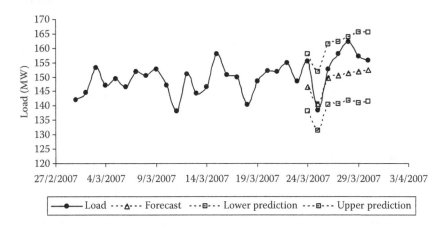

FIGURE 3.6 Bus 1 daily load peak forecasting, seven days ahead, with prediction intervals.

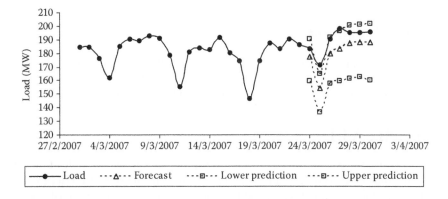

FIGURE 3.7 Bus 2 daily load peak forecasting, seven days ahead, with prediction intervals.

Results include prediction intervals with 90% of confidence degree. The removal of distributional assumptions usually generates wider but more reliable prediction intervals. For bus 1, the following forecasting mean absolute percentage errors (MAPEs) have been obtained: 3.4, 3.6, 3.6, 3.7, 3.8, 3.8, and 3.8 for one, two, . . ., and seven days ahead, respectively. For bus 2, the MAPEs are the following: 4.0, 4.3, 4.7, 4.9, 5.2, 5.4, and 5.6 for one, two, . . ., and seven days ahead, respectively.

3.7 Conclusions

This chapter has presented efficient methodologies for implementing autonomous neural-network-based short-term load forecasters. Autonomous means that both input selection and model structure identification are performed in an adaptive and automatic mode. The proposed methodologies provide reliable forecasts with very little intervention from the user. They seem to be the answer for dealing with the large-scale bus load forecasting problem, in which the particular dynamics of each load series does not allow manually tuned solutions.

The input representation problem when applying neural network models to forecast electric loads has been overlooked for a long time. This chapter has explained recent developments for selecting inputs to MLP-based short-term load forecasters. An appropriate measure of information is presented for input preselection. Extensions of Bayesian methods are proposed for completing the ANN design. Motivations for choosing these techniques include data-driven nature, automatic application, and suitability for MLP models.

According to the prescribed approach, the MLP with the largest evidence is the selected one. A possibility for making the forecasts even more robust is to combine a certain number of MLPs, taking into account the corresponding evidences. Support vector regression has shown great potential for STLF [22]. However, one drawback of such a method is the difficulty in estimating the learning parameters [12]. The application of Bayesian inference to overcome this problem looks very promising. Based on the advantages of the framework presented in this chapter, it seems worthwhile to pursue the recommended direction for the next generation of STLF tools.

Acknowledgments

This work was supported by the Brazilian Research Council (CNPq) and by the State of Rio de Janeiro Research Foundation (FAPERJ). The authors would like to thank Mr Sílvio Michel de Rocco, from COPEL, for making the bus load data available.

Appendix A3.1

Wavelet analysis uses a prototype function called mother wavelet ($g(t)$). This function has null mean and sharply drops in an oscillatory way, as in Figure A3.1.1. Data are represented by superposition of scaled and translated versions of the prespecified mother wavelet. The "continuous wavelet transform" (CWT) of a given signal $x(t)$, with respect to $g(t)$, is defined in Equation A3.1.1, where a and b are the scale and translation factors, respectively.

$$\text{CWT}(a,b) = \frac{1}{\sqrt{a}} \int_{-\infty}^{+\infty} x(t)g\left(\frac{t-b}{a}\right)dt \qquad (A3.1.1)$$

A CWT(a,b) coefficient, at a particular scale and translation, represents how well the original signal $x(t)$ and the scaled/translated mother wavelet match. Thus, the set of all wavelet coefficients CWT(a,b), associated with a particular signal $x(t)$, is the wavelet representation of the signal with respect to the mother wavelet $g(t)$. As the CWT is achieved by continuously scaling and translating the mother wavelet, substantial redundant information is generated. Therefore, instead of doing that, the mother wavelet can be scaled and translated using certain scales and positions based on powers of two. This scheme is more efficient and just as accurate as the CWT. It is known as the "discrete wavelet transform" (DWT), defined as

$$\text{DWT}(m,k) = \frac{1}{\sqrt{a_0^m}} \sum_n x(n)g\left(\frac{k - nb_0 a_0^m}{a_0^m}\right) \qquad (A3.1.2)$$

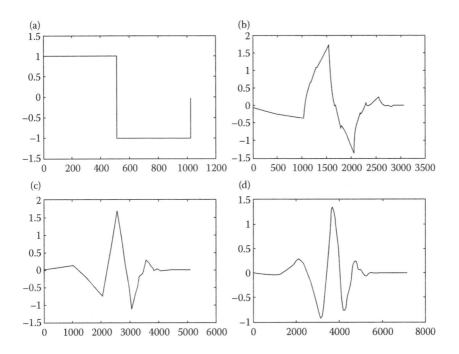

FIGURE A3.1.1 Examples of mother wavelets: (a) Haar, (b) Daubechies of order 2, (c) Daubechies of order 3, and (d) Daubechies of order 4.

The scaling and translation parameters a and b, in Equation A3.1.1, are functions of the integer variable m ($a = a_0^m$ and $b = nb_0 a_0^m$). In Equation A3.1.2, k is an integer variable that refers to a particular point of the input signal and n is the discrete time index.

A fast DWT procedure called pyramidal algorithm uses filters to obtain "approximations" and "details" from a given signal. An approximation is a low-frequency representation of the original signal, whereas a detail is the difference between two successive approximations. An approximation holds the general trend of the original signal, whereas a detail depicts high-frequency components of it. The pyramidal algorithm presents two stages: decomposition (analysis) and reconstruction (synthesis). The former calculates the "fast wavelet transform" (FWT), whereas the latter computes the "inverse fast wavelet transform" (IFWT). Multiresolution can be obtained by using a filter bank composed of L, H, L', and H', as shown in Figure A3.1.2. The low- and high-pass decomposition filters (L and H) together with their corresponding reconstruction filters (L' and H') are based on mother wavelets.

Starting from signal S (i.e., $x(n)$), two sets of coefficients are produced by the FWT: approximation coefficients cA_1 and detail coefficients cD_1. This decomposition is obtained by convolving S with the low-pass filter L for approximation, and with the high-pass filter H for detail, followed by downsampling, that is, by throwing away every other coefficient. Conversely, starting from cA_1 and cD_1, the IFWT reconstructs S by

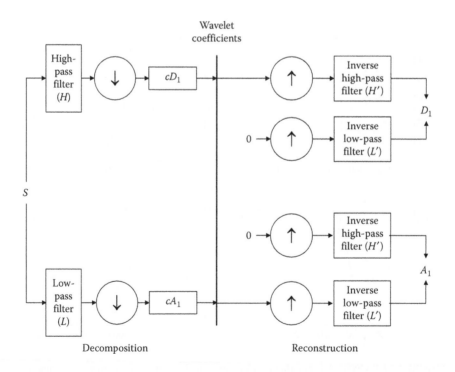

FIGURE A3.1.2   Single-resolution analysis via Mallat's algorithm ($S = A_1 + D_1$).

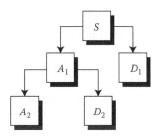

FIGURE A3.1.3 Multiple-level decomposition scheme ($S = A_2 + D_2 + D_1$).

inverting the decomposition stage. Inversion is achieved by inserting zeros between the wavelet coefficients (upsampling) and by convolving the resulting signal with the reconstruction filters L' and H' (Figure A3.1.2). Note that a multilevel decomposition process can be achieved according to Figure A3.1.3. The successive approximations are decomposed so that S is broken down into lower-resolution components.

References

1. A.D. Papalexopoulos and T.C. Hesterberg, A regression-based approach to short-term system load forecasting, *IEEE Transactions on Power Systems*, 5(4), 1535–1547, 1990.
2. A.S. Debs, Short-term load forecasting, Chapter 9, in *Modern Power System Control and Operation*. Boston, MA: Kluwer Academic Publishers, pp. 335–358, 1988.
3. D.C. Park, M.A. El-Sharkawi, and R.J. Marks, II, An adaptively trained neural network, *IEEE Transactions on Neural Networks*, 2(3), 334–345, 1991.
4. H. Mori and H. Kobayashi, Optimal fuzzy inference for short-term load forecasting, *Proceedings of the Power Industry Computer Application Conference*, Salt Lake City, USA, pp. 312–318, 1995.
5. A.G. Bakirtzis, J.B. Theocharis, S.J. Kiartzis, and K.J. Satsios, Short term load forecasting using fuzzy neural networks, *IEEE Transactions on Power Systems*, 10(3), 1518–1524, 1995.
6. A. Khotanzad, R. Afkhami-Rohani, and D. Maratukulam, ANNSTLF—artificial neural network short-term load forecaster—generation three, *IEEE Transactions on Power Systems*, 13(4), 1413–1422, 1998.
7. R.P. Brent, Fast training algorithms for multi-layer neural nets, *IEEE Transactions on Neural Networks*, 2(3), 346–354, 1991.
8. A.P. Alves da Silva and L.S. Moulin, Confidence intervals for neural network based short-term load forecasting, *IEEE Transactions on Power Systems*, 15(4), 1191–1196, 2000.
9. T. Iizaka, T. Matsui, and Y. Fukuyama, A novel daily peak load forecasting method using analyzable structured neural network, *Proceedings of the IEEE PES Transmission and Distribution Conference and Exhibition*, Yokohama, Japan, Vol. 1, pp. 394–399, 2002.

10. S. Mallat, A theory for multiresolution signal decomposition—the wavelet representation, *IEEE Transactions on Pattern Analysis and Machine Intelligence*, 11(7), 674–693, 1989.
11. I. Drezga and S. Rahman, Input variable selection for ANN-based short-term load forecasting, *IEEE Transactions on Power Systems*, 13(4), 1238–1244, 1998.
12. V.H. Ferreira and A.P. Alves da Silva, Toward estimating autonomous neural network-based electric load forecasters, *IEEE Transactions on Power Systems*, 22(4), 1554–1562, 2007.
13. A.J.R. Reis and A.P. Alves da Silva, Feature extraction via multi-resolution analysis for short-term load forecasting, *IEEE Transactions on Power Systems*, 20(1), 189–198, 2005.
14. M. Misiti, Y. Misiti, G. Oppenheim, and J.-M. Poggi, *Wavelet Toolbox Manual— User's Guide*, The Math Works Inc., USA, 1996.
15. R. Batitti, Using mutual information for selecting features in supervised neural net learning, *IEEE Transactions on Neural Networks*, 5(4), 537–550, 1994.
16. A.K.C. Wong and D.K.Y. Chiu, An event-covering method for effective probabilistic inference, *Pattern Recognition*, 20(2), 245–255, 1987.
17. A.P. Alves da Silva, V.H. Quintana, and G.K.H. Pang, A probabilistic associative memory and its application to signal processing in electrical power systems, *Engineering Applications of Artificial Intelligence*, 5(4), 309–318, 1992.
18. A.P. Alves da Silva, V.H. Ferreira, and R.M.G. Velasquez, Input space to neural network based load forecasters, *International Journal of Forecasting*, 24(4), 616–629, 2008.
19. D.J.C. Mackay, Bayesian methods for adaptive models, PhD dissertation, California Institute of Technology, Pasadena, USA, 1992.
20. C.M. Bishop, *Neural Networks for Pattern Recognition*. Oxford, UK: Oxford University Press, 1995.
21. I.T. Nabney, *NETLAB: Algorithms for Pattern Recognition*. London/Berlin/ Heidelberg: Springer-Verlag, 2002.
22. B.-J. Chen, M.-W. Chang, and C.-J. Lin, Load forecasting using support vector machines: A study on EUNITE competition 2001, *IEEE Transactions on Power Systems*, 19(4), 1821–1830, 2004.

4

Short-Term Electricity Price Forecasting

Nima Amjady

4.1 Introduction

Electricity price is an important signal for all participants of the electricity market and the motive behind most of their activities. Price forecast plays a major role in today's power markets and is key input data for market participants. Companies that trade in electricity markets make extensive use of price forecast techniques either to bid or to hedge against volatility. However, despite the importance of electricity price prediction, it is a complex signal for forecasting. Electrical energy cannot be considerably stored and power system frequency stability requires constant balance between generation and load. On short timescales, most users of electricity are unaware of or indifferent to its price [1]. Moreover, on a short timescale, transmission bottlenecks may prevent free exchange among different regions. These facts lead to extreme price volatility or even price spikes in the electricity market; for example, the price spikes of the PJM (Pennsylvania–New Jersey–Maryland) and California markets in 1999 and 2000, respectively [1,2]. Besides, volatility in fuel price, load uncertainty, fluctuations in hydro-electricity production, generation uncertainty (outages), and behavior of market partici-pants also contribute to electricity price uncertainty [3].

The electricity price signal features extreme jumps of magnitudes rarely seen in financial markets that also occur at greater frequency [4]. Amjady and Hemmati [5] have discussed how the uncertainty of hourly loads and some other stochastic signals, such as equipment outages and fuel prices, are combined resulting in a higher level of uncertainty in the electricity price. For instance, the electricity price data of the PJM market, a well-established electricity market in the United States, are shown in Figure 4.1. High variability and volatility of the electricity price time series can be seen from this figure. Volatility of a signal is the measure of its change over a given period of time [3]. Two price spike thresholds of 150 and 200 $/MWh are also indicated by dash-dot and dashed lines in Figure 4.1, respectively [4]. It is observed that many price spikes have occurred with both the price spike thresholds, further illustrating the volatile behavior of the electricity price time series.

Additionally, electricity price is a nonlinear time variant mapping function of its input features. It nonlinearly changes with respect to variations in the inputs. For instance, load demand is an important driver for electricity price. However, load variations in low- and high-load levels have different impacts on the electricity price. Moreover, its time-variant nature is related, for instance, to discrete changes in participants' strategies (e.g., agents decide to switch from a conservative behavior to a more aggressive or risky one) or to

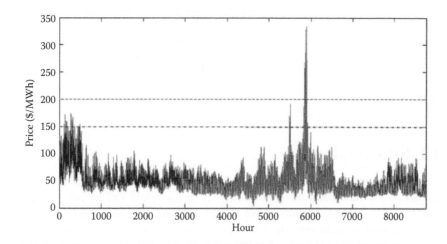

FIGURE 4.1   Price data of day-ahead electricity market of PJM in 2006.

changes in market regulations. A discussion about the other characteristics of electricity price time series such as multiple seasonality (e.g., daily and weekly periodicities), high-frequency changes, and high percentages of unusual prices (outliers) can be found in Reference [5].

4.2 Overview of Electricity Price Forecast Methods

Importance and complexity of electricity price forecast motivate many research works in the area. For short-term price forecast, considered in this chapter, forecast step is usually from a fraction of an hour (e.g., 5, 15, and 30 min) to an hour. Its forecast horizon can be from 1 h ahead to 1 week ahead. However, the most common forecast horizon for short-term price forecast is the next day, used in day-ahead electricity markets.

Due to the diverse nature of short-term electricity price forecast methods, a classification of these techniques can give a better insight about them. However, one point should be mentioned here before proceeding to the classification of the price forecast methods. An essential characteristic of the electricity markets is the pricing mechanism, which can be uniform or pay-as-bid pricing. Under the uniform pricing structure, the marginal bid block sets the market clearing price (MCP). In the presence of congestion in the power system, locational marginal price (LMP), which is the marginal cost of each bus [3], should be considered instead of MCP. However, in the pay-as-bid (discriminatory) pricing structure, every winning block gets its bid price as its income. The pricing mechanism can affect the competition, efficiency, consumer surplus, and total revenue of the players in the electricity markets. More details about this matter can be found in Reference [6]. This chapter focuses on electricity price forecast in the uniform pricing structure, which is the most commonly accepted structure of electricity markets around the world. A discussion about price forecast in the pay-as-bid (discriminatory) pricing structure can be found in Reference [7].

4.2.1 Classification of Price Forecast Methods

A lot of forecast methods have been proposed for prediction of MCP or LMP of electricity markets in the literature. Some of these methods are basically the short-term load forecasting (STLF) methods. However, electricity prices are usually more volatile than hourly loads and so short-term price forecasting is more complex than STLF.

In general, electricity price forecast methods can be divided into two main categories. Methods of the first category try to directly predict electricity price by analyzing the electricity market dynamics and effective parameters on the market price, such as production costs and strategic behavior of market participants. An important group of these methods is based on the game and auction theories. Another group consists of fundamental or structural models, based on traditional cost models, which have been developed for centralized systems and adapted to liberalized markets [8]. Methods of the second category try to forecast MCP without analyzing in detail the underlying physical processes. These methods, based on the black box models, analyze price evolution by means of statistical data. Methods of the second category, such as those based on the time-series techniques and neural networks, are more commonly used for electricity price forecast than those of the first category, due to greater flexibility, less input data required, and greater adaptability to market participants' conditions.

A similar and more detailed classification of electricity price forecast methods has been presented by Weron [9], wherein the methods are divided into six classes including production cost (or cost-based) models, equilibrium (or game theoretic) approaches, fundamental (or structural) methods, quantitative (or stochastic, econometric, reduced form) models, statistical approaches, and artificial intelligence-based techniques. Among these methods, artificial intelligence-based techniques have received more attention in recent years because of their high ability to tackle nonlinear complex input/output mapping functions with limited available data. For instance, Guo and Luh [10] have discussed how neural networks are universal approximators and can approximate any continuous function.

A complete review of all electricity price forecast methods is beyond the scope of this chapter. In the next sections, artificial intelligence-based methods will be analyzed and evaluated.

4.2.2 Structure and Components of Artificial Intelligence-Based Price Forecast Methods

Structure of an artificial intelligence-based electricity price forecast strategy can be depicted as shown in Figure 4.2. At first, a data model is constructed based on the available

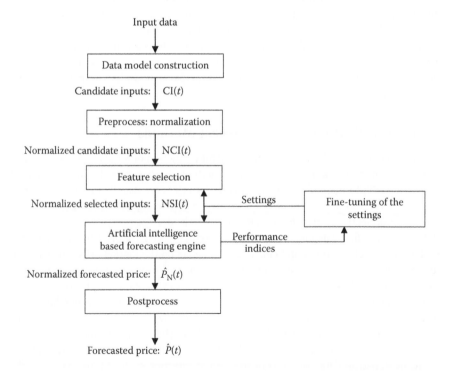

FIGURE 4.2 Structure of an artificial intelligence-based electricity price forecast strategy.

input data and engineering judgment. The candidate inputs of the data model, denoted by CI(t) in the figure, have different ranges and should be normalized. Then, the normalized candidate inputs NCI(t) are processed by a feature selection component to select the most informative features among them. The artificial intelligence-based forecasting engine (e.g., a neural network, combination of neural networks, or combination of neural networks with other artificial intelligence systems) is fed by the normalized selected inputs NSI(t).

Both feature selection and the artificial intelligence-based forecasting engine usually have some settings, such as the number of neurons in the hidden layer(s) of the multilayer perceptron (MLP) neural network. Performance of these components is dependent on the fine-tuning of these settings and different methods, such as cross-validation techniques and search procedures, have been proposed for this purpose. The output of the forecasting engine, denoted by $\hat{P}_N(t)$ in Figure 4.2, will be in the normalized form. The normalized forecast price $\hat{P}_N(t)$ should be returned to the actual range, shown by $\hat{P}(t)$ in the figure, by the inverse transform in the postprocess. Throughout this chapter, forecast values are shown by a circumflex accent.

The above components will be described in more detail in the next sections and different alternatives for each component will be presented.

4.3 Input Data Preparation for Electricity Price Forecast

The first step of the setup phase for an artificial intelligence-based electricity price forecast strategy is the input data preparation. For this purpose, the data model of the forecast process should be constructed and appropriately refined. The refined data model determines the input features of the artificial intelligence-based forecasting engine. In other words, the forecasting engine should learn and extract the input/output mapping function of the forecast process in the form of NSI(t)$\rightarrow \hat{P}_N(t)$. For the learning phase, the forecasting engine requires training samples that are constructed based on the historical data. Input and output features of these training samples are NSI(t) and $\hat{P}_N(t)$ in the previous time intervals, respectively.

4.3.1 Formation of Candidate Input Set

Electricity price is a nonlinear function of many candidate inputs including its past values as well as past and forecast values of the exogenous variables (such as load demand and available generation). These candidate inputs constitute auto-regression and cross-regression of the electricity price forecast model, respectively. Moreover, in addition to the time-domain candidate inputs, the frequency domain may also contain some useful information for electricity price forecast. Thus, some candidate inputs from the frequency domain may also be considered in the data model of the forecast process. Consequently, two kinds of data model can be considered for electricity price prediction: a nonhybrid data model only including time domain candidate inputs and a hybrid data model containing both time- and frequency-domain candidate features.

4.3.1.1 Nonhybrid Data Model

A large set of candidate inputs including lagged features as much as possible should be considered for electricity price prediction so that no informative candidate feature, which can be effective for the forecast process, is missed. The electricity price signal has a short-run trend characteristic (dependency on the previous neighboring hours' values) as well as daily periodicity behavior (dependency on the values of the same hour in the previous days) and weekly periodicity behavior (dependency on the values of the same hour in the previous weeks). Taking into account these characteristics, data models including the lagged values of the price and exogenous variables up to, at least, 200 h ago plus the predicted values of the exogenous variables (provided that they are available) have been suggested for the electricity price forecast process by Amjady and Keynia [2,11]. For instance, based on this suggestion and only considering the load demand and available generation as the exogenous variables, the following nonhybrid data model can be constructed [2]:

$$
\mathrm{CI}(t) = \begin{cases} P(t-1),\ldots, P(t-200),\ \hat{L}\ (t),\ L(t-1),\ldots,L(t-200), \\ \hat{G}(t),\ G(t-1),\ldots,G(t-200) \end{cases} \tag{4.1}
$$

where P, L, and G indicate electricity price, load demand, and available generation, respectively. The forecast features of $\hat{L}(t)$ and $\hat{G}(t)$ may be either obtained from separate prediction processes or provided by the market operator.

4.3.1.2 Hybrid Data Model

The hybrid data models contain the information content of both time- and frequency-domain candidate features for the electricity price forecast process. Mathematical transformations are applied to signals to obtain further information that is not readily available in the raw time-domain signal. Fourier transform gives the spectral contents of the signal, but it gives no information regarding the time in which those spectral components appear. Thus, the conventional Fourier analysis is suited for dealing with frequencies that do not evolve with time, that is, stationary signals [12]. On the other hand, it is well known that electricity price has several nonstationary characteristics and should be considered as a nonstationary signal [11,13]. Short-time Fourier transform provides the time information by computing different Fourier transforms for consecutive time intervals and putting them together. Consecutive time intervals of the signal are obtained by truncating the signal using a sliding windowing function. However, short-time Fourier transform gives a fixed resolution at all times. Low-frequency estimates require long windows, whereas high-frequency ones need small windows.

Wavelet analysis overcomes the limitations of the Fourier analysis methods by using functions that retain a useful compromise between time location and frequency information. Wavelets are located in time, despite trigonometric functions of the Fourier analysis. Moreover, they have a window that automatically adapts itself to give the appropriate resolution. These characteristics enable wavelets to analyze many nonstationary signals [12].

The one-dimensional continuous wavelet transform $W(a,b)$ of signal $f(x)$ with respect to a wavelet $\phi(x)$ is given by

$$W(a,b) = \frac{1}{\sqrt{a}} \int\limits_{-\infty}^{+\infty} f(x)\phi\left(\frac{x-b}{a}\right) dx \qquad (4.2)$$

where the scale factor a controls the spread of the wavelet and shift parameter b determines its central position. $\phi(x)$ is also called the mother wavelet. Daubechies wavelet of order 4 (db4) [11], Daubechies wavelet of order 5 (db5) [14], Mexican hat [15], and Morlet wavelet [16] are some well-known mother wavelet functions that have been used in previous electricity price forecast research works. Daubechies wavelets of orders 2–10 (db2–db10), Mexican hat, and Morlet wavelets are shown in Figures 4.3, 4.4, and 4.5, respectively.

Fourier analysis consists of breaking up a signal into sine waves of various frequencies. Similarly, wavelet analysis is the breaking up of a signal into scaled and shifted versions of the mother wavelet. A $W(a,b)$ coefficient represents how well the signal $f(x)$ and the scaled/shifted mother wavelet match. Thus, the set of all wavelet coefficients

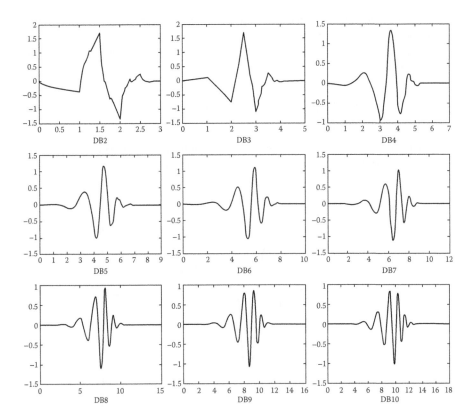

FIGURE 4.3 Daubechies wavelets from db2 to db10.

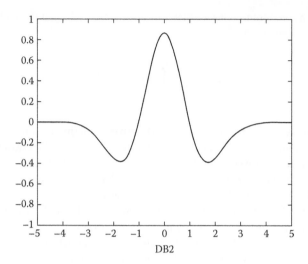

FIGURE 4.4 Mexican hat wavelet.

$W(a,b)$, associated with a particular signal, is the wavelet representation of the signal with respect to the mother wavelet.

Similarly, discrete wavelet transform (DWT) for signal $f_{(t)}$, with discrete time t, can be written as follows:

$$W(m,n) = 2^{-(m/2)} \sum_{t=0}^{T-1} f(t)\phi\left(\frac{t - n.2^m}{2^m}\right) \tag{4.3}$$

where T is the length of signal $f_{(t)}$. The scaling and shifting parameters are functions of the integer variables m and n ($a = 2^m$, and $b = n.2^m$). As electricity price time series

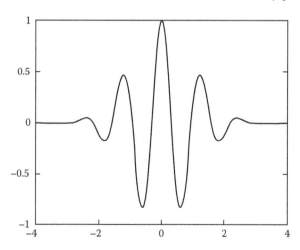

FIGURE 4.5 Morlet wavelet.

usually represents a discrete one-dimensional signal, the DWT of Equation 4.3 is used in the associated electricity price forecast research works. A computationally efficient algorithm to implement DWT using quadrature mirror filters was developed by Mallat [17,18]. This algorithm has two stages: decomposition and reconstruction. In first stage, the original signal is passed through two complementary filters and emerges as two signals: approximation (general trend or low-frequency component) and detail (high-frequency component). Each of these signals has the same number of data points, and these are downsampled by two to get the DWT coefficients. This decomposition can be iterated and successive approximations can be decomposed to many lower-resolution components. For instance, DWT decomposition with three decomposition levels is shown in Figure 4.6. Approximation $A3$ and three detail components $D1$, $D2$, and $D3$ are obtained from this decomposition process. In the second stage, these components can be reassembled into the original signal. Thus, wavelet decomposition involves filtering and downsampling, and wavelet reconstruction involves upsampling and filtering. Mathematical details of Mallat's algorithm can be found in References [17,18]. For instance, the price time series of the PJM electricity market in October and November 2008 along with its wavelet components $A3$, $D3$, $D2$, and $D1$ are shown in Figure 4.7. It can be seen that the price time series has both low- and high-frequency components (fast and slow changes in the time domain). However, approximation $A3$ is the low-frequency component and the detail subseries contain the high-frequency content of the price signal. $D1$ is the highest-frequency component. $D2$ is a lower-frequency component than $D1$, and $D3$ contains lower-frequency variations than $D1$ and $D2$. In this way, DWT can decompose the price signal to its components with different frequency contents.

DWT can be used in electricity price forecasting in three different ways. The first alternative is constructing a data model for the forecast process including candidate inputs from the original price time series and from wavelet domain subseries (e.g., $A3$, $D1$, $D2$, and $D3$ in Figure 4.6) [4,12]. In the second alternative, the electricity price time series is decomposed to wavelet components. Then, each component is separately predicted by a forecasting engine. Finally, the predictions of the components are returned to the original domain by the inverse transform to construct the price forecast. For

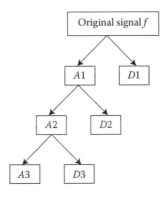

FIGURE 4.6 Multilevel decomposition process with three decomposition levels. A and D stand for approximation and detail, respectively ($f = A3 + D3 + D2 + D1$).

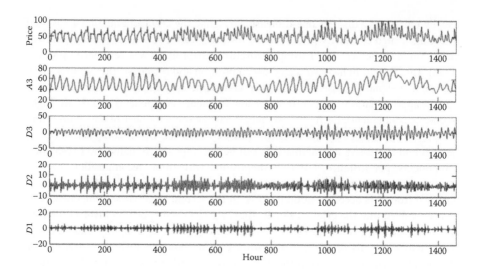

FIGURE 4.7   Electricity price time series and its wavelet components A3, D3, D2, and D1.

instance, this approach has been followed by Amjady and Keynia [11] and Conejo et al. [14]. The third option is merging wavelets with neural networks in the forecasting engine known as the wavelet neural network (WNN). WNN is a special case of single hidden layer feed-forward neural network with wavelets as the activation function of the hidden neurons [15,16].

Based on the first alternative, the hybrid data models can be constructed for electricity price forecast. For instance, a hybrid data model can be constructed from the nonhybrid model of Equation 4.1 based on DWT decomposition with three decomposition levels (Figure 4.6) as follows:

$$
CI(t) = \begin{cases}
P(t-1),\dots,P(t-200),\hat{L}(t),L(t-1),\dots,L(t-200),\hat{G}(t),\\
G(t-1),\dots,G(t-200),A3(t-1),\dots,A3(t-200),D1(t-1),\dots,\\
D1(t-200),D2(t-1),\dots,D2(t-200),D3(t-1),\dots,D3(t-200)
\end{cases} \quad (4.4)
$$

where $A3(.)$, $D1(.)$, $D2(.)$, and $D3(.)$ are the wavelet components of the signal $P(.)$.

It is noted that the given explanations about the data model are general guidelines. However, the data model of an electricity price forecast process should be constructed keeping in mind the specific characteristics of the price signal and associated electricity market. For instance, wind power sources, as a kind of nonpollutant renewable energy, are rapidly grown in many countries around the world. Considerable amounts of wind power generation are currently seen in some power systems and it is expected that wind energy will become an important component in the supply mix to meet the growing demand for electric energy in the near future. However, wind power sources have considerable differences with conventional generators. Wind is primarily an energy resource and not a capacity resource [19]. Thus, wind power generation may not be considered in the available generation capacity $G(.)$, and lagged and forecast values of wind power generation (as an exogenous variable) may be required to be taken into account in the data model of the electricity price forecast for electricity markets with high wind power integration.

4.3.2 Single-Stage Feature Selection Techniques

Feature selection is a well-known process in data mining that seeks optimal or suboptimal subsets of the original features by preserving the main information carried by the collected complete data, to facilitate future analysis for high-dimensional problems. The best subset contains the least number of dimensions that most contribute to accuracy; the remaining, unimportant dimensions are discarded. Feature selection usually plays a fundamental role in data mining, information retrieval, and, more generally, machine learning tasks for a variety of reasons, including facilitating data interpretation, reducing measurement and storage requirements, defying the curse of dimensionality, and, more importantly, improving generalization performance.

From the discussions in the previous sections, it can be seen that the data models of the electricity price forecast process usually have many candidate inputs such as the set of candidate inputs shown in Equations 4.1 and 4.4. However, such a large set of inputs is not directly applicable to a forecasting engine. Moreover, it may include ineffective features that complicate the extraction of the input/output mapping function of the prediction process for the forecasting engine and degrade its performance. Thus, the set of candidate inputs CI(t) should be refined by a feature selection technique such that a minimum subset of the most informative features is selected and the other unimportant candidates are filtered out.

Feature selection algorithms can be classified into wrappers and filters. Wrappers search through the space of feature subsets using the estimated accuracy from a learning algorithm (here, the forecasting engine) as the measure of the goodness for a particular feature subset. A common choice for performing the evaluation in a wrapper method is cross-validation. Cross-validation techniques will be discussed in Section 4.5. The wrapper methods are usually restricted by the time complexity of the learning algorithm. When the number of candidate features is large, which is the case for electricity price forecast considering its data models, the wrappers may become prohibitively expensive to run, as too many subsets of the candidate inputs need to be constructed and evaluated by the forecasting engine. Thus, wrapper methods will not be discussed for the feature selection of electricity price forecast in this chapter. On the other hand, filter methods use criteria not involving any learning machine. In other words, filter methods select a subset of features independent of the learning algorithm. In these methods, features are selected based on intrinsic characteristics, which determine their relevance to the target. In this chapter, we focus on filter methods. Thus, hereafter, by "feature selection," we mean "filter-type feature selection."

Two kinds of feature selection methods are described in this chapter: single-stage and two-stage feature selection techniques. Single-stage methods evaluate relevancy of candidate inputs with the target variable (forecast feature) and select the most relevant candidates for the prediction process. Three data mining criteria, including correlation coefficient, mutual information, and RELIEF weights, will be presented in the next sections. Two-stage feature selection techniques evaluate both relevancy and redundancy (redundant information) of candidate inputs to select a subset of the most relevant and least redundant candidates. These techniques are usually constructed by combining or extending single-stage methods.

It is noted that candidate inputs of the electricity price forecast process (such as the lagged prices and loads) have different ranges. Thus, to avoid the masking effect, candidate inputs should be first normalized. Suppose that a candidate input x is in the range of $[x_{min}, x_{max}]$. This range can be obtained from the historical data. The normalized form of x, denoted by x_n, within the range $[a,b]$, can be obtained from the following linear transformation:

$$x_n = a + (x - x_{min}) \cdot \frac{b - a}{x_{max} - x_{min}} \qquad (4.5)$$

Two common ranges for the normalized variables are $[a,b] = [0,1]$ and $[a,b] = [-1,1]$. An advantage of the above linear transformation is that it does not distort the distribution of data. Each candidate input should be separately normalized based on its own minimum and maximum values. After the normalization process, all normalized candidate inputs will be in the range $[a,b]$. The feature selection is performed by the normalized candidate inputs NCI(t), as shown in Figure 4.2. Finally, the inverse transform in the form of

$$x = x_{min} + (x_n - a) \cdot \frac{x_{max} - x_{min}}{b - a} \qquad (4.6)$$

is executed in the postprocess (Figure 4.2) to transform the normalized forecast price $\hat{P}_N(t)$ to $\hat{P}(t)$ in the actual range.

4.3.2.1 Correlation Analysis

Correlation analysis is a single-stage feature selection technique that has been used for feature selection of electricity price forecast in some research works [13,15]. The correlation coefficient between two random variables x and y, denoted by $\rho_{x,y}$, can be computed from the following relation:

$$\rho_{x,y} = \frac{cov(x,y)}{\sigma_x \cdot \sigma_y} \qquad (4.7)$$

In Equation 4.7, $cov(x,y) = E\big((x - \mu_x).(y - \mu_y)\big)$ is the covariance of x and y, where $E(.)$ is the expected value operator and μ_x and μ_y represent expected values of x and y, respectively. Moreover, $\sigma_x = \sqrt{E\big((x - \mu_x)^2\big)}$ and $\sigma_y = \sqrt{E\big((y - \mu_y)^2\big)}$ are the standard deviations of x and y, respectively. In Equation 4.7, $\rho_{x,y} = 1$ corresponds to a perfect linear correlation whereas an intermediate value describes partial correlations, and $\rho_{x,y} = 0$

represents no correlation at all. In other words, higher $\rho_{x,y}$ value results in more similarity between x and y. To use correlation analysis to select candidate inputs for the electricity price forecast process, the correlation coefficient between each candidate input and target variable (price of the next hour) is computed based on the historical data. Then, candidate features with a correlation coefficient greater than a threshold TH1 are selected as the relevant inputs and the remaining candidates with a correlation coefficient less than TH1 (less relevant features) are filtered out. Threshold TH1 is a set point or adjustable parameter of the feature selection technique. Fine-tuning of the adjustable parameters of a price prediction strategy (including those of the feature selection component and forecasting engine) will be discussed in Section 4.5.

Despite its simplicity, the correlation coefficient is a linear measure of cross-information and so a linear criterion for selecting candidate features. On the other hand, electricity price is a nonlinear mapping function of its inputs, as discussed in Section 4.1. Consequently, correlation analysis may not correctly evaluate the nonlinear dependencies of the electricity price signal and thus may not correctly evaluate the actual information value of candidate inputs for the forecast process. Therefore, this feature selection technique is not further discussed here. Instead, two nonlinear feature selection criteria, including mutual information and modified RELIEF, will be presented in the next two sections.

4.3.2.2 Mutual Information

A recently developed feature selection method is mutual information, which is based on the entropy concept. The individual entropy $H(x)$ of a continuous random variable x with probability distribution $P(x)$ is defined as follows [20,21]:

$$H(x) = -\int P(x)\log_2\big(P(x)\big)\mathrm{d}x \tag{4.8}$$

For a discrete random variable x with n values, $H(x)$ is computed as follows:

$$H(x) = -\sum_{i=1}^{n} P(x_i)\log_2\big(P(x_i)\big) \tag{4.9}$$

where $P(x_i)$ indicates probability of the value x_i. Strictly, electricity price and its candidate inputs (such as the load demand, available generation, and wavelet components) are continuous random variables. However, in practice, we have a few samples of each of them based on the historical data. Thus, for the sake of simplicity, we treat electricity price and its candidate inputs as discrete random variables to introduce the mutual information-based feature selection technique. In this way, the integral operations required to compute mutual information (e.g., the integral in Equation 4.8) are reduced to summations (e.g., the summation in Equation 4.9).

The entropy is often considered a measure of uncertainty. For instance, suppose that the random variable x represents the occurrence of an event such that x has two values of

$x_1 = 0$ (not occurred) and $x_2 = 1$ (occurred). If we know without any uncertainty that the event has occurred [i.e., $P(x_1) = 0$ and $P(x_2) = 1$] or has not occurred [i.e., $P(x_1) = 1$ and $P(x_2) = 0$], then the individual entropy $H(x)$ becomes zero. On the other hand, if there is high uncertainty about the occurrence/nonoccurrence of the event [i.e., $P(x_1) = 0.5$ and $P(x_2) = 0.5$] then the individual entropy $H(x)$ becomes 1. This value represents the highest possible uncertainty about the random variable x.

The joint entropy $H(x,y)$ of two discrete random variables x and y with n and m values, respectively, and joint probability distribution $P(x,y)$ is defined as follows:

$$H(x, y) = -\sum_{i=1}^{n} \sum_{j=1}^{m} P(x_i, y_j) \log_2 \left(P(x_i, y_j) \right) \tag{4.10}$$

where $H(x,y)$ represents the total entropy of the random variables x and y. When the random variable x is known and y is not, the remaining uncertainty of y is measured by the conditional entropy:

$$H(y/x) = \sum_{i=1}^{n} P(x_i) H(y/x = x_i) = -\sum_{i=1}^{n} P(x_i) \sum_{j=1}^{m} P(y_j/x_i) \log_2 \left(P(y_j/x_i) \right)$$
$$= -\sum_{i=1}^{n} \sum_{j=1}^{m} P(x_i, y_j) \log_2 \left(P(y_j/x_i) \right) \tag{4.11}$$

In other words, the conditional entropy $H(y/x)$ indicates the remaining uncertainty of y after observing x. Mathematically, the joint entropy and conditional entropy have the following relationship, which is known as the chain rule [21]:

$$H(x, y) = H(x) + H(y/x) = H(y) + H(x/y) \tag{4.12}$$

In other words, the total entropy of the random variables x and y is the individual entropy of x plus the remaining entropy of y for a given x.

In data mining, the information found commonly in two random variables x and y is of importance. This is known as the mutual information between the two variables, denoted by $MI(x,y)$, which is defined as follows:

$$MI(x, y) = \sum_{i=1}^{n} \sum_{j=1}^{m} P(x_i, y_j) \log_2 \left(\frac{P(x_i, y_j)}{P(x_i) P(y_j)} \right) \tag{4.13}$$

If the mutual information between two random variables is large, the two variables are closely related and vice versa. If the mutual information becomes zero, the two random variables are totally unrelated or independent. The mutual information has the following relationships with the individual entropy, joint entropy, and conditional entropy:

$$MI(x, y) = H(x) - H(x/y) \tag{4.14}$$

$$MI(x, y) = H(y) - H(y/x) \tag{4.15}$$

$$MI(x, y) = H(x) + H(y) - H(x, y) \tag{4.16}$$

$$MI(x, y) = MI(y, x) \tag{4.17}$$

$$MI(x, x) = H(x) \tag{4.18}$$

Figure 4.8 gives a better insight into the above relationships. On the basis of Equation 4.14, the mutual information $MI(x,y)$ measures how much the uncertainty of x is reduced if y has been observed. A similar conclusion is obtained from Equation 4.15 for y by observing x. If x and y are independent, their mutual information becomes zero; that is, observing y does not reduce the uncertainty of x and vice versa. In this case, $H(x,y) = H(x) + H(y)$ based on Equation 4.16. As the two random variables x and y have no common information, their joint entropy becomes the sum of their individual entropies. On the other hand, if the two random variables x and y become completely related, which occurs when $x = y$, then, according to Equation 4.18, the mutual information between these two variables reaches its maximum value $H(x)$.

To use the mutual information technique for the feature selection of electricity price forecast, assume that the set of normalized candidate inputs includes $y_1, y_2, \ldots, y_N$. For instance, these candidate inputs can include time-domain features as shown in

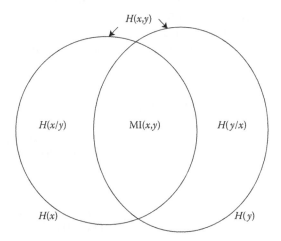

FIGURE 4.8 Illustration of individual entropies, conditional entropies, joint entropy, and mutual information for two random variables x and y.

Equation 4.1 for a nonhybrid data model and can include both time- and wavelet-domain features as shown in Equation 4.4 for a hybrid data model. The candidate input y_m ($1 \leq m \leq N$) owning more mutual information MI(x,y_m) with the target variable x (the forecast feature or the next-hour normalized price) is a better candidate input feature, as by considering y_m the uncertainty of x reduces more than using other candidate inputs. Consequently, the mutual information MI(x,y_m) can assign a value to each candidate input y_m for forecasting the target variable x. In other words, we can rank candidate inputs based on their mutual information with the target variable or their information value for the forecast process. Then, candidate inputs with the mutual information MI(x,y_m) greater than the relevancy threshold TH1 are selected as relevant features for price forecast and other less relevant candidates are filtered out. It is noted that the appropriate values of threshold TH1 for the mutual information technique can be different from its appropriate values for correlation analysis. In general, the threshold(s) of each feature selection method should be specifically fine-tuned for it.

An important issue for the implementation of the mutual information-based feature selection technique is efficiently constructing the individual and joint probability distributions shown in Equation 4.13. Even considering the nonhybrid data model of Equation 4.1, 602 mutual information values, by the number of candidate inputs, should be computed in each feature selection phase. Thus, 603 individual probability distribution functions (for 602 candidate inputs and one output feature) and 602 joint probability distribution functions are required. Consequently, a computationally efficient method should be adopted to construct these probability distributions. One solution is to incorporate data quantization as an initial processing step for feature selection. A binary quantization approach has been proposed for this purpose by Amjady and Keynia [2], which is a good compromise between computation burden and estimation accuracy. In this approach, the median of the normalized candidate inputs and the output feature is first computed. Half the number of values of each normalized variable is more than its median which is rounded to 1 and the other half is less than it which is rounded to 0. After this process, a binomial distribution is obtained for each candidate input and the target variable. On the basis of the constructed binomial distributions, MI(x,y) in Equation 4.13 can be approximated as follows:

$$
\begin{aligned}
\mathrm{MI}(x, y) &= \sum_{i=1}^{2} \sum_{j=1}^{2} P(x_i, y_j) \log_2 \left[\frac{P(x_i, y_j)}{P(x_i) \times P(y_j)} \right] \\
&= P(x = 0, y = 0) \times \log_2 \left[\frac{P(x = 0, y = 0)}{P(x = 0) \times P(y = 0)} \right] \\
&\quad + P(x = 0, y = 1) \times \log_2 \left[\frac{P(x = 0, y = 1)}{P(x = 0) \times P(y = 1)} \right] \\
&\quad + P(x = 1, y = 0) \times \log_2 \left[\frac{P(x = 1, y = 0)}{P(x = 1) \times P(y = 0)} \right] \\
&\quad + P(x = 1, y = 1) \times \log_2 \left[\frac{P(x = 1, y = 1)}{P(x = 1) \times P(y = 1)} \right]
\end{aligned}
\tag{4.19}
$$

Each of the individual and joint probabilities in the above equation can be easily computed on the basis of the historical data. Another alternative is tristate quantization, in which each of the input and output features is quantized into three states at the positions $\mu \pm \delta$ (μ is the mean value and δ is the standard deviation of the feature): it takes -1 if it is less than $\mu - \delta$, $+1$ if larger than $\mu + \delta$, and 0 if otherwise [22]. For the tristate quantization, Equation 4.13 becomes as follows:

$$
\begin{aligned}
\text{MI}(x, y) = \sum_{i=1}^{3} \sum_{j=1}^{3} & P(x_i, y_j) \log_2 \left[\frac{P(x_i, y_j)}{P(x_i) \times P(y_j)} \right] \\
= P(x_i = -1, y_j = -1) & \log_2 \left[\frac{P(x_i = -1, y_j = -1)}{P(x_i = -1) \times P(y_j = -1)} \right] \\
+ P(x_i = -1, y_j = 0) & \log_2 \left[\frac{P(x_i = -1, y_j = 0)}{P(x_i = -1) \times P(y_j = 0)} \right] \\
+ P(x_i = -1, y_j = 1) & \log_2 \left[\frac{P(x_i = -1, y_j = 1)}{P(x_i = -1) \times P(y_j = 1)} \right] \\
+ P(x_i = 0, y_j = -1) & \log_2 \left[\frac{P(x_i = 0, y_j = -1)}{P(x_i = 0) \times P(y_j = -1)} \right] \\
+ P(x_i = 0, y_j = 0) & \log_2 \left[\frac{P(x_i = 0, y_j = 0)}{P(x_i = 0) \times P(y_j = 0)} \right] \\
+ P(x_i = 0, y_j = 1) & \log_2 \left[\frac{P(x_i = 0, y_j = 1)}{P(x_i = 0) \times P(y_j = 1)} \right] \\
+ P(x_i = 1, y_j = -1) & \log_2 \left[\frac{P(x_i = 1, y_j = -1)}{P(x_i = 1) \times P(y_j = -1)} \right] \\
+ P(x_i = 1, y_j = 0) & \log_2 \left[\frac{P(x_i = 1, y_j = 0)}{P(x_i = 1) \times P(y_j = 0)} \right] \\
+ P(x_i = 1, y_j = 1) & \log_2 \left[\frac{P(x_i = 1, y_j = 1)}{P(x_i = 1) \times P(y_j = 1)} \right]
\end{aligned}
\tag{4.20}
$$

Similarly, the individual and joint probability values of Equation 4.20 can be obtained from the historical data. Quantization with more than three states can also be considered. However, with more states, less historical data can be assigned to each state. Thus, the construction of the individual and joint probability distributions required for the computation of mutual information values becomes more complex and even inaccurate. For applications where it is unclear how to properly quantize the input/output features, an alternative solution is to use a density estimation method (e.g., Parzen windows) to approximate $\text{MI}(x,y)$ [22].

4.3.2.3 Modified RELIEF

The RELIEF algorithm is a nonlinear feature selection method that uses instance-based learning to assign a relevance weight to each feature. The key idea of RELIEF is to iteratively estimate feature weights according to their ability to discriminate between neighboring patterns [23]. In other words, RELIEF strives to reinforce similarities between instances of the same class and simultaneously decrease similarities between instances of opposite classes where similarity is defined by proximity in the feature space [24].

In each iteration, the RELIEF algorithm works by randomly selecting a training sample TS from the training set. Then, two nearest neighbors for the selected sample TS are found, one from the same class (called the nearest hit or NH) and the other from a different class (called the nearest miss or NM). A feature's weight is updated according to how well its values distinguish the selected sample TS from its nearest hit NH(TS) and its nearest miss NM(TS). A feature receives a high weight if it largely varies between samples from opposite classes [e.g., TS and NM(TS)] and has close values for samples of the same class [e.g., TS and NH(TS)].

To find the nearest neighbors of the selected sample, we should have a distance measure. The Euclidean distance can be used for this purpose. For instance, the Euclidean distance between TS and NH(TS) is defined as follows:

$$D\big(\text{TS},\text{NH(TS)}\big) = \big\|\text{TS} - \text{NH(TS)}\big\| = \sqrt{\big(\text{TS} - \text{NH(TS)}\big).\big(\text{TS} - \text{NH(TS)}\big)}$$

$$= \sqrt{\big\|\text{TS}\big\|^2 + \big\|\text{NH(TS)}\big\|^2 - 2\,\text{TS.NH(TS)}} \qquad (4.21)$$

where $D(\text{TS},\text{NH(TS)})$ is the Euclidean distance between the vectors of TS and NH(TS); the dot sign indicates the inner product of two vectors; $\|.\|$ represents the Euclidean norm. In the RELIEF algorithm, the relevance weights of all features are first initialized to zero. Then, all feature weights are updated on the basis of the randomly selected sample TS in each iteration as follows:

$$W_i = W_i + \big|\text{TS}_{(i)} - \text{NM}_{(i)}\big(\text{TS}\big)\big| - \big|\text{TS}_{(i)} - \text{NH}_{(i)}\big(\text{TS}\big)\big|, \quad i = 1,2,\ldots,\text{DIM(NCI)} \qquad (4.22)$$

where W_i is the relevance weight of ith feature; $\text{TS}_{(i)}$, $\text{NM}_{(i)}(\text{TS})$, and $\text{NH}_{(i)}(\text{TS})$ indicate the ith feature of TS, NM(TS) and NH(TS), respectively; DIM(NCI) represents the dimension (number of features) of the set of normalized candidate inputs NCI. If $\text{TS}_{(i)}$ and $\text{NM}_{(i)}(\text{TS})$ have different values, the ith candidate input discriminates two samples from different classes. This is desirable and so W_i increases according to Equation 4.22. On the other hand, if $\text{TS}_{(i)}$ and $\text{NH}_{(i)}(\text{TS})$ have different values, the ith candidate input separates two samples from the same class. This is inconsistent and so W_i decreases according to Equation 4.22. In other words, a relevant feature should largely change between dissimilar samples and slightly change (or does not change in the ideal case) between similar samples. The cycle of randomly selecting the sample TS, finding its nearest hit NH(TS) and nearest miss NM(TS), and updating the feature weights W_i is repeated until the maximum number of iterations (which is a user-defined parameter) is reached. Then, the RELIEF algorithm is terminated and the candidate inputs are ranked according to the weight values finally obtained.

Using the RELIEF algorithm for the feature selection of electricity price forecasting has an inconsistency. RELIEF can only be used for two class problems, which can be seen from its formulation. However, for electricity price forecasting, the output feature is a continuous variable and so the nearest hit and nearest miss cannot be found. To remedy this inconsistency, a modification is proposed for the RELIEF algorithm here. The resultant algorithm is called modified RELIEF. Assume that x represents the continuous output feature and $\bar{x}$ indicates its mean. Construct the auxiliary variable $z = x - \bar{x}$. Considering $\bar{z} = 0$, the training samples of the training set can be categorized into two classes based on the positive and negative values of z. Thus, the RELIEF algorithm can be applied for the feature selection of the problem. For instance, if the auxiliary variable z corresponding to the output feature x of the randomly selected training sample TS becomes positive, the NH(TS) and NM(TS) should be the nearest samples to TS with the positive and negative values of z, respectively. To compute the auxiliary variable z for electricity price forecasting, the output feature x can be selected as the electricity price $P(t)$ or normalized electricity price $P_N(t)$, as the linear normalization introduced in Equation 4.5 does not change the distribution of $P(t)$. The auxiliary variable z is only used to determine NH(TS) and NM(TS) for the randomly selected sample TS. However, it is noted that the feature weights are computed by means of Equation 4.22 for the normalized candidate inputs. After determining the relevance weights for all normalized candidate inputs, the features with W_i greater than the relevancy threshold TH1 are selected as relevant features for the price forecast process.

Amjady et al. [25] present an extended form of the modified RELIEF algorithm in which k nearest misses and hits are used instead of one nearest miss and hit. Moreover, that version of the modified RELIEF algorithm simultaneously considers all training samples to determine the relevance weights, whereas the RELIEF algorithm takes into account a few randomly selected training samples such that only one of them is used in each iteration.

4.3.3 Two-Stage Feature Selection Techniques

Single-stage feature selection techniques only consider relevancy of the candidate inputs with the target variable to select the informative features for the machine learning process. Here, the aim of the machine learning process is extraction of the input/output mapping function of the electricity price forecast. Although the relevancy of a candidate input is a necessary condition for selecting the candidate, it may not be a sufficient condition. In feature selection, it has been recognized that the combinations of individually good features do not necessarily lead to good classification performance. In other words, "the m best features are not the best m features" [22]. This is because the individually relevant features may have redundant information, which decreases the whole information content of the set of relevant features. For electricity price forecasting that many candidate inputs are successive lagged features [such as $P(t-1)$, $P(t-2)$, etc.] and some candidate inputs have close dependencies [such as $P(t-k)$ and $L(t-k)$], evaluating the redundant information between candidate features is an important issue for selecting the most informative set of inputs. Thus, feature selection methods that can evaluate both relevancy and redundancy of the candidate features are required for electricity price forecasting. Two-stage feature selection techniques are presented here for this purpose. These feature selection methods consist of irrelevancy and redundancy filters to filter out irrelevant and redundant candidate inputs. On the basis of the filter methods previously presented, three two-stage feature selection techniques, including two-stage correlation analysis, two-stage mutual information, and modified RELIEF combined with mutual information, will be presented in the next sections.

4.3.3.1 Two-Stage Correlation Analysis

Two-stage correlation analysis can evaluate the linear independency of the candidate inputs in addition to their correlation with the target variable and so it can filter out redundant information and collinear candidate inputs in addition to uncorrelated candidates. The structure of this feature selection technique is shown in Figure 4.9. The first stage of this feature selection technique includes an irrelevancy filter based on correlation analysis. Relevant candidate inputs with a correlation coefficient with the target variable more than TH1 pass this filter. The remaining uncorrelated candidate features with the correlation coefficient less than TH1 are filtered out. Note that

$$\text{Subset of relevant candidate inputs} \cup \text{Subset of irrelevant candidate inputs} = \text{NCI}(t)$$

The performance of the first stage or irrelevancy filter has been described in Section 4.3.2.1. In the second stage of the two-stage feature selection technique, a redundancy filter analyzes the "subset of relevant candidate inputs" that pass the irrelevancy filter. A cross-correlation analysis is performed in the redundancy filter of the two-stage correlation analysis. If the correlation coefficient between any two candidate features is smaller than a prespecified threshold TH2, then both features are retained; else, only the feature with the larger correlation coefficient with the target variable is retained, whereas the other is not considered any further [26,27]. The redundancy filtering process is continued until no redundant feature is found among the subset of relevant candidate inputs.

The union of the filtered and retained features of the redundancy filter is the outcome of the irrelevancy filter. The threshold TH2 is a degree of freedom for the two-stage

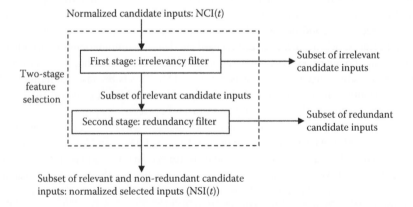

FIGURE 4.9   Structure of a two-stage feature selection technique.

correlation analysis similar to TH1. The two-stage feature selection techniques with two filters have two threshold set points. The outcome of the redundancy filter is the final result of the two-stage feature selection technique including the normalized selected inputs NSI(*t*), as shown in Figure 4.9. These inputs are applied to the forecasting engine (Figure 4.2). The filtered candidate inputs in the first and second stages comprise all the filtered features of this feature selection technique.

4.3.3.2 Two-Stage Mutual Information

Deficiency of the correlation analysis for the feature selection of electricity price forecasting has been described in Section 4.3.2.1. A more efficient two-stage feature selection technique can be implemented on the basis of the mutual information criterion. Two-stage mutual information follows the same logic of two-stage correlation analysis. However, the measure of mutual information is used in this feature selection technique instead of correlation coefficient. The first stage implements a mutual information-based irrelevancy filter, the same as described in Section 4.3.2.2. Then, a mutual information-based redundancy filtering process is performed on the candidate inputs that pass the first stage.

Higher value of mutual information between two selected features of the first stage y_k and y_m, that is, higher MI(y_k,y_m), means more common information between y_k and y_m. Thus, these features have a higher level of redundancy. Therefore, the following redundancy criterion RC(.) is defined to measure the redundancy of each feature, $y_k \in$ Subset of relevant candidate inputs, with the other features of this subset:

$$RC(y_k) = \max_{y_m \in \text{Subset of relevant candidate inputs} - \{y_k\}} \left(MI(y_k, y_m) \right) \qquad (4.23)$$

It is seen that y_m belongs to the subset of relevant candidate inputs–$\{y_k\}$ in Equation 4.23 as each feature is fully redundant with itself; thus, we should exclude y_k from this subset. We can rank the candidate features of the subset according to the redundancy measure of Equation 4.23 such that a higher value of RC(y_k) means y_k is a more redundant feature or equivalently a less informative candidate input. If RC(y_k) becomes greater than the redundancy threshold TH2, y_k is considered as a redundant candidate input and so between this candidate and its competitor, one feature should be filtered out. For instance, suppose that y_k has the highest redundancy (mutual information) with y_c among all features of the subset of relevant candidate inputs–$\{y_k\}$. Hence, the following relationship can be written:

$$\arg\max_{y_m \in \text{Subset of relevant candidate inputs} - \{y_k\}} \left(MI(y_k, y_m) \right) = y_c \qquad (4.24)$$

In such a case, y_c is called the competitor of y_k. If MI(y_k,y_c) > TH2, between y_k and its competitor y_c, one feature should be filtered out. For this purpose, the relevancy factors of these features including MI(y_k,x) and MI(y_c,x) are considered (*x* indicates the target variable or forecast feature). The feature with the lower relevancy factor (less relevant feature or less effective feature for the forecast process) is removed. The redundancy filtering process is repeated for all features of the subset of relevant candidate inputs until the no-redundancy measure of Equation 4.23 becomes greater than TH2. The features of this subset that pass the redundancy filter, having RC(.) less than TH2, comprise the selected features of the two-stage mutual information technique, including relevant and nonredundant candidate inputs.

4.3.3.3 Modified RELIEF and Mutual Information

The first stage or irrelevancy filter of this feature selection technique is modified RELIEF, as presented in Section 4.3.2.3. The second stage includes a mutual information-based redundancy filter as described in the previous section.

The thresholds TH1 and TH2 have different effects on the finally selected features of a two-stage feature selection technique. By increasing TH1, more candidate inputs are considered as irrelevant features and filtered out in the first stage. In other words, the rejection band of the irrelevancy filter increases leading to less features in the subset of relevant candidate inputs. On the other hand, higher values of TH2 broaden the pass band of the redundancy filter, as more candidate inputs with higher levels of redundant information are allowed to pass the redundancy filter. Thus, the performance of a two-stage feature selection technique is dependent on the fine-tuning of these settings. Some techniques for this purpose will be introduced in Section 4.5.

Although two-stage feature selection techniques use single-stage methods as the irrelevancy filter, the values of the threshold TH1 used for two-stage and single-stage feature selection techniques are different. For instance, values of the threshold TH1 used for single-stage mutual information techniques and for the irrelevancy filter of two-stage mutual information methods are different. Single-stage methods are only based on one filter. Thus, higher values of the threshold TH1 (narrower pass bands) may be selected for these methods such that a reasonable number of selected inputs, which can be applicable to the forecasting engine, are obtained. On the other hand, the irrelevancy filter of a two-stage feature selection method should have lower values of the threshold TH1 (wider pass bands), such that a sufficiently large subset of relevant candidate inputs is obtained for the second filter and the number of features finally selected does not become too low.

Some representative results comparing the single-stage and two-stage feature selection methods are presented in Tables 4.1 through 4.4 to give a better insight into

TABLE 4.1   MAPE Values (%) Obtained by Different Feature Selection Methods for Day-Ahead Price Forecast of the Four Test Weeks of the PJM Electricity Market in 2010

Test Week	Correlation Analysis	Mutual Information	Modified RELIEF	Two-Stage Correlation Analysis	Two-Stage Mutual Information	Modified RELIEF and Mutual Information
Winter	7.04	6.49	6.54	6.29	5.61	5.75
Spring	7.54	6.73	6.71	6.27	5.75	5.84
Summer	7.91	7.26	7.42	6.76	6.21	6.32
Fall	7.38	6.89	7.13	6.67	5.52	5.31
Average	7.47	6.84	6.95	6.50	5.77	5.81

TABLE 4.2 AMAPE Values (%) Obtained by Different Feature Selection Methods for Day-Ahead Price Forecast of the Four Test Weeks of the PJM Electricity Market in 2010

Test Week	Correlation Analysis	Mutual Information	Modified RELIEF	Two-Stage Correlation Analysis	Two-Stage Mutual Information	Modified RELIEF and Mutual Information
Winter	6.89	6.26	6.33	6.02	5.39	5.57
Spring	7.38	6.55	6.58	6.11	5.67	5.71
Summer	7.74	7.02	7.22	6.58	6.12	6.24
Fall	7.19	6.62	7.00	6.52	5.25	5.26
Average	7.30	6.61	6.78	6.31	5.61	5.69

TABLE 4.3 VAR Values ($\times 10^{-4}$) Obtained by Different Feature Selection Methods for Day-Ahead Price Forecast of the Four Test Weeks of the PJM Electricity Market in 2010

Test Week	Correlation Analysis	Mutual Information	Modified RELIEF	Two-Stage Correlation Analysis	Two-Stage Mutual Information	Modified RELIEF and Mutual Information
Winter	52.15	38.54	41.84	35.53	30.18	30.72
Spring	54.28	43.94	45.08	32.98	31.28	30.19
Summer	55.42	45.37	46.53	33.95	30.53	31.38
Fall	53.07	41.64	44.78	33.12	31.59	32.25
Average	53.73	42.37	44.56	33.89	30.89	31.13

their performance. The real effectiveness of a feature selection method can be evaluated on the basis of the quality of the forecast results obtained using that method. Thus, in this numerical experiment, the artificial intelligence-based electricity price forecast strategy, shown in Figure 4.2, is separately implemented with each of the presented feature selection methods, and the results obtained are reported in Tables 4.1 through 4.4. For the sake of a fair comparison, the other parts of the strategy are kept unchanged for all applied feature selection techniques. The nonhybrid data model of Equation 4.1 is used for all feature selection methods shown in Tables 4.1 through 4.4. Moreover, the linear transformation of Equation 4.5 with $[a,b] = [0,1]$ and its inverse transform Equation 4.6 are used for the preprocess (normalization) and postprocess blocks of Figure 4.2, respectively. Furthermore, cascaded neural networks are used as the

TABLE 4.4 AVAR Values ($\times 10^{-4}$) Obtained by Different Feature Selection Methods for Day-Ahead Price Forecast of the Four Test Weeks of the PJM Electricity Market in 2010

Test Week	Correlation Analysis	Mutual Information	Modified RELIEF	Two-Stage Correlation Analysis	Two-Stage Mutual Information	Modified RELIEF and Mutual Information
Winter	52.24	38.29	42.67	33.81	30.47	31.28
Spring	53.35	44.34	44.38	32.66	30.32	30.09
Summer	54.62	44.61	45.57	34.52	30.49	31.33
Fall	53.72	41.29	45.59	33.34	32.04	30.86
Average	53.48	42.13	44.55	33.58	30.83	30.89

forecasting engine and line search procedure for fine-tuning the settings for all methods in Tables 4.1 through 4.4. These two components will be introduced in Sections 4.4.2.1 and 4.5.2, respectively.

This numerical experiment is performed by the real data of the PJM electricity market in 2010. Four test weeks corresponding to four seasons of 2010 (including the fourth weeks of February, May, August, and November) are considered here, indicated in the first column of Tables 4.1 through 4.4. This is to represent the whole year in the numerical experiment. The historical data of 50 days prior to each forecast day including $50 \times 24 = 1200$ h samples is used for both the feature selection technique and the forecasting engine of the strategy (Figure 4.2) in the day-ahead electricity price forecast of this numerical experiment. For instance, the individual and joint probability values of the mutual information, discussed in Section 4.3.2.2, are computed based on this historical data.

The forecasting accuracy is the main concern for power engineers. The first error criterion used in this numerical experiment, measuring the forecasting accuracy, is mean absolute percentage error (MAPE), defined as follows:

$$\text{MAPE} = \frac{1}{\text{NH}} \sum_{t=1}^{\text{NH}} \frac{|P(t) - \hat{P}(t)|}{P(t)} \qquad (4.25)$$

where NH indicates the number of hours in the test period. In this numerical experiment with weekly test periods, NH = 168. In this equation, $P(t)$ and $\hat{P}(t)$ represent the actual and forecast values of electricity price for hour t of the test period. It is noted that the length of the forecasting horizon for the day-ahead hourly electricity price prediction of this numerical experiment is 24 h or 24 forecast steps. In other words, the forecasting engine proceeds by 24 steps in each prediction stage. Then, the historical data are updated at the end of each day and the price forecast of the next day is performed by the updated data. However, each test period in Tables 4.1 through 4.4 is 1 week, to give a better evaluation of the prediction performance over a longer period. The prediction results for each test week are obtained from seven successive day-ahead forecast processes.

As the electricity price could be close to zero, the MAPE could be large even if the price forecast has a small deviation from the actual value. To remedy this problem, another version of MAPE, denoted here by AMAPE, is defined as follows:

$$\text{AMAPE} = \frac{1}{\text{NH}} \sum_{t=1}^{\text{NH}} \frac{|P(t) - \hat{P}(t)|}{P_{\text{ave}}} \qquad (4.26)$$

where P_{ave} is as follows:

$$P_{\text{ave}} = \frac{1}{\text{NH}} \sum_{t=1}^{\text{NH}} P(t) \qquad (4.27)$$

In other words, the average of actual prices or P_{ave} is used for the denominator of AMAPE, instead of $P(t)$, to avoid the adverse effect of prices close to zero. Results obtained for MAPE and AMAPE in this numerical experiment are reported in Tables 4.1 and 4.2, respectively.

In addition to accuracy, the stability of forecast results is another important issue for a prediction process. A stable forecast process should have low uncertainty, which is usually measured by the statistical dispersion of the forecast errors. For instance, the variance of electricity price forecast errors has been used as an index of uncertainty for the prediction results in several studies [1,2,14,28]. Lower values of the error variance mean less statistical dispersion of the forecast errors leading to a more stable electricity price prediction process. Consistent with the MAPE and AMAPE definitions in Equations 4.25 and 4.26, variance (VAR) and alternative variance (AVAR) for the forecast errors are defined as follows:

$$VAR = \frac{1}{NH} \sum_{t=1}^{NH} \left(\frac{|P(t) - \hat{P}(t)|}{P(t)} - MAPE \right)^2 \tag{4.28}$$

$$AVAR = \frac{1}{NH} \sum_{t=1}^{NH} \left(\frac{|P(t) - \hat{P}(t)|}{P_{ave}} - AMAPE \right)^2 \tag{4.29}$$

To better illustrate the concepts of forecast accuracy and forecast stability, two kinds of prediction errors with normal distribution are shown in Figure 4.10. The first error series, shown by a black solid line, has mean absolute error 0.9 and standard deviation

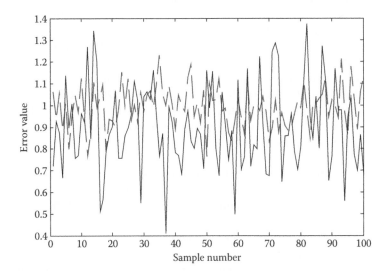

FIGURE 4.10 Forecast error series with mean absolute error 0.9 and standard deviation 0.2 (black, solid line) and forecast error series with mean absolute error 1 and standard deviation 0.1 (gray, dashed line).

0.2, while the second one, shown by a gray dashed line, has mean absolute error 1 and standard deviation 0.1. In Figure 4.10, although the first error series has better accuracy in terms of mean absolute error, larger oscillations are seen in its error curve. Thus, the results of the first forecast method may have more uncertainty leading to a more unstable prediction process. The second forecast method with more stable results may be preferable for some electricity market participants to design the bidding strategy and risk analysis.

From Tables 4.1 through 4.4 it can be seen that, among the single-stage feature selection techniques, mutual information and modified RELIEF lead to both better price forecast accuracy (lower values of MAPE and AMAPE) and better price forecast stability (lower values of VAR and AVAR) than correlation analysis. Mutual information and modified RELIEF can better evaluate the impact of different candidate inputs on the target variable in a nonlinear forecast process than correlation analysis. MAPE, AMAPE, VAR, and AVAR values of mutual information are close to those of the modified RELIEF. Tables 4.1 through 4.4 also show that the two-stage feature selection methods perform better than the single-stage techniques. Even the two-stage correlation analysis leads to better forecast accuracy and stability than the single-stage techniques of mutual information and modified RELIEF revealing the importance of considering redundant information in the feature selection processes. However, the nonlinear feature selection methods of two-stage mutual information and modified RELIEF combined with mutual information have lower values of MAPE, AMAPE, VAR, and AVAR than the linear feature selection technique of two-stage correlation analysis. The performance measures of the two-stage mutual information are close to those of the modified RELIEF combined with mutual information.

From Tables 4.1 and 4.2, it can be seen that the AMAPE values follow the same trend of the MAPE values. However, the AMAPE values are slightly lower than their corresponding MAPE values due to the softening effect of AMAPE that makes softer sharp errors. Similar conclusions can be drawn comparing the VAR and AVAR values in Tables 4.3 and 4.4.

To better illustrate the performance of the feature selection process, sample results, including the selected features for the forecast day May 23, 2010 of the PJM electricity market (a test day from the spring test week), are shown in Table 4.5. For the sake of conciseness, this more detailed analysis is only performed for the most accurate feature selection method in Tables 4.1 through 4.4, that is, the two-stage mutual information technique.

In the first stage of the two-stage mutual information method, 60 features are selected by the irrelevancy filter. Considering 602 candidate inputs in the nonhybrid data model of Equation 4.1, the filtering ratio of this filter becomes 602/60 = 10.03. The second stage or redundancy filter removes redundant features among the 60 selected candidate inputs of the first stage; 23 features out of the 60 candidate inputs can pass the redundancy filter. These 23 features and their ranks are shown in Table 4.5. The ranks of the finally selected candidate inputs are determined on the basis of their relevancy factors or mutual information with the target variable. The filtering ratio of the irrelevancy filter is 60/23 = 2.61. Although the filtering ratio of the irrelevancy filter is greater than that of the redundancy filter, the second stage is more competitive than the first. The initial set

TABLE 4.5 Results Obtained from the Two-Stage Mutual Information Technique (Finally Selected Candidate Inputs) for the Day-Ahead Price Forecast of the PJM Electricity Market on May 23, 2010

Rank	Selected Feature
1	$P(t-1)$
2	$\hat{L}(t)$
3	$L(t-1)$
4	$P(t-2)$
5	$L(t-168)$
6	$P(t-168)$
7	$L(t-24)$
8	$P(t-24)$
9	$P(t-23)$
10	$P(t-25)$
11	$L(t-167)$
12	$P(t-192)$
13	$P(t-144)$
14	$P(t-169)$
15	$L(t-144)$
16	$L(t-169)$
17	$P(t-48)$
18	$P(t-120)$
19	$L(t-23)$
20	$L(t-25)$
21	$G(t-1)$
22	$G(t-24)$
23	$L(t-72)$

of candidate inputs or CI(t) in Equation 4.1 usually includes many ineffective features that are filtered out by the irrelevancy filter. However, the 60 selected candidate inputs of the first stage are relevant features. In other words, in the second stage, the competition is among more qualified candidate inputs that pass the first qualification and so selecting informative features among them is a more critical task. By combining the filtering ratios of the two stages, the filtering ratio of the whole two-stage mutual information technique for this test case is determined as $10.03 \times 2.61 = 26.17$, which can also be obtained from $602/23 = 26.17$. It is seen that the feature selection process can effectively filter out the ineffective candidate features and reduce the size of the input set for the forecasting engine.

From these results another important point can also be seen. Although the two-stage feature selection methods usually perform better than the single-stage techniques, they require more computation effort. For instance, the single-stage feature selection technique of mutual information requires the computation of 602 mutual information values between each candidate input and target variable for this test case. The two-stage mutual information method also requires this computation for its first stage. Additionally, mutual information values between all pairs of the selected candidate inputs of the first stage by the number of

$$C^2_{\text{Number of relevant features}} = C^2_{60} = \frac{60!}{(60-2)! \cdot 2!} = \frac{60 \times 59}{2} = 1770 \qquad (4.30)$$

should also be computed for the redundancy filter. In other words, the computation burden of the second stage is about $1770/602 = 2.94$ times more than the computation burden of the first stage for this test case. Thus, for the forecast processes with serious limits on the computation time, the single-stage methods may be preferable.

Among the 23 selected features in Table 4.5, 11 candidate inputs are from the autoregression part of the data model (past values of the electricity price signal) and 12 features are from the cross-regression part (past and forecast values of the exogenous variables of load and available generation). These results illustrate the importance of modeling electricity price signal as a multivariate forecast process as more than half of the selected features are from the external variables in this test case. This information content is not accessible in the univariate electricity price forecast processes only relying on the history of price. Among the 12 selected candidate inputs of the exogenous variables, 10 load features and 2 available generation features are seen, as load is a more important price driver than available generation.

Short-run trend, daily periodicity, and weekly periodicity, characteristics of the electricity price time series have been discussed in Section 4.3.1.1. The effect of these characteristics can be observed from the results reported in Table 4.5. Price, load, and available generation of the previous neighboring hours [including $P(t-1)$, $P(t-2)$, $L(t-1)$, and $G(t-1)$] are among the selected features of this table representing the effect of short-run trend characteristic of the price signal. Moreover, the features of the same hour in the previous day [including $P(t-24)$, $L(t-24)$, and $G(t-24)$] are selected, indicating the daily periodicity behavior of the electricity price time series. Some features of the same hour a few days ago [such as $P(t-48) = P(t-2 \times 24)$, $L(t-72) = L(t-3 \times 24)$] are also seen among the selected features representing the multiperiod effect of the daily periodicity characteristic. Furthermore, the features of the same hour in the previous week [including $P(t-168)$ and $L(t-168)$] are selected by the two-stage mutual information technique, indicating the effect of weekly periodicity. The price time series largely takes these periodic behaviors from its most important driver, that is, the load time series [29]. Finally, the combined effects of these characteristics can also be seen from the selected candidate inputs in Table 4.5. For instance, the combined effect of short-run trend and daily periodicity behavior results in the selection of some neighboring hours around the same hour in the previous day, such as the selection of $P(t-23) = P(t-24+1)$, $P(t-25) = P(t-24-1)$, $L(t-23) = L(t-24+1)$, and $L(t-25) = L(t-24-1)$. Similarly, the combined effect of short-run trend and weekly periodicity behavior results in the selection of $L(t-167) = L(t-168+1)$ and $L(t-169) = L(t-168-1)$ and the combined effect of daily and weekly periodicity characteristics leads to the selection of $P(t-144) = P(t-168+24)$ and $P(t-192) = P(t-168-24)$.

It should be noted that the previous discussion only presents general guidelines for the feature selection of electricity price forecast. However, this is not to say that the selected candidate inputs of Table 4.5 are the best subset of features for any electricity price forecast process. Price drivers are dependent on the electricity market and so the selected features for price forecast may change from an electricity market to another one. Even the selected candidate inputs may change from one forecast day to another due to the time-variant behavior of the input/output mapping function of the electricity price signal, discussed in Section 4.1. Hence, it is better to separately perform the feature selection analysis for each electricity market and each forecast horizon (e.g., each forecast day) by the latest available data.

4.4 Electricity Price Forecast Engines

The forecasting engine is fed by the feature selection technique such that the selected candidate features are considered as the inputs of the forecasting engine. Thus, the mission of the forecasting engine is constructing an input/output mapping function for the electricity price prediction process in the form of NSI(t)➔$\hat{P}_N(t)$ as shown in Figure 4.2. In this section, some efficient price forecast engines are presented, which can be categorized as nonhybrid and hybrid methods. These two categories will be introduced in the next sections.

4.4.1 Nonhybrid Price Forecast Engines

Here, we call the electricity price forecast methods that only contain a single prediction technique (e.g., a time series technique, an expert system, a neural network, etc.) as the nonhybrid price forecast engines. These forecast engines are usually simpler than the hybrid forecast methods, but their performance may be lower. Among the nonhybrid forecasting engines, time-series techniques and neural networks, used more than the other methods for electricity price prediction, have been introduced here.

4.4.1.1 Time-Series Techniques

Although time-series techniques usually are not taken into account as artificial intelligence-based forecasting engines, they are discussed here as these methods are used in many electricity price prediction research works. Among the time-series methods, auto-regressive integrated moving average (ARIMA), dynamic regression, and transfer function models have gained more attention for electricity price prediction due to their easy implementation and relatively good performance. For instance, ARIMA, dynamic regression, and transfer function methods have been used for electricity price forecasting in several studies [14,28,30–33]. In the following, ARIMA time series and its variants are detailed. Dynamic regression and transfer function models have similar structures.

An auto-regressive moving average (ARMA) model for prediction of $x(t)$ can be expressed as follows [30]:

$$\phi(B) \cdot x(t) = c + \theta(B) \cdot \varepsilon(t) \tag{4.31}$$

where $x(t)$ is the value of the target variable (here, normalized electricity price) at time t; c is a constant term; $\varepsilon(t)$ is the error term at time t. Usually, $\varepsilon(t)$ is assumed to be an independently and identically distributed normal random variable with zero mean and constant variance σ_ε^2, that is, a Gaussian white noise process [32,33]. In Equation 4.31, $\phi(B)$ and $\theta(B)$ are polynomial functions of the backshift operator B as follows:

$$\phi(B) = 1 - \sum_{k=1}^{p} \phi_k \cdot B^k \tag{4.32}$$

$$\theta(B) = 1 - \sum_{k=1}^{q} \theta_k \cdot B^k \tag{4.33}$$

Note that

$$B^k \cdot x(t) = x(t-k), \quad B^k \cdot \varepsilon(t) = \varepsilon(t-k) \tag{4.34}$$

In Equation 4.31, $\phi(B).x(t)$ constitutes the auto-regressive (AR) part and $\theta(B).\varepsilon(t)$ is the moving average (MA) part, both of which together construct the ARMA model. On the basis of Equations 4.32 and 4.33, the degree of the AR and MA parts is p and q, respectively. Thus, these parts are denoted as AR(p) and MA(q), respectively, and so the ARMA model of Equation 4.31 is referred to as ARMA(p,q). The ARMA model of Equation 4.31 is a univariate forecast model. An effective extension to the ARMA model is the multivariate ARMA or ARMAX [34]. The ARMAX model explicitly includes the effect of exogenous variables, such as load and available generation, for electricity price forecast. In the ARMAX, a delay polynomial, similar to $\phi(B)$ and $\theta(B)$, is also considered for each exogenous variable. The free parameters of the ARMA(p,q) model include c, ϕ_k ($1 \le k \le p$), and θ_k ($1 \le k \le q$). For the ARMAX model, the coefficients of the delay polynomials of the exogenous variables (similar to ϕ_k and θ_k) should also be considered. These free parameters are estimated based on the historical data, for instance, by the least-squares approach or its variants [35]. It is noted that only the lagged features of the target variable and the exogenous variables that have been selected by the feature selection technique, that is, the selected candidate inputs, have a nonzero coefficient in the corresponding delay polynomials.

Another well-known extension to the ARMA model is ARIMA. Consider the dth-order differenced series $z(t)$ constructed from the target variable series $x(t)$ as follows:

$$z(t) = (1 - B)^d \cdot x(t) \tag{4.35}$$

The ARMA(p,q) model for the differenced series $z(t)$ is referred to as the ARIMA(p,q,d) model for the original series $x(t)$ [33]. Thus, the ARIMA model can be represented as follows:

$$\phi(B) \cdot (1 - B)^d \cdot x(t) = c + \theta(B) \cdot \varepsilon(t) \tag{4.36}$$

Nonstationary behavior in a time series arises from instability in the mean and variance of the series. The differentiation of the ARIMA model is usually used to deal with the nonstationary behavior induced by the variable mean. To stabilize the variance, several transformations, such as Box–Cox transformation [31], can be applied to the originals series.

Similar to ARMAX, a multivariate ARIMA model can be constructed by considering a delay polynomial for each exogenous variable in the ARIMA model. The ARMA and ARIMA models are nonseasonal models. To take into account the periodic behaviors of the target variable, the seasonal ARMA model

$$\phi(B) \cdot \Phi(B^s) \cdot x(t) = c + \theta(B) \cdot \Theta(B^s) \cdot \varepsilon(t) \tag{4.37}$$

and the seasonal ARIMA model

$$\phi(B) \cdot \Phi(B^s) \cdot (1 - B)^d \cdot (1 - B^s)^D \cdot x(t) = c + \theta(B) \cdot \Theta(B^s) \cdot \varepsilon(t) \tag{4.38}$$

can be used. In Equations 4.37 and 4.38, the delay polynomials $\phi(B)$ and $\theta(B)$ are as defined in Equations 4.32 and 4.33, respectively; $\Phi(B^s)$ and $\Theta(B^s)$ are polynomial functions of B^s instead of B. Similar to Equation 4.34, $B^s \cdot x(t) = x(t - s)$. The order s represents the corresponding seasonality. For instance, to model daily and weekly periodicity characteristics of the hourly price time series, $s = 24$ and $s = 168$ should be taken into account, respectively. In the above models, $\phi(B)$ and $\theta(B)$ are referred to as nonseasonal parts and $\Phi(B^s)$ and $\Theta(B^s)$ are called seasonal parts. Considering P and Q as the orders of the delay polynomials $\Phi(B^s)$ and $\Theta(B^s)$, as the functions of B^s, the ARMA model of Equation 4.37 is shown by ARMA$(p,q)(P,Q)_s$ and the ARIMA model of Equation 4.38 is illustrated by ARIMA$(p,q,d)(P,Q,D)_s$.

As previously described, both the ARMA and ARIMA models are dependent on the homoskedastic assumption of constant variance σ_ε^2 for the error term $\varepsilon(t)$. On the other hand, the forecasting models that are not based on the constant variance assumption are called heteroskedastic. Garcia et al. [36] present a heteroskedastic time-series technique, known as the generalized auto-regressive conditional heteroskedastic (GARCH) model, which can consider time dependent error variance and also model serial correlation in it.

4.4.1.2 Multilayer Perceptron Neural Network

The electricity price forecast process NSI(t)➔$\hat{P}_N(t)$, shown in Figure 4.2, is a nonlinear input/output mapping function. Theoretically, neural networks can represent any nonlinear function. In general, a neural network is a computer information processing system that simulates the function of the human brain. A discussion about biological neurons and modeling their performance in the neural networks (more accurately, artificial neural networks) can be found in Reference [3]. Among different neural networks, the MLP has received more attention for electricity price forecasting due to its flexibility and efficient learning algorithms developed for it. In this section, the MLP neural network and some of its commonly used training mechanisms are described.

The structure of an MLP neural network is shown in Figure 4.11. Observe from this figure that the MLP structure has one input layer, a few hidden layers, and one output layer. Each layer is composed of parallel processing units (also called cells, nodes, and neurons) that act in parallel and usually perform similar processing tasks (i.e., they have similar activation functions) [37]. The first layer of the MLP or input layer receives information (m inputs in Figure 4.11) from the outside world. For electricity price forecasting, these inputs are those selected by the feature selection technique or the candidate features in NSI(t). The last layer or the output layer gives the outputs of the neural network (p outputs in Figure 4.11) to the outside world after the incoming information is processed by the network. Between input and output layers of the MLP structure, there are hidden layers. The hidden layers only connect to each other or to the input and output layers. These layers do not have any connection to the outside world and so are called hidden layers. According to Kolmogorov's theorem [27], an MLP neural network can solve a problem by using one hidden layer, provided it has the proper number of neurons. Thus, one hidden layer may be used in the MLP structure ($n = 1$), but the number of its neurons must properly be selected. Hereafter, we focus on this MLP structure with a single hidden layer. This choice has the advantage of having a minimum number of parameters to be adjusted numerically, as the number of nodes of each hidden layer is an adjustable parameter for the MLP neural network.

In the MLP structure, each hidden layer neuron is connected by weights (Figure 4.11) to all neurons of the input and output layers. For instance, consider the connection between two neurons in the input and hidden layers. The output of the input layer node is multiplied by the corresponding weight and its weighted form is received at the input of the hidden layer node. In the fully connected structure of the MLP, each neuron in the hidden and output layers receives a vector of weighted outputs generated by the neurons of the previous layer. These neurons sum their received weighted inputs. Then, the result is applied to a nonlinear function called activation function of the neuron. For instance, the sigmoid and the hyperbolic tangent are two well-known activation functions of the hidden and output layer nodes of the MLP neural network [38]. The output of the activation function is the output of the neuron. Each input layer node has one input as

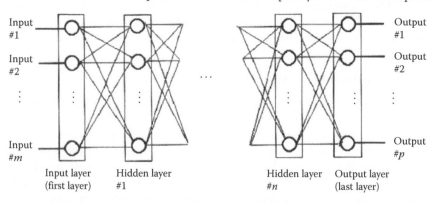

FIGURE 4.11 Structure of a multilayer perceptron (MLP) neural network (narrow lines illustrate the weights of the MLP).

shown in Figure 4.11. Moreover, these nodes usually do not perform any processing task. In other words, their activation functions are identity function $[I(x) = x]$. The input layer nodes only propagate inputs of the MLP to the next hidden layer. The nodes of the hidden and output layers process their inputs and generate their outputs until the outputs of the MLP (the outputs of the last layer shown in Figure 4.11) are produced. This is known as forward propagation in the MLP neural network. For this reason, MLP sometimes is considered as a feed-forward neural network.

From the above explanations, it can be seen that, by combining nonlinear activation functions, MLP constructs an input/output mapping function to map its inputs to the outputs. In other words, MLP constructs complex nonlinear mapping functions by combining simpler ones. From the linear algebra viewpoint, the activation functions of MLP act as the elements of the basis set. Thus, more efficient activation functions enhance the capability of MLP to construct a larger set of nonlinear functions. This is still a matter of research. For instance, WNN is a kind of MLP that uses mother wavelets as the activation functions of the hidden layer nodes [15,16]. The weights of the MLP are the coefficients of the combination. It can be said that the weights are the degrees of freedom of the neural network. The role of weights for an MLP neural network is similar to the role of free parameters $(c, \phi_K, \text{ and } \theta_K)$ for an ARMA time-series model.

MLP adapts its weights based on training samples taken from different operating points of the problem (here, electricity price forecast process). Indeed, this characteristic gives a learning ability to MLP, as, by adjusting its weights, MLP can construct a specific mapping function for the problem under consideration. The learning algorithm of the neural network performs this task. Usually, the weights of MLP are randomly initialized. In other words, the learning algorithm begins from a random initial point and determines the weights of the MLP so that the error of mapping the inputs of the training samples to their outputs is minimized with the expectation that a low error is also obtained for the unseen test samples. For instance, for day-ahead price prediction, these test samples can be 24-h forecast samples of the next day. Various learning algorithms have been presented for the MLP neural network in the literature. For instance, many of these learning algorithms and different activation functions are available in the neural network tool box of MATLAB® software package [39]. For electricity price forecasting, LM (Levenberg–Marquardt), BFGS (Broyden, Fletcher, Goldfarb, Shanno), and BR (Bayesian regularization) are three commonly used training mechanisms. A description of these learning algorithms can be found in References [26,40]. After training, the weights of the MLP are determined and the trained neural network can be used for price forecasting of the next time intervals.

An effective method to cope with the time-variant behavior of the input/output mapping function of the electricity price signal is adaptive training in which the weights of the MLP neural network are regularly updated based on the latest available data. For instance, for day-ahead price forecasting, the training phase of the MLP can be repeated each day considering the newly obtained data of that day.

Day-ahead electricity price prediction by the MLP neural network can be implemented by two approaches including direct forecasting and iterative forecasting [11]. In direct forecasting, the number of output layer nodes (p in Figure 4.11) is equal to the length of the forecasting horizon. For instance, for hourly price prediction of the next day, $p = 24$

in direct forecasting. By means of this approach, the future values of the electricity price signal are directly predicted from the MLP outputs. On the other hand, the neural network has one output node in the iterative forecasting method. Multiperiod price forecast, for example, prediction of electricity price for the next 24 h, is reached through recursion, that is, by feeding input variables with the neural network's outputs. For instance, predicted price for the first hour is used as $P(t-1)$ for the price forecast of the second hour provided that $P(t-1)$ is among the selected candidate inputs of NSI(t). Iterative forecasting may lead to cumulative error for the last time intervals of the forecasting horizon. On the other hand, in direct forecasting, the MLP neural network should learn the mapping function between the inputs and much more outputs (e.g., 24 outputs). This complicates the learning task of the MLP and may decrease its training efficiency. For this reason, iterative forecasting is used more than direct forecasting in the electricity price prediction research works using neural network-based forecasting engines.

4.4.2 Hybrid Price Forecast Engines

Hybrid forecast engines are constructed by hybridizing single prediction methods, such as combining neural networks with each other or combining neural networks with other forecast techniques. By combining forecast capabilities of different methods, hybrid price forecast engines have the potential to attain better prediction performance than nonhybrid forecast methods. However, effectiveness of a hybrid forecast engine is dependent on its building blocks, structure, and data flow. It cannot be said that any hybrid forecast method has prediction performance superior to that of a nonhybrid forecast technique. Even a poor design could lead to inferior performance of a hybrid forecast engine compared with its building blocks. For instance, Catalão et al. [41] have presented a hybrid price forecast method that is composed of wavelet transform, particle swarm optimization, and adaptive network-based fuzzy inference system. It has been shown that this hybrid method outperforms several other time series and neural network-based forecast engines for the price prediction of a well-known case study from the electricity market of mainland Spain. On the other hand, the prediction results of different forecast methods used in the third Makridakis competition (M3-Competition) have been analyzed by Makridakis and Hibon [42] and it is concluded that "Statistically sophisticated or complex methods do not necessarily produce more accurate forecasts than simpler ones" (p. 458). Thus, care must be taken in the design of the hybrid forecast methods including their building blocks, structure, and data flow.

In this section, three hybrid electricity price forecast engines are introduced. Also, a numerical comparison between these hybrid forecast engines and some nonhybrid electricity price prediction methods is presented.

4.4.2.1 Cascaded Neural Networks

The structure of cascaded neural networks (CNNs) is shown in Figure 4.12. This forecasting engine includes a set of forecasters arranged in sequence. Each forecaster is an MLP, although other kinds of neural networks can be used as well. Observe in Figure 4.12 that the number of cascaded forecasters is the number of time intervals of the forecasting horizon, that is, NH. On the basis of the terminology adopted in this chapter, $\hat{P}_N(t)$

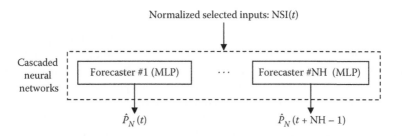

FIGURE 4.12 Structure of cascaded neural networks.

represents the predicted electricity price for the next time interval and so $\hat{P}_N(t + \text{NH} - 1)$ indicates the predicted price for the last time interval of the forecasting horizon. For day-ahead hourly price forecasting, $\text{NH} = 24$ and $\hat{P}_N(t), \ldots, \hat{P}_N(t + \text{NH} - 1) = \hat{P}_N(t + 23)$ represent the predicted prices for the next 24 h.

In CNNs, each MLP neural network is trained to learn the input/output mapping function between the electricity price of one time interval and its corresponding selected inputs or NSI(t). In other words, the price time series of each time interval is separately modeled and predicted in this forecasting engine. CNNs can be considered a compromise between direct forecasting and iterative forecasting. Each MLP of a CNN has one output node similar to iterative forecasting. The single output of each MLP provides the price forecast of its corresponding time interval, as shown in Figure 4.12. At the same time, the predictions of the future values of the electricity price signal are directly obtained from the outputs of the MLP neural networks. If forecaster #k of the CNN [predicting $\hat{P}_N(t + k - 1)$] requires $\hat{P}_N(t), \ldots, \hat{P}_N(t + k - 2)$ related to the previous time intervals of the forecasting horizon (such as the previous hours of the forecast day), these predictions are provided by its previous forecasters, that is, forecasters #1, ..., #$k - 1$. For instance, assume that the price of the previous time interval is among the selected inputs of NSI(t). Then, forecaster #2 should obtain this feature from the predicted price of forecaster #1, despite iterative forecasting based on a single MLP that the forecaster obtains this feature from its own previously predicted price.

In CNNs, the training samples of each MLP neural network reduce by 1/NH. Thus, the training efficiency of the forecasters may decrease. Although we can enlarge the training set by considering a longer training period, it means that old historical data are included in the training set. Considering time variant behavior of the electricity price signal, old historical data may have poor relevance or even be misleading for the training of the forecasting engine. Despite this deficiency, CNNs can have better prediction performance than a single MLP. CNNs can better model the periodic behavior of the electricity price signal. For instance, a CNN with 24 forecasters can effectively model the daily periodicity characteristic of the hourly price time series. Garcia-Martos et al. [43] have discussed how the 24 price time series constructed in this way are more homogenous. This simplifies the learning task of the forecasters. Moreover, each neural network of the CNN should predict one step ahead despite a single MLP that should predict the price values of all time intervals of the forecasting horizon (e.g., 24 h ahead).

4.4.2.2 Hybrid Neural Networks

As described in the previous section, electricity price is a nonlinear mapping function of many input variables. It is very hard for a single neural network to correctly learn the impact of all of these inputs on electricity price. Thus, enhancing the learning capability of a neural network-based forecasting engine can significantly improve its price forecast performance. A well-designed combination of different neural networks can potentially enhance their learning capability in modeling a complex process. For example, some parallel and cascaded structures for combining neural networks with improved price prediction performance have been proposed in various studies [8,10,44]. Moreover, CNNs, introduced in the previous section, are a cascaded structure of neural networks. However, these structures share the input data among their building blocks. Thus, the extracted knowledge of a block is not really shared with other blocks. In the hybrid neural network (HNN), on the other hand, a knowledge transfer procedure is envisioned from one neural network to another. For instance, the architecture of HNN, including three neural networks, is shown in Figure 4.13 [25]. These neural networks, denoted by NN1, NN2, and NN3 in the figure, have the same MLP structure.

Figure 4.13 shows that each neural network of the HNN transfers two sets of information to the next neural network. The first set includes the values obtained for the weights. As described in Section 4.4.1.2, usually, the weights of an MLP neural network are randomly initialized. Then, the MLP stores its extracted knowledge during the learning process in its weights. Thus, each neural network of the HNN transfers its obtained knowledge to the next one. Consequently, the next neural network can begin its learning process from the point that the previous one terminated (instead of beginning from a random point). Only the first neural network should begin with an initial set of random

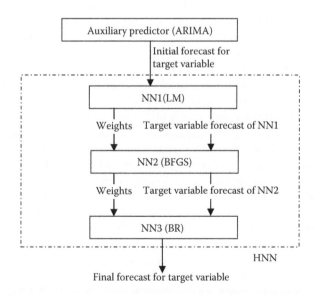

FIGURE 4.13 Architecture of a hybrid neural network and its auxiliary predictor.

values for the weights. As all neural networks of the HNN have the same MLP structure, these weights can be directly used by the next neural network and then it can increase the obtained knowledge of the previous one. Additionally, by selecting a variety of MLP learning algorithms for the neural networks, the HNN can benefit from wider learning capability. The three neural networks of the HNN in Figure 4.13 have LM, BFGS, and BR learning algorithms. Further discussions justifying these choices can be found in Reference [25].

The second set of results transferred from a neural network of the HNN to the next one is target variable forecast. This is due to the fact that the predicted price values, obtained from the other prediction methods, can be useful information for the price forecast of a forecaster. Thus, the obtained price forecast of each neural network of the HNN is used by the next one. In this way, each neural network of the HNN has an input close to its output feature, which enhances the training efficiency of the neural network to learn the behaviors and patterns of the target variable. In other words, each neural network of the HNN uses both the obtained knowledge and forecast result of its previous one. For the first neural network of the HNN, an auxiliary predictor provides the target variable forecast. This auxiliary predictor is chosen as the ARIMA time series for the HNN in a study by Amjady et al. [25] used for electricity price forecast (shown in Figure 4.13) and as the radial basis function (RBF) neural network for the HNN in another study by Amjady et al. [45] used for wind power prediction.

The auxiliary predictor is fed by the selected candidate inputs of the feature selection part, that is, NSI(t). The first neural network of the HNN is supplied with the selected candidate inputs plus the price forecast provided by the auxiliary predictor. NN2 of the HNN uses the same input features of NN1 except that the price forecast of the auxiliary predictor is replaced by the price forecast of NN1, as the price forecast of each block usually has better accuracy than the previous one and so can be a better choice for the input feature of the next block. Similarly, the input features of NN3 of the HNN include the selected candidate inputs plus the price forecast of NN2.

After the setup process for the auxiliary predictor and neural networks of the HNN, this forecast engine can predict price values for the time intervals of the forecasting horizon. MLP blocks of the HNN have one output node and so multiperiod price forecast is achieved through iterative forecasting described in Section 4.4.1.2. If we can use more accurate forecasts for initial time intervals, error propagation decreases and better predictions for all later intervals can be obtained. In other words, the problem of cumulative error for iterative forecasting is mitigated. The HNN has the potential to do that. For this purpose, each neural network of the HNN and also the auxiliary predictor, instead of recursive forecasting the price values of the whole time intervals of the forecasting horizon (e.g., 24 h ahead), only predicts the next-hour price and transfers it to the following block until the final price prediction of the HNN for that time interval is obtained by NN3 (as shown in Figure 4.13). This final price forecast, owning the least prediction error in the entire system, is used in the auxiliary predictor and neural networks of the HNN to predict its next-hour price. This cycle is repeated until the price values of the whole time intervals are predicted.

4.4.2.3 Hybrid Neuroevolutionary System

The structure of a hybrid neuroevolutionary system (HNES) including three neuroevolutionary blocks is shown in Figure 4.14 [26]. Observe that HNES is composed of neuroevolutionary blocks whereas building blocks of the HNN are the MLP neural networks. The neuroevolutionary blocks of the HNES have the same MLP structure, similar to the neural networks of the HNN. However, they are trained by a combination of neural network learning and evolutionary algorithms, whereas the training mechanisms of the neural networks of the HNN only include learning algorithms. This is the main difference between HNN and HNES.

As described in Section 4.4.1.2, the learning algorithm of an MLP neural network tries to optimize its weight values such that the error of the training samples or training error is minimized. For electricity price prediction, near-optimum solutions of this optimization problem are usually close to each other in the solution space. Each potential solution of the optimization problem includes all the weight values of the MLP neural network. At the end of the training phase, the learning algorithm of the MLP may find one of these near-optimum solutions, whereas the better ones might be in its vicinity and remain unseen for the neural network, as the learning algorithms usually search the solution space in a special direction (like the steepest descent). Hence, this provides the motivation to search around the final solution of the learning algorithm in various directions as much as possible to find a better solution. An evolutionary algorithm (EA) can be a suitable candidate for what is required. All EA parts of the HNES, denoted by

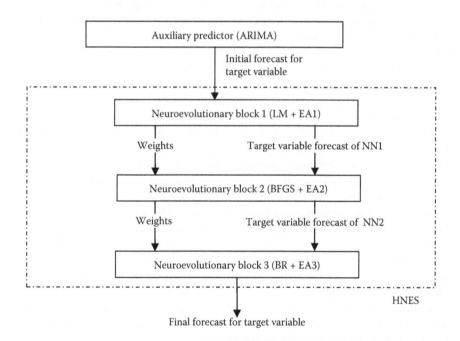

FIGURE 4.14 Architecture of a hybrid neuroevolutionary system and its auxiliary predictor.

EA1, EA2, and EA3 in Figure 4.14, have the same evolution mechanism that can be described as follows:

$$\Delta W_{i(n+1)} = m.\Delta W_{i(n)} + (1 - m).g.W_{i(n)} \tag{4.39}$$

$$W_{i(n+1)} = W_{i(n)} + \Delta W_{i(n+1)} \tag{4.40}$$

where W_i represents the ith element of the vector W containing all the weights of the MLP neural network; ΔW_i indicates change of W_i; subscripts n and $n + 1$ represent two successive generations (parent and child, respectively) of the EA; g is a small random number separately generated for each weight; and m is the momentum constant. Use of the momentum can smooth the search path decreasing sudden changes. For electricity price forecasting, $m = 0.5$ and selection of g in the range of $(0,0.1)$ for all generations of the EA could be appropriate choices. In other words, a uniform search without using localizing techniques (such as a hill climbing operator [46]) is recommended for the EA parts of the HNES. Nonuniform searches are suitable when there is one optimum point in the solution space and it is desired that the stochastic search technique, like EA, converges to it. However, in the optimization problem of the neural network training, there are several optimum solutions and each one may be better than the other.

To start the EA evolution, $W_{i(0)}$ is initialized as the obtained value from the MLP learning algorithm for the weight W_i and $\Delta W_{i(0)}$ is set to zero. In each cycle, the EA updates all weights W_i by means of Equations 4.39 and 4.40. Then, the error function of the MLP is evaluated for the solution of the new generation, that is, $W_{(n+1)}$. This error function can be the training error or the validation error of the MLP neural network, which will be introduced in the next section. If the child $W_{(n+1)}$ has less error function value than its parent $W_{(n)}$, the parent is replaced by the child, otherwise the parent is restored and the next cycle of the EA is executed. So, at the end of the EA, the best obtained solution with the lowest error function value among all generations will be selected.

The EA is executed after each learning algorithm of the HNES to enhance the training efficiency as much as possible. This matter is shown in Figure 4.14 for the three neuroevolutionary components of the HNES as "neuroevolutionary block 1 (LM + EA1)," "neuroevolutionary block 2 (BFGS + EA2)," and "neuroevolutionary block 3 (BR + EA3)," respectively. So, the whole training phase of the HNES can be represented as follows:

$$W_{\text{Initial}} \xrightarrow{\text{LM}} W(\text{LM}) \xrightarrow{\text{EA1}} W(\text{NE1}) \xrightarrow{\text{BFGS}} W(\text{BFGS})$$
$$\xrightarrow{\text{EA2}} W(\text{NE2}) \xrightarrow{\text{BR}} W(\text{BR}) \xrightarrow{\text{EA3}} W(\text{NE3}) \tag{4.41}$$

where the vector W_{Initial} includes the initial values for the weights. W_{Initial} is obtained from random initialization. The LM learning algorithm of the neuroevolutionary block 1 of the HNES, shown in Figure 4.14, begins from W_{Initial}. When the LM algorithm terminates, the values obtained for the weights, denoted by $W(\text{LM})$ in Equation 4.41, are given to its corresponding EA or EA1. Termination conditions for the neural network learning algorithms will be introduced in the next section. Then, EA1 further optimizes

W(LM) based on the described evolution mechanism. The stopping condition for the EA can be the maximum number of generations or it can be stopped when no better solution is obtained after a few successive generations. The weight values obtained by EA1, denoted by W(NE1) in Equation 4.41, represent the weights of the neuroevolutionary block 1. In other words, the weights of W(NE1) are loaded into the MLP structure of this block. At this stage, the training phase of the neuroevolutionary block 1 is terminated. Then, W(NE1) is passed to the neuroevolutionary block 2 to begin its training process. The BFGS learning algorithm of this block begins from the weight values of W(NE1) and its obtained solution, denoted by W(BFGS), is given to EA2 for further optimization. Similarly, the weight values obtained by EA2, that is, W(NE2), indicate the weights of the neuroevolutionary block 2 and are loaded into its MLP structure. W(NE2) is also given as initial values to the BR learning algorithm of the neuroevolutionary block 3. In the same way, the weight values obtained by the BR, that is, W(BR), are given to EA3 and its obtained solution W(NE3) represents the weight vector for the MLP structure of the neuroevolutionary block 3.

When all neuroevolutionary blocks of the HNES are trained, its training process is terminated and it is ready for forecasting the future values of the electricity price signal. Multiperiod price forecast of the HNES is obtained by means of iterative forecasting, described for the HNN in the previous section, to decrease the cumulative error.

Sample price prediction results for the forecast engines presented in this section are shown in Tables 4.6 and 4.7. The same test weeks as in Tables 4.1 through 4.4 are also considered in this numerical experiment. Also, the MAPE and VAR values reported in Tables 4.6 and 4.7 are as defined in Equations 4.25 and 4.28, respectively. For the sake of a fair comparison, all price forecast engines of Tables 4.6 and 4.7 have the same

- Nonhybrid data model of Equation 4.1.
- Linear transformation of Equation 4.5 with $[a,b] = [0,1]$ and its inverse transform Equation 4.6 as the preprocess (normalization) and postprocess blocks, respectively (Figure 4.2).
- Two-stage mutual information as the feature selection technique.
- Line search procedure for the fine tuning of the settings (this procedure will be introduced in Section 4.5.2).

Moreover, the historical data of 50 days prior to each forecast day, including $50 \times 24 = 1200$ h samples, are used for the setup of each forecast strategy of this numerical experiment.

TABLE 4.6 MAPE Values (%) Obtained by Different Forecast Engines for Day-Ahead Price Forecast of the Four Test Weeks of the PJM Electricity Market in 2010

Test Week	ARMAX	RBF	MLP + BR	MLP + BFGS	MLP + LM	CNN	HNN	HNES
Winter	9.48	7.69	7.24	7.11	6.42	5.61	4.82	4.77
Spring	8.82	6.93	6.85	6.97	6.14	5.75	5.12	5.28
Summer	9.31	7.79	7.83	7.86	6.85	6.21	5.67	5.38
Fall	9.76	7.95	7.79	7.69	6.67	5.52	4.91	4.59
Average	9.34	7.59	7.43	7.41	6.52	5.77	5.13	5.01

TABLE 4.7 VAR Values (×10⁻⁴) Obtained by Different Forecast Engines for Day-Ahead Price Forecast of the Four Test Weeks of the PJM Electricity Market in 2010

Test Week	ARMAX	RBF	MLP + BR	MLP + BFGS	MLP + LM	CNN	HNN	HNES
Winter	36.43	33.74	34.21	32.13	32.12	30.18	26.81	27.29
Spring	36.88	34.88	34.93	32.78	31.73	31.28	26.79	26.05
Summer	37.32	34.83	34.17	33.82	31.52	30.53	27.12	26.07
Fall	38.07	34.78	34.64	33.77	32.19	31.59	28.81	27.45
Average	37.17	34.56	34.49	33.12	31.89	30.89	27.38	26.71

Tables 4.6 and 4.7 show that the multivariate ARMA or ARMAX has the poorest results among all forecast engines of these tables in terms of both price prediction accuracy and stability. This is due to the fact that ARMAX is a linear forecast model, whereas electricity price is a nonlinear mapping function of its input variables. The nonlinear price forecast methods of neural networks including RBF, MLP trained by BR (MLP + BR), MLP trained by BFGS (MLP + BFGS), and MLP trained by LM (ML + PLM) lead to lower MAPE and VAR values. Among these neural network approaches, MLP + LM leads to better price forecast results due to greater effectiveness of this learning algorithm.

For day-ahead hourly electricity price forecasting of this numerical experiment, the CNN includes 24 cascaded forecasters such that each forecaster is an MLP + LM. The CNN, based on its hourly partitioning mechanism, can better model the daily periodicity characteristic of the hourly price signal and benefit from more specific training processes for its forecasters. Observe from Tables 4.6 and 4.7 that the CNN has better price forecast accuracy and stability than the single time-series and neural network approaches. It is noted that the training time of the CNN is not 24 times more than a single MLP + LM. Although the CNN has 24 MLP + LM neural networks, each one of them is trained by $1/NH = 1/24$ of the whole training samples. In other words, each neural network of the CNN should learn $1/24 \times 1200 = 50$ training samples (less training samples, but more specific ones, for each neural network). Thus, the increase in the training time of the CNN with respect to a single MLP + LM is much lower than 24 times (in this numerical experiment, it is about two times).

From Tables 4.6 and 4.7, it can be seen that the HNN has better MAPE and VAR values than the previous forecast engines. Although BFGS and BR may result in slightly less price forecast accuracy than LM for a single MLP, by combining the learning ability of its neural networks the HNN can reach a higher level of learning capability, which is a key issue for modeling the complex forecast process of the electricity price signal. The solution space for the optimization problem of MLP neural network training can be considered as a vector space, where its dimensions are the weights of the neural network. A learning algorithm for the MLP in this vector space can be taken into account as a search mechanism that searches for a point with minimum error function value. Now, if a search mechanism is saturated, another one may still be able to proceed, especially when it is equipped with a more efficient initial point. Moreover, transferring price forecast results from the auxiliary predictor to the first MLP of the HNN and between its neural networks is another important issue for the enhanced price prediction performance of the HNN. If an MLP neural network has an initial forecast following the trend of the target signal, it can easier learn the behavior of the signal [47]. In this case, the MLP must learn the difference between the two trajectories instead of the global values of the target trajectory. In the HNN, each MLP neural network has such an initial forecast. Moreover, the accuracy of this initial forecast increases from the first neural network to the last one.

Finally, the HNES overall has the best price prediction performance in terms of both price forecast accuracy and stability among all forecast engines of Tables 4.6 and 4.7. The average MAPE and VAR values of the HNES are lower than those of all other methods, illustrated in the last rows of Tables 4.6 and 4.7, respectively. Adding the local search ability of the EA parts to the HNES further enhances its learning capability with respect to the HNN.

4.5 Fine-Tuning of the Adjustable Parameters

In the previous sections, it has been seen that both the feature selection methods and forecast engines have adjustable parameters and their efficiency is dependent on the fine-tuning of these parameters. Moreover, the effectiveness of the training process of a neural network-based forecast engine for electricity price prediction is usually dependent on its termination mechanism. Premature termination can result in the incomplete learning of the neural network. On the other hand, a large number of training iterations may lead to the overfitting problem. In this section, efficient solution methods for these two problems (the fine-tuning of the adjustable parameters and effectively terminating the training process of the neural network) are presented.

4.5.1 Cross-Validation Techniques

To present cross-validation techniques, the error function of a neural network should be first introduced. The classical choice for the error function of a neural network is the error of the training samples or training error. However, only minimizing the training error of a neural network in the training phase may lead to the overfitting problem in which the neural network begins to memorize the training samples instead of learning them. When overfitting occurs in a neural network, the training error continues to decrease and it seems that the training process progresses, whereas in fact the generalization capability of the neural network degrades and loses its prediction capability for unseen forecast samples (generalization is a measure of how well the neural network performs on the actual problem once training is complete [48]). For electricity price prediction, in which price is a time-variant signal and its functional relationships vary with time, the problem becomes more serious. Although it seems that the neural network learns the training data well, it may encounter large prediction errors in the prediction phase.

To remedy this problem, the generalization performance of the neural network should also be monitored along its training phase. In other words, to have a correct measure of the price prediction capability of the forecast method, its error for the unseen part of the signal should be evaluated. However, as forecast error is not available in the training phase, validation error is used as an approximation of it. For neural network learning, validation samples are a subset of the training set that is not used for the training of the neural network (optimization of its weights) and retained unseen for the neural network. Thus, the error of validation samples or validation error can give an estimate of the neural network error for unseen forecast samples (e.g., electricity price values of 24 h ahead). Compared with the training error, validation error can better measure generalization capability of a neural network and avoid the overfitting problem [48].

Cross-validation techniques monitor the validation error of the neural network, in addition to its training error, along the learning process. These techniques can solve the two problems mentioned at the beginning of this section for neural network-based forecast engines.

After constructing the training set (including, for instance, the 1200-h samples of the 50 days prior to the forecast day), it is partitioned into the subset of training samples (used for the optimization of the neural network weights) and the subset of validation samples (retained unseen for the neural network). The learning algorithm trains the neural network to minimize the error of the training samples or training error. After each training iteration, both the training error and the validation error of the neural network are evaluated. For instance, training and validation errors of an MLP neural network for electricity price forecasting can be measured in terms of MAPE and AMAPE defined in Equations 4.25 and 4.26, respectively. To measure training/validation error by means of MAPE or AMAPE, NH should be the number of hours in the training/validation period (i.e., the number of training/validation samples) instead of the number of hours in the test period. Moreover, $P(t)$ and $\hat{P}(t)$ should be the actual and predicted values of electricity price, respectively, for hour t of the training/validation period.

Usually, the validation error of a neural network decreases during the initial iterations of the training phase, as does the training error. However, when the validation error starts to increase, the generalization performance of the neural network begins to degrade, indicating the occurrence of the overfitting problem. Thus, the training process of the neural network should be terminated at that iteration. This termination mechanism is also known as the early stopping condition [49]. In other words, the training iteration owning the lowest validation error brings the final results (the weight values) of the neural network training, wherein it is expected that the generalization capability of the neural network is maximized (its forecast error is minimized). In this way, cross-validation techniques, based on the validation error and early stopping condition, can effectively terminate the training process of the neural network, and thus the second problem is solved.

To better illustrate this matter, sample results of leave-one-out (LOO) cross-validation are shown in Figure 4.15 (this cross-validation method will be introduced in the subsequent paragraphs). The results shown in this figure are obtained for the day-ahead price forecast of the PJM electricity market for a forecast day of the summer test week of Tables 4.1 through 4.4 (August 20, 2010). It can be seen that the training error (heavy gray curve) continuously decreases. However, the validation error decreases up to a point (training iteration 12) and after that it increases due to the overfitting effect. Thus, this training process should be terminated in training iteration 12 with the lowest value of the validation error MAPE (0.0761% or 7.61%). Moreover, it is observed that the trend

of the test error or forecast error (light gray curve) is similar to the trend of the validation error curve such that after training iteration 12, the test error approximately begins
to increase. This observation confirms that training iteration 12 is also the best termination point for the learning algorithm based on the test error.

To solve the first problem, the cross-validation technique is executed by different sets
of adjustable parameters. More accurately, by different sets of the adjustable parameters,
the feature selection process and the training process of the neural network based forecast engine are carried out. Each time, the training process is terminated according to
the early stopping condition. In other words, each execution of the cross-validation
technique results in a training error curve and validation error curve (like the heavy
gray and black curves in Figure 4.15, respectively) and the validation error obtained
(e.g., MAPE = 0.0761 in Figure 4.15) is recorded. These results are denoted by the performance indices in Figure 4.2. The set of adjustable parameters leading to the minimum
value of the validation errors recorded among the executions of the cross-validation
technique is selected. We expect that these values of adjustable parameters also lead to
the minimum test error based on the available data.

The only remaining part of the cross-validation technique is selection of the subsets
of validation samples and training samples among the training set. Two common
approaches for this purpose are k-fold cross-validation and LOO cross-validation. In the
k-fold cross-validation technique, we randomly divide the training set into k subsets.
Using each subset as the validation samples, the setup process of the price forecast strategy (including the execution of the feature selection phase and training process of the
forecast engine) is carried out by the rest of the training set. The validation subset is
retained unseen for the feature selection technique and forecast engine to simulate the
situation of test samples. The performance of the forecast strategy is determined by the

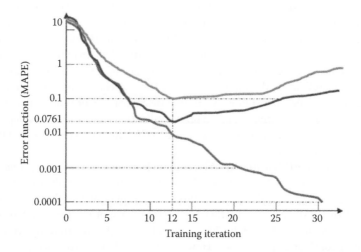

FIGURE 4.15 Sample results for training error (heavy gray), validation error (black), and test
error (light gray) obtained by the leave-one-out cross-validation for day-ahead price forecast of
the PJM electricity market for a forecast day of the summer test week (August 20, 2010).

average of the k validation errors. The combination of adjustable parameters leading to the best performance (the least value for the average of the k validation errors) is selected. For instance, the application of a tenfold cross-validation technique in electricity price spike prediction [4] and mid-term load forecast [50] has been presented in the literature.

In the LOO cross-validation technique, one subset of the training set is used as the validation samples and the remaining part of the training set is used for the setup of the price prediction strategy. For instance, for day-ahead price forecasting, the hourly samples of 1 day of the training period (e.g., 1 day out of the 50 days) are retained as the validation samples and the hourly samples of the remaining days (e.g., 49 days out of the 50 days) are used for the feature selection and training process of the forecast engine. As LOO cross-validation only uses one validation subset, its validation samples should be as similar as possible to the forecast samples so that the validation error can give a reasonable estimate of the prediction error. Considering the daily periodicity characteristic of the electricity price signal (described in Section 4.3.1.1), the hourly samples related to the day before the forecast day can be a suitable choice for the validation subset. However, other choices such as the hourly samples of the same day in the previous week (based on the weekly periodicity characteristic) may also be considered as the validation subset. Both the early stopping condition and the fine-tuning of adjustable parameters are implemented based on the single validation subset.

Numerical results obtained by fourfold cross-validation ($k = 4$), tenfold cross-validation (k = 10), and LOO for the four test weeks of Tables 4.1 through 4.4 are shown in Tables 4.8 and 4.9. The MAPE and VAR values reported in Tables 4.6 and 4.7 are as defined in Equations 4.25 and 4.28, respectively. All cross-validation methods of Tables 4.8 and 4.9 have:

- The nonhybrid data model of Equation 4.1.
- The linear transformation of Equation 4.5 with $[a,b] = [0,1]$ and its inverse transform Equation 4.6 as the preprocess (normalization) and postprocess blocks, respectively.
- Two-stage mutual information as the feature selection technique.
- HNES as the forecast engine.
- The historical data of 50 days prior to each forecast day including $50 \times 24 = 1200$ h samples as the training set.

TABLE 4.8 MAPE Values (%) Obtained by Different Cross-Validation Techniques for Day-Ahead Price Forecast of the Four Test Weeks of the PJM Electricity Market in 2010

Test Week	LOO	Fourfold	Tenfold
Winter	4.67	4.61	4.41
Spring	5.24	5.21	4.88
Summer	5.57	5.14	5.13
Fall	4.78	4.48	4.31
Average	5.07	4.86	4.68

TABLE 4.9   VAR Values (×10⁻⁴) Obtained by Different Cross-Validation Techniques for Day-Ahead Price Forecast of the Four Test Weeks of the PJM Electricity Market in 2010

Test Week	LOO	Fourfold	Tenfold
Winter	27.42	26.65	26.00
Spring	26.41	26.01	25.42
Summer	26.32	25.74	25.48
Fall	27.90	26.38	26.21
Average	27.01	26.19	25.78

Tables 4.8 and 4.9 show that the MAPE and VAR values obtained by the fourfold and tenfold cross-validation methods are slightly lower than those obtained by the LOO cross-validation technique. The fourfold and tenfold cross-validations can more accurately evaluate the effectiveness of each combination of adjustable parameters based on more validation subsets compared with the LOO cross-validation. Similarly, the tenfold cross-validation leads to slightly better MAPE and VAR results than the fourfold cross-validation. However, the computation burden of the tenfold and fourfold cross-validations (especially the tenfold cross-validation) is much more than that of the LOO cross-validation. It can roughly be said that the computation effort required for a k-fold cross-validation technique is k times more than that required for an LOO cross-validation method. Thus, considering computation time limits for short-term price forecasting applications, LOO cross-validation might become the preferable choice.

4.5.2 Search Procedures

Search procedures are a more systematic approach for fine-tuning adjustable parameters than cross-validation techniques. Suppose that $(SP_1, \ldots, SP_{NS})$ represent the set points or adjustable parameters of the electricity price forecast strategy, by the number of NS, including those of the feature selection technique and forecast engine. For instance, the price forecast strategy used in the previous numerical experiment includes two set points for the feature selection part (the thresholds TH1 and TH2 for the two-stage mutual information) and at least one set point for the HNES forecast engine (the number of nodes in the single hidden layer of the MLP neural networks of the HNES). Thus, NS for this electricity price forecast strategy becomes 3. Cross-validation techniques randomly change these set points within their allowable ranges (obtained from the engineering judgment) until the combination of adjustable parameters leading to the best performance (the minimum validation error) is found. Variation steps for these parameters can be selected based on the allowable computation time. Smaller variation steps results in higher resolutions of the search process of the cross-validation technique and finding the appropriate values of the adjustable parameters with more accuracy. At the same time, the computation burden will increase. For instance, the thresholds TH1 and TH2 for the normalized mutual information (normalized with respect to their maximum values [2]) may be varied in the range of [0.3,0.8] and the allowable range for the number of nodes in the single hidden layer of the MLP neural networks of the HNES may be [10,30]. If the variation step for TH1 and TH2 is selected as 0.1 and for the number of nodes of the hidden layer as 2, the solution space of the optimization problem will have $6 \times 6 \times 11 = 396$ points. By making these variation steps smaller/larger a more dense/coarse grid can be constructed. The cross-validation technique examines some of these points (or all of them) on the basis of the permissible computation time and the best found combination of adjustable parameters is selected.

Despite cross-validation techniques, search procedures do not change the adjustable parameters simultaneously. These procedures divide the adjustable parameters into subsets. Each time, the parameters of only one subset are changed, whereas the parameters of the other subsets are kept constant. After finding the appropriate values for the set points of one subset, the next subset is processed. In this way, we can better control the variation of the parameters. The simplest form of the search procedure is line search in which only one parameter is changed each time. In other words, we search along one dimension of the solution space (or along a line) each time. After fine-tuning all adjustable parameters one by one, a cycle of the line search procedure is terminated. Then, the next cycle begins with the values obtained in the previous cycle. Only the first cycle begins with the randomly chosen values for the adjustable parameters within their allowable ranges. When the maximum number of cycles is reached or when there is no change between the values of adjustable parameters in two successive cycles, the line search procedure can be terminated. More details about the search procedures can be found in Reference [2].

It is noted that the search procedures can be implemented on the basis of the LOO or *k*-fold mechanisms like the cross-validation techniques. Indeed, these procedures have different search mechanisms with respect to cross-validation methods.

Results obtained by the line search procedure with the LOO mechanism for the numerical experiment described in Tables 4.8 and 4.9 are reported in Tables 4.10 and 4.11, respectively, and are compared with the results obtained by the LOO cross-validation method. From Tables 4.10 and 4.11, it can be observed that the line search procedure leads to slightly more accurate and more stable price forecast results than the LOO cross-validation technique, whereas their computation times for this numerical experiment are approximately the same. This is due to a more effective search mechanism of the line search procedure.

Up to now, all components of an artificial intelligence-based electricity price forecast strategy, shown in Figure 4.2, have been introduced. To also give a graphical view about

TABLE 4.10 MAPE Values (%) Obtained by the Line Search Procedure and Cross-Validation Technique, Both with the Leave-One-Out (LOO) Mechanism, for Day-Ahead Price Forecast of the Four Test Weeks of the PJM Electricity Market in 2010

Test Week	Line Search Procedure	LOO Cross-Validation
Winter	4.77	4.67
Spring	5.28	5.24
Summer	5.38	5.57
Fall	4.59	4.78
Average	5.01	5.07

TABLE 4.11 VAR Values (×10⁻⁴) Obtained by the Line Search Procedure and Cross-Validation Technique, Both with the Leave-One-Out (LOO) Mechanism, for Day-Ahead Price Forecast of the Four Test Weeks of the PJM Electricity Market in 2010

Test Week	Line Search Procedure	LOO Cross-Validation
Winter	27.29	27.42
Spring	26.05	26.41
Summer	26.07	26.32
Fall	27.45	27.90
Average	26.71	27.01

the overall effectiveness of the presented components, the electricity price prediction results of the forecast strategy are shown in Figures 4.16 through 4.19 for the four test weeks, respectively. The forecast strategy has the same data model, preprocess and postprocess blocks, feature selection technique, forecast engine and training set of the previous numerical experiment. Moreover, it has early stopping condition as the termination mechanism for the training process of the neural networks of the HNES (forecast engine) and line search procedure for fine-tuning adjustable parameters, on the basis of the LOO mechanism. From Figures 4.16 through 4.19, volatile behavior and irregular patterns of the electricity price signal can be observed. Despite this, the forecasts generated using artificial intelligence, data mining, and optimization techniques present good accuracy with reasonable computation burden. In all figures, the forecast curve accurately follows the actual curve and the error curve overall has small values.

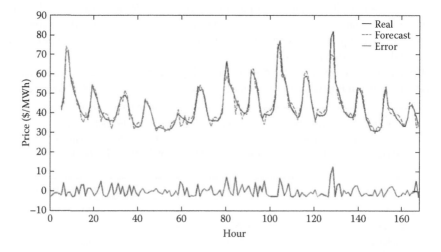

FIGURE 4.16 Results obtained by the artificial intelligence-based electricity price forecast strategy for day-ahead price forecast of the winter test week of the PJM electricity market.

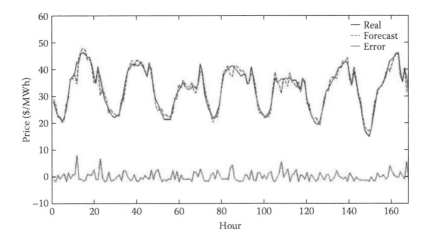

FIGURE 4.17 Results obtained by the artificial intelligence-based electricity price forecast strategy for day-ahead price forecast of the spring test week of the PJM electricity market.

4.6 Price Spike Forecast

Price spikes are a distinctive characteristic of the electricity price signal. Electricity price spikes are important for market participants. For instance, they can bring serious economic damages to customers. In Section 4.1, a discussion about some effective factors on the generation of price spikes has been presented. In this section, at first, the concept of electricity price spike is more accurately introduced. Then, two key forecast processes of price spikes, including price spike occurrence prediction and price spike

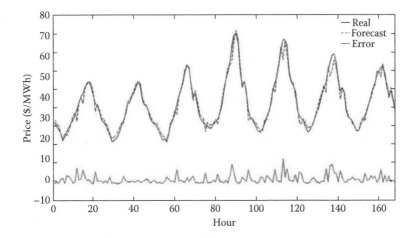

FIGURE 4.18 Results obtained by the artificial intelligence-based electricity price forecast strategy for day-ahead price forecast of the summer test week of the PJM electricity market.

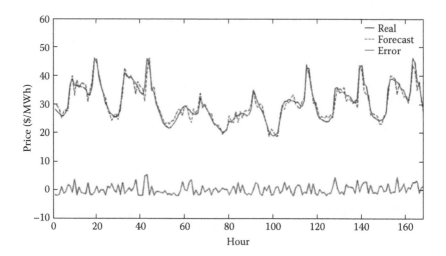

FIGURE 4.19 Results obtained by the artificial intelligence-based electricity price forecast strategy for day-ahead price forecast of the fall test week of the PJM electricity market.

value prediction, are discussed and solution methods for these two forecast processes are presented.

4.6.1 What is a Price Spike?

A price spike can be generally defined as an abnormal price value, which has significant difference with respect to the expected value of the signal. On the basis of this definition, price spikes may be classified into three categories [51,52]:

1. Abnormal high price: A price that is significantly higher than its expected value.
2. Abnormal jump price: If the absolute value of difference between electricity price values in two successive time intervals is greater than a jump threshold JTH, that is, we have

$$|P(t) - P(t-1)| > \text{JTH} \tag{4.42}$$

   then $P(t)$ is defined as a price spike of abnormal jump price type.
3. Negative price: A price value lower than zero is defined as a negative price.

An abnormal high price is the most common form of electricity price spikes [52]. Moreover, most research in the area of price spike forecasting focuses on the abnormal high price-type price spikes [4,51–54]. Thus, hereafter, we also focus on this type of price spike. To analyze these price spikes, at first, we must determine how high the prices should be in order to be considered as spikes. In other words, a price spike threshold should be chosen to discriminate price spikes from normal prices. Here, nonspike prices are considered as normal prices. The threshold of $\mu_p + 2\sigma_p$ has been proposed for this purpose [51–53], where μ_p and σ_p indicate mean and standard deviation of historical market prices, respectively. On the basis of this criterion, different price spike thresholds can be inferred for different electricity markets. In Figure 4.1, two price spike thresholds of 150 and 200 \$/MWh, proposed by Amjady and Keynia [4,54] for the price time series of the PJM electricity market, are shown. A discussion about selection of price spike thresholds can be found in References [4,9]. After selecting the threshold, historical data for price spikes can be collected. Thus, the setup process of the price spike forecaster can be executed by the collected data. For price spike forecasting, both price spike occurrence and value should be predicted. Prediction methods for these two forecast processes will be presented in the next two sections.

4.6.2 Prediction of Price Spike Occurrence

The artificial intelligence-based electricity price forecast strategy shown in Figure 4.2 can also be used for price spike occurrence prediction. However, it should be noted that the target variable $P(t)$ for each time interval t is a continuous real-valued variable. On the other hand, the forecast process of price spike occurrence prediction is a binary classification task wherein the output feature is a binary-valued variable indicating whether or not a price spike in the corresponding time interval occurs.

The previously presented forecast engines are efficient estimators of continuous variables such as electricity price value. Although an estimator can also be used for classification tasks (e.g., by rounding its estimated output to the closest class label), it is better to use specifically designed classifiers for this purpose. For instance, a probabilistic neural network (PNN) is an efficient classifier that can be used for price spike occurrence prediction. Details of the PNN classifier can be found in Reference [55]. An advantage of the PNN is its rapid training, as the learning phase of the PNN is done in one pass of each training sample rather than in several of them.

In addition to using a classifier as the forecasting engine, it is recommended that a hybrid data model including both time- and frequency-domain candidate features, such as the data model shown in Equation 4.4, is used for price spike occurrence prediction. A price spike contains sudden changes in the time domain or equivalently high-frequency components in the frequency domain. Thus, frequency domain candidate inputs can contain important information for price spike occurrence prediction. Moreover, as price spikes usually constitute a small portion of the whole electricity price time series (which can be seen, for instance, from Figure 4.1), calendar indicators may also include useful information to enhance the discrimination capability of a classifier to predict price spike occurrence. For instance, seasonal, daily, and hourly calendar indicators can be added to the hybrid data model encoding 4 seasons of a year, 7 days of a week (public holidays may be treated like weekends), and 24 h of a day, respectively.

Another useful candidate input for price spike occurrence prediction is the existence feature. It is a binary variable for each time interval t indicating whether or not a price spike occurs in 24 h before t. Zhao et al. [51] have discussed how spikes tend to occur together over a short period of time and this period can be several hours but no longer than a day. From Figure 4.1 too it can be seen that price spikes are usually close to each other. However, a binary existence variable cannot bring whole related information about consecutive price spikes. For instance, the probability of spike occurrence reduces when the closest price spike becomes farther. Hence, the existence candidate input has been improved in the study by Amjady and Keynia [4] as follows:

$$\text{Existence}(t) = \begin{cases} k & \text{if the closest price spike occurs at } t - k \text{ such that } 1 \leq k \leq 24 \\ 25 & \text{if no price spike occurs in the period } [t - 24, t - 1] \end{cases} \quad (4.43)$$

In this way, the existence feature can also measure the distance with the closest price spike.

On the basis of the above explanations, a price spike occurrence prediction strategy can be constructed as follows:

- The hybrid data model of Equation 4.4 plus the seasonal, daily, and hourly calendar indicators and the existence feature defined in Equation 4.43.
- The linear transformation of Equation 4.5 with $[a,b] = [0,1]$ and its inverse transform Equation 4.6 as the preprocess (normalization) and postprocess blocks, respectively.
- Two-stage mutual information as the feature selection technique.
- PNN as the forecast engine.
- The historical data of 1 year prior to each forecast day including $365 \times 24 = 8760$ h samples as the training set.
- A search procedure or cross-validation technique for fine-tuning adjustable parameters.

Results obtained from this prediction strategy for price spike occurrence forecasting of day-ahead electricity market of PJM in 2006 have been presented by Amjady and Keynia [4] and compared with the results of some other price spike occurrence prediction strategies. Despite the previous numerical experiments in this chapter (reported in Tables 4.1 through 4.4 and 4.6 through 4.11) that consider four specific test weeks, the whole price data of the PJM electricity market in 2006 have been considered in the study by Amjady and Keynia [4], as mentioned. This is due to the fact that a large number of test samples ($4 \times 168 = 792$) can be generated for price forecast by considering four test weeks. However, only a small number of price spikes may be found in four test weeks and so the price spike prediction performance of the forecast strategy may not be correctly evaluated by the price data of four test weeks. Similarly, the historical data of 1 year prior to each forecast day has been considered as the training set here (instead of 50 days prior to each forecast day) to have a sufficient number of price spikes among the training samples.

The price spike occurrence prediction results in the study by Amjady and Keynia [4] have been presented for two electricity price spike thresholds of 150 and 200 \$/MWh, indicated in Figure 4.1. In the day-ahead electricity market of PJM in 2006, the number of price spikes with a threshold of 150 \$/MWh, that is, 73, is more than the number of price spikes with a threshold of 200 \$/MWh, that is, 25. This can also be observed in Figure 4.1.

A well-known criterion for evaluating the performance of classifiers is classification accuracy defined as follows:

$$\text{Classifier accuracy (\%)} = \frac{\text{Number of correctly classified test samples}}{\text{Number of test samples}} \times 100 \qquad (4.44)$$

However, for price spike occurrence prediction, this criterion usually is not informative because the data of the problem are seriously imbalanced [51]. For instance, for the day-ahead electricity market of PJM in 2006, $73/8760 = 0.83\%$ and $25/8760 = 0.29\%$ out of all test samples are price spikes with the two price spike thresholds of 150 and 200 \$/MWh, respectively. Thus, even if the price spike occurrence prediction strategy misclassifies all price spikes as normal prices, high classification accuracies of 99.17% and 99.71% can still be obtained for the thresholds of 150 and 200 \$/MWh, respectively. Thus, new criteria are required to correctly evaluate price spike occurrence prediction performance of a forecast strategy. Two criteria are defined for this purpose by Zhao et al. [51,53]:

$$\text{Spike prediction accuracy (\%)} = \frac{\text{Number of correctly predicted spikes}}{\text{Number of spikes}} \times 100 \qquad (4.45)$$

$$\text{Spike prediction confidence (\%)} = \frac{\text{Number of correctly predicted spikes}}{\text{Number of predicted spikes}} \times 100 \qquad (4.46)$$

Spike prediction accuracy measures the ability to correctly predict spikes. At the same time, the classifier may misclassify some normal prices as price spikes. The spike prediction confidence can measure these errors of the price spike occurrence prediction strategy. The number of correctly predicted price spikes, the number of incorrect predictions (normal prices predicted as price spikes), spike prediction accuracy, and spike prediction confidence for the day-ahead electricity market of PJM in 2006 with the two price spike thresholds of 150 and 200 \$/MWh are shown in Table 4.12. These results have been obtained by the price spike occurrence forecasting strategy of Amjady and Keynia [4] outlined above.

For instance, for the price spike threshold of 150 \$/MWh, the spike prediction accuracy of the strategy becomes $71/73 = 97.3\%$ and its spike prediction confidence becomes $71/(71 + 10) = 87.7\%$. It is seen that despite high volatility of price spikes, both reasonable accuracy and confidence can be obtained for price spike occurrence prediction by using appropriately designed forecast strategies. From Table 4.12, it can also be observed that better price spike occurrence prediction accuracy and confidence are obtained for the price spike threshold of 150 \$/MWh compared with the threshold of 200 \$/MWh.

TABLE 4.12 Results Obtained for Price Spike Occurrence Prediction of the Day-Ahead Electricity Market of PJM in 2006 with Two Price Spike Thresholds of 150 and 200 \$/MWh

Price spike threshold	150	200
Number of spikes	73	25
Number of correct predictions	71	23
Number of incorrect predictions	10	3
Spike prediction accuracy (%)	97.3	92.0
Spike prediction confidence (%)	87.7	88.5

More spikes can be found with the lower price spike threshold of 150 \$/MWh. Thus, the feature selection process and the forecast engine's training process can be carried out more effectively with more spiky training samples.

4.6.3 Prediction of Price Spike Value

The prediction of price spike value in addition to price spike occurrence can produce a more informative forecast for price spike. After discriminating normal prices from price spikes by means of the price spike occurrence predictor, two estimators can be specifically trained to forecast price spike values and normal price values; thus, their training efficiency enhances compared with a single estimator that should learn the behaviors of both price spikes and normal prices. In other words, a complex forecast process is decomposed to two more specific ones, which can be implemented more effectively.

Results obtained from such a prediction strategy for price spike value prediction and normal price value prediction of the day-ahead electricity market of PJM in 2008 have been presented by Amjady and Keynia [54]. The price spike occurrence prediction is as described in the previous section and its two estimators (used for price spike value prediction and normal price value prediction) are as described in Section 4.5.2 to generate the forecast results of Figures 4.16 through 4.19. Moreover, a closed-loop prediction mechanism is also proposed by Amjady and Keynia [54] to combine the classifier and the two estimators with the aim of generating consistent forecasts for price spike occurrence, price spike value, and normal price value. In other words, when the price spike occurrence predictor forecasts the status of the electricity price for a future time interval as spike/normal (and so the corresponding estimator is activated), the forecast generated for the price value should be greater/smaller than the price spike threshold. The conventional open-loop forecast methodologies cannot guarantee this consistency and may generate inconsistent predictions for price spike occurrence and value, as these forecast methods have no feedback from their outputs (generated predictions). On the other hand, the closed-loop prediction strategy, using the concept of closed-loop control systems, can remedy this problem.

In engineering applications, closed-loop control systems, based on the feedback from the output, can change their inputs to modify their outputs, a characteristic that is not seen in the open-loop control systems. Similarly, the closed-loop prediction strategy takes the feedback from the output. If the prediction generated for price spike occurrence and the forecast produced for the value of price spike/normal price are inconsistent, the inputs of the price spike occurrence predictor are changed accordingly on the basis of the price value prediction by the activated estimator. In this way, the price spike occurrence predictor can modify its output on the basis of the closed-loop operation. This cycle is continued until consistently more accurate forecasts for price spike occurrence, price spike value, and normal price value are obtained. More details about the closed-loop price spike prediction strategy can be found in Reference [54].

4.7 Conclusions

The electricity price signal is a nonlinear time-variant mapping function of many input variables. Its time series usually represents volatile behavior in the form of sudden changes, irregular patterns, outliers, and even price spikes. To predict the future behavior of this signal, a qualified data model should be first constructed including the possible price drivers as much as possible. In addition to time-domain features, frequency-domain candidate inputs may also be considered to further enrich the data model. The constructed data model should be refined by means of data mining techniques and information theoretic criteria to select a minimum subset of the most informative features for the forecast engine. The prediction method should be able to extract the mapping function between the selected inputs and the target variable. Learning capability of the forecast engine is a key issue in this regard. As much as a more informative set of inputs is selected and the forecast engine can better learn the impact of the selected inputs on the target variable, the price prediction performance of the forecast strategy enhances.

Feature selection methods and forecast engines usually have a few adjustable parameters and their performance is dependent on fine-tuning these settings. These parameters nonlinearly affect the performance of the forecast strategy and may interact with each other. Thus, selecting appropriate values for the settings is a hard task for the user and automatically fine-tuning these parameters is required. Designing effective cross-validation techniques and search procedures that can optimize the values of adjustable parameters with a reasonable computation burden is another important issue for the efficiency of the price forecast strategy. It is noted that an effective prediction strategy should be able to produce both accurate and stable price forecasts.

Prediction of price spikes, which have more volatility than normal prices, is a more complex forecast process. For price spike forecasting, we should predict both price spike occurrence and value. Moreover, these predictions should be consistent. Combining classifiers with estimators and using closed prediction mechanisms can present potential solutions for these requirements.

References

1. N. Amjady, Day-ahead price forecasting of electricity markets by a new fuzzy neural network, *IEEE Transactions on Power Systems*, 21(2), 887–896, 2006.
2. N. Amjady and F. Keynia, Day-ahead price forecasting of electricity markets by mutual information technique and cascaded neuro-evolutionary algorithm, *IEEE Transactions on Power Systems*, 24(1), 306–318, 2009.
3. M. Shahidehpour, H. Yamin, and Z. Li, *Market Operations in Electric Power Systems*, New York: Wiley, 2002.
4. N. Amjady and F. Keynia, Electricity market price spike analysis by a hybrid data model and feature selection technique, *Electric Power Systems Research*, 80(3), 318–327, 2010.
5. N. Amjady and M. Hemmati, Energy price forecasting—problems and proposals for such predictions, *IEEE Power and Energy Magazine*, 4(2), 20–29, 2006.
6. Y. S. Son, R. Baldick, K. Lee, and S. Siddiqi, Short-term electricity market auction game analysis: Uniform and pay-as-bid pricing, *IEEE Transactions on Power Systems*, 19(4), 1990–1998, 2004.

7. N. Bigdeli, K. Afshar, and N. Amjady, Market data analysis and short-term price forecasting in the Iran electricity market with pay-as-bid payment mechanism, *Electric Power Systems Research*, 79(6), 888–898, 2009.

8. A. M. Gonzalez, A. M. San Roque, and J. G. Gonzalez, Modeling and forecasting electricity prices with input/output hidden markov models, *IEEE Transactions on Power Systems*, 20(2), 13–24, 2005.

9. R. Weron, *Modeling and Forecasting Electricity Loads and Prices: A Statistical Approach*, Chichester: Wiley, 2006.

10. J. J. Guo and P. B. Luh, Improving market clearing price prediction by using a committee machine of neural networks, *IEEE Transactions on Power Systems*, 19(4), 1867–1876, 2004.

11. N. Amjady and F. Keynia, Day ahead price forecasting of electricity markets by a mixed data model and hybrid forecast method, *International Journal of Electric Power and Energy Systems*, 30(9), 533–546, 2008.

12. A. J. Rocha Reis and A. P. Alves da Silva, Feature extraction via multiresolution analysis for short-term load forecasting, *IEEE Transactions on Power Systems*, 20(1), 189–198, 2005.

13. S. Fan, C. Mao, and L. Chen, Next-day electricity-price forecasting using a hybrid network, *IET Generation, Transmission & Distribution*, 1(1), 176–182, 2007.

14. A. J. Conejo, M. A. Plazas, R. Espinola, and A. B. Molina, Day-ahead electricity price forecasting using the wavelet transform and ARIMA models, *IEEE Transactions on Power Systems*, 20(2), 1035–1042, 2005.

15. N. M. Pindoriya, S. N. Singh, and S. K. Singh, An adaptive wavelet neural network-based energy price forecasting in electricity markets, *IEEE Transactions on Power Systems*, 23(3), 1423–1432, 2008.

16. L. Wu and M. Shahidehpour, A hybrid model for day-ahead price forecasting, *IEEE Transactions on Power Systems*, 25(3), 1519–1530, 2010.

17. S. Mallat, A theory for multiresolution signal decomposition-the wavelet representation, *IEEE Transactions on Pattern Analysis and Machine Intelligence*, 11(7), 674–693, 1989.

18. S. Mallat, *A Wavelet Tour of Signal Processing*, New York: Academic Press, 1998.

19. M. Milligan, K. Porter, E. DeMeo, P. Denholm, H. Holttinen, B. Kirby, N. Miller et al., Wind power myths debunked, *IEEE Power and Energy Magazine*, 7(6), 89–99, 2009.

20. N. Kwak and C. H. Choi, Input feature selection for classification problems, *IEEE Transactions on Neural Networks*, 13(1), 143–159, 2002.

21. N. Kwak and C. H. Choi, Input feature selection by mutual information based on Parzen window, *IEEE Transactions on Pattern Analysis and Machine Intelligence*, 24(12), 1667–1671, 2002.

22. H. Peng, F. Long, and C. Ding, Feature selection based on mutual information: Criteria of max-dependency, max-relevance and min-redundancy, *IEEE Transactions on Pattern Analysis and Machine Intelligence*, 27(8), 1226–1238, 2005.

23. Y. Sun, Iterative RELIEF for feature weighting: Algorithms, theories, and applications, *IEEE Transactions on Pattern Analysis and Machine Intelligence*, 29(6), 1035–1051, 2007.

24. N. Amjady and A. Daraeepour, Design of input vector for day-ahead price forecasting of electricity markets, *Journal of Expert Systems with Applications*, 36(10), 12281–12294, 2009.

25. N. Amjady, A. Daraeepour, and F. Keynia, Day-ahead electricity price forecasting by modified RELIEF algorithm and hybrid neural network, *IET Generation, Transmission & Distribution*, 4(3), 432–444, 2010.

26. N. Amjady and F. Keynia, Application of a new hybrid neuro-evolutionary system for day-ahead price forecasting of electricity markets, *Applied Soft Computing*, 10(3), 784–792, 2010.

27. G. J. Tsekouras, N. D. Hatziargyriou, and E. N. Dialynas, An optimized adaptive neural network for annual midterm energy forecasting, *IEEE Transactions on Power Systems*, 21(1), 385–391, 2006.

28. F. J. Nogales, J. Contreras, A. J. Conejo, and R. Espinola, Forecasting next-day electricity prices by time series models, *IEEE Transactions on Power Systems*, 17(2), 342–348, 2002.

29. P. Mandal, T. Senjyu, N. Urasaki, T. Funabashi, and A. K. Srivastava, A novel approach to forecast electricity price for PJM using neural network and similar days method, *IEEE Transactions on Power Systems*, 22(4), 2058–2065, 2007.

30. A. J. Conejo, J. Contreras, R. Espinola, and M. A. Plazas, Forecasting electricity prices for a day-ahead pool-based electric energy market, *International Journal of Forecasting*, 21(3), 435–462, 2005.

31. F. J. Nogales and A. J. Conejo, Electricity price forecasting through transfer function models, *Journal of the Operational Research Society*, 57(4), 350–356, 2006.

32. J. Contreras, R. Espinola, F. J. Nogales, and A. J. Conejo, ARIMA models to predict next-day electricity prices, *IEEE Transactions on Power Systems*, 18(3), 1014–1020, 2003.

33. H. Zareipour, C. A. Canizares, K. Bhattacharya, and J. Thomson, Application of public-domain market information to forecast Ontario's wholesale electricity prices, *IEEE Transactions on Power Systems*, 21(4), 1707–1717, 2006.

34. N. Amjady, Generation adequacy assessment of power systems by time series and fuzzy neural network, *IEEE Transactions on Power Systems*, 21(3), 1340–1349, 2006.

35. S. L. Marple, *Digital Spectral Analysis with Applications*, Englewood Cliffs, NJ: Prentice-Hall, 1987.

36. R. C. Garcia, J. Contreras, M. V. Akkeren, and J. B. C. Garcia, A GARCH forecasting model to predict day-ahead electricity prices, *IEEE Transactions on Power Systems*, 20(2), 867–874, 2005.

37. N. Amjady, *Introduction to Intelligent Systems*. Semnan, Iran: Semnan University Press, 2002.

38. N. Amjady and H. Hemmati, Day-ahead price forecasting of electricity markets by a hybrid intelligent system, *European Transactions on Electrical Power*, 19(1), 89–102, 2009.

39. MathWorks Inc., *Neural Networks Toolbox User's Guide*, 2005.

40. J. P. S. Catalão, S. J. P. S. Mariano, V. M. F. Mendes, and L. A. F. M. Ferreira, Short-term electricity prices forecasting in a competitive market: A neural network approach, *Electric Power Systems Research*, 77(10), 1297–1304, 2007.

41. J. P. S. Catalão, H. M. I. Pousinho, and V. M. F. Mendes, Hybrid wavelet-PSO-ANFIS approach for short-term electricity prices forecasting, *IEEE Transactions on Power Systems*, 26(1), 137–144, 2011.

42. S. Makridakis and M. Hibon, The M3-Competition: Results, conclusions and implications, *International Journal of Forecasting*, 16(4), 451–476, 2000.

43. C. Garcia-Martos, J. Rodriguez, and M. J. Sanchez, Mixed models for short-run forecasting of electricity prices: Application for the Spanish market, *IEEE Transactions on Power Systems*, 22(2), 544–551, 2007.

44. L. Zhang, P. B. Luh, and K. Kasiviswanathan, Energy clearing price prediction and confidence interval estimation with cascaded neural network, *IEEE Transactions on Power Systems*, 18(1), 99–105, 2003.

45. N. Amjady, F. Keynia, and H. Zareipour, Wind power prediction by a new forecast engine composed of modified hybrid neural networks and enhanced particle swarm optimization, *IEEE Transactions on Sustainable Energy*, 2(3), 265–276, 2011.

46. I. G. Damousis, A. G. Bakirtzis, and P. S. Dokopoulos, A solution to the unit commitment problem using integer-coded genetic algorithm, *IEEE Transactions on Power Systems*, 19(2), 1165–1172, 2004.

47. N. Amjady, Short-term bus load forecasting of power systems by a new hybrid method, *IEEE Transactions on Power Systems*, 22(1), 333–341, 2007.

48. D. R. Hush and B. G. Horne, Progress in supervised neural networks, *IEEE Signal Processing Magazine*, 10(1), 8–39, 1993.

49. N. Amjady and A. Daraeepour, Mid-term demand prediction of electrical power systems using a new hybrid forecast technique, *IEEE Transactions on Power Systems*, 26(2), 755–765, 2011.

50. B.-J. Chen, M.-W. Chang, and C.-J. Lin, Load forecasting using support vector machines: A study on EUNITE competition 2001, *IEEE Transactions on Power Systems*, 19(4), 1821–1830, 2004.

51. J. H. Zhao, Z. Y. Dong, X. Li, and K. P. Wong, A framework for electricity price spike analysis with advanced data mining methods, *IEEE Transactions on Power Systems*, 22(1), 376–385, 2007.

52. X. Lu, Z. Y. Dong, and X. Li, Electricity market price spike forecast with data mining techniques, *Electric Power Systems Research*, 73(1), 19–29, 2005.

53. J. H. Zhao, Z. Y. Dong, and X. Li, Electricity market price spike forecasting and decision making, *IET Generation, Transmission & Distribution*, 1(4), 647–654, 2007.

54. N. Amjady and F. Keynia, A new prediction strategy for price spike forecasting of day-ahead electricity markets, *Applied Soft Computing*, 11(6), 4246–4256, 2011.

55. Y. Wang, L. Li, J. Ni, and S. Huang, Feature selection using tabu search with long-term memories and probabilistic neural networks, *Pattern Recognition Letters*, 30(7), 661–670, 2009.

5

Short-Term Wind Power Forecasting

Gregor Giebel

Michael Denhard

5.1 Introduction

To optimize the scheduling of grid components, such as the power plant types explained in the introduction to this book during the unit commitment process, information is needed on the future development. The expected load is a major contributor to the unit commitment process, but load can be predicted quite well. In systems with sizeable penetrations of renewable energy, the uncertainty stemming from their production can easily be the largest uncertainty of the next-day profile. Not all renewable energy plants are equal in this respect: tidal power, for example, is very well predictable, biomass- or biogas-fired plants can even be scheduled, as the primary resource can easily be stored, and hydropower plants with storage dams are ideal to schedule as well. Hydropower plants on the other hand have issues on much longer timescales than the unit commitment time scale: for example, in Portugal, the variability of hydro resource on an annual basis is much larger than the variability of wind power, both averaged over the whole country [1]. Even solar power can be very well predicted half of the day—at night, when photovoltaic and concentrated solar power plants without heat storage are not producing energy. For wind power and solar power during the day, the situation is not so easy.

5.1.1 Timescales

Let us first look at the timescales involved. For wind power, there are two timescales clearly outside the unit commitment or power system realm that are still drawing attention: the forecast a few seconds ahead allowing the control system to pitch the blades properly into the wind, and the wind resource for the next 20 years, needed for citing studies possibly including climate change effects. The former can lead to a higher aerodynamic efficiency of the turbine, but can reasonably only be achieved by measurements of the incoming wind field. As a ring of measurements around the wind turbine is prohibitively expensive, the only possibility for this is a forward-looking lidar (light detection and ranging, essentially a laser beam being reflected and Doppler shifted on aerosols flying with the wind speed) or similar remote sensing instrument, mounted on the nacelle and being able to turn into the wind with the turbine [2,3]. The latter is a standard technique since at least the mid-1980s and is not within the realm of this book.

This leads to two timescales that are important for the power system (see also Chapter 1): unit commitment and scheduling, and the market timescale. The unit commitment timescale is usually given by the largest block in the system, often a large coal-fired power station needing 8–12 h from cold to produce the first power. The other timescale is the one that has in recent years received most attention, because it is where energy companies make money: the market timescale. In Europe, like in the case of NordPool for the Nordic countries or the German electricity exchanges, this mostly means next-day forecasting, to have a good forecast at 1100 h local time for the next full day, that is, a forecast with 13–37 h horizon. Whereas the market timescale is a relatively new phenomenon, only brought about through the deregulation process in the last decade or so, the scheduling problematic has been on the radar since the first paper on wind power forecasting by Brown et al. in 1984 [4]. In retrospect, it is surprising how complete the paper already was, using a transformation to a Gaussian distribution of the wind speeds, forecasting with an autoregressive process, upscaling with the power law (but discussing the potential benefit of using the log law), and then predicting power using a measured power curve. Additionally, the removal of seasonal and diurnal swings in the autoregressive components is discussed, alongside prediction intervals and probability forecasts. Noteworthy is also that their work was sponsored by Bonneville Power Administration, which much later entered the forecasting business again as a sponsor, this time with a special emphasis on ramps prediction [5,6].

Forecasting more than a few days ahead can be important for scheduled maintenance on the power system, either to replace a conventional generator at a time when enough wind power is expected or to do maintenance on a power line in dependency of the loading state. As this is not connected to the unit commitment problem, a larger discussion of the topic can be found in References [7,8].

5.1.2 Wind Power Forecasting Methods

Although forecasting for the scheduling horizon is sometimes attempted without taking resort to numerical weather prediction (NWP) (i.e., the type of weather model outputs we see every evening in the news), for the next day it is paramount to take into account the predicted wind fields from NWP for the power conversion model. The first attempt at that was made by McCarthy [9] for the Central California wind resource area. It was run in the summers of 1985–1987 on an HP 41CX programmable calculator, using meteorological observations and local upper air observations. The program was built around a climatological study of the site and had a forecast horizon of 24 h. It forecast daily average wind speeds with better skill than either persistence or climatology alone. However, this model required one to manually input the calculation data into the calculator, and upscaling it to a whole nation would have been quite difficult. The father of the modern NWP-based forecast was Landberg [10,11], who developed a short-term prediction model, now known as Prediktor, based on physical reasoning similar to the Wind Atlas Analysis and Application Program (WAsP) methodology developed for the European Wind Atlas [12]. The idea is to use the wind speed and direction from an NWP, then transform this wind to the local site, then use the power curve, and finally modify this with the PARK efficiency. A model output statistics (MOS) module could calibrate the forecast using real measurements. Landberg used the Danish or Risø version for all the parts in the model: the high-resolution local area model (HIRLAM) of the Danish Meteorological Institute (DMI) as NWP input, the WAsP model from Risø to convert the wind to local conditions, and the Risø PARK model to account for the lower output in a wind park due to wake effects. The site assessment regarding roughness is done as input for WAsP. Either a roughness rose or a roughness map is needed. From this, WAsP determines an average roughness at hub height. Only one WAsP correction matrix is used, which could be too little for a larger wind farm [13]. Prediktor was also used in the generic supervisory control and data acquisition (SCADA) system *CleverFarm* for maintenance scheduling [14].

Prediktor is mentioned here as an example for a physical model, which tries to establish a physical connection between the coarse NWP results and the actual measurements. Another way to do it is with a statistical model, directly establishing a connection between the input data and the measurements. Most models currently in use involve a mix of both methods.

5.1.3 Typical Results

The reason why next-day forecasts are usually done with the use of NWP data is the far better accuracy those models achieve in comparison with time-series models. The verification of model performance is dependent on the error type. Models can be good at one particular error and bad at another. The typical behavior of the error function for models using time-series approaches or NWP is shown in Figure 5.1 for the case of Prediktor applied to an older Danish wind farm in the mid-1990s (the farm has been repowered since), using root mean square erros (RMSE) as the error measure.

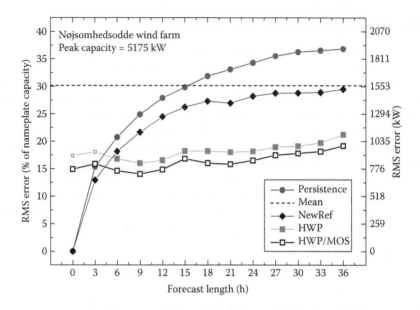

FIGURE 5.1 Root mean square (RMS) error for different forecast lengths and different prediction methods. The wind farm is the old Nøjsomheds Odde farm (before repowering) with an installed capacity of 5175 kW. NewRef refers to the New Reference Model [1]. HWP/MOS refers to the HWP approach (HIRLAM/WAsP/PARK, nowadays called Prediktor) coupled with an MOS model. MOS, model output statistics; HIRLAM, high-resolution local area modeling; WAsP, Wind Atlas Analysis and Application Program.

Figure 5.1 highlights a number of issues. Persistence is the model most frequently used to compare against the performance of a forecasting model. It is one of the simplest prediction models, second only to predicting the mean value for all times (a climatology prediction). In this model, the forecast for all times ahead is set to the value it has now. Hence, by definition, the error for zero time steps ahead is zero. For short prediction horizons (e.g., a few minutes or hours), this model is the benchmark all other prediction models have to beat. This is because the dominant timescales of large synoptic scale changes in the atmosphere are in the order of days (at least in Europe, where the penetration of wind power is still the highest). It takes in the order of days for a low-pressure system to cross the continent. As the pressure systems are the driving force for the wind, the rest of the atmosphere undergoes periodicity on the same timescales. High-pressure systems can be more stationary, but these are typically not associated with high winds, and therefore not so important. Mesoscale features (fronts, low-pressure troughs, large thunderstorms, mesoscale cellular convection, gravity waves, etc.) operate on timescales of hours, and have reasonable predictability using mesoscale models. To predict much better than persistence for short horizons using the same input, that is, online measurements of the predictand, is only possible with some effort.

One can see that persistence beats the NWP-based model easily for short prediction horizons (*ca.* 3–6 h). However, for forecasting horizons beyond *ca.* 15 h, even forecasting with the climatological mean (the dashed line) is better. This is not surprising, as it can be shown theoretically [15] that the mean square error of forecasting by mean value is half that of the mean square error of a completely decorrelated time-series with the same statistical properties (which is similar to persistence for very long horizons).

After about 4 h the quality of the "raw" NWP model output [marked HIRLAM/ WAsP/PARK (HWP), full squares] is better than persistence even without any postprocessing. The quality of the New Reference Model [15] (essentially persistence with a trend toward the mean of the time-series) is reached after 5 h. The relatively small slope of the line is a sign of the relatively poor quality of the assessment of the initial state of the atmosphere by the NWP, but of the good quality of the predictive equations used in the model from that initial state. The first two points in the HWP line are fairly theoretical; owing to the data assimilation and calculating time of HIRLAM (~4 h), these cannot be used for practical applications and could be regarded as hindcasting. The improvement attained by using a simple linear MOS (the line marked HWP/MOS, the model now known as Prediktor, open squares) is quite pronounced.

One line of results is missing in Figure 5.1 (for reasons of sharper distinction between time-series analysis methods and NWP methods): a result for current statistical methods using both NWP and online data as input. That line would of course be a horizon-dependent weighting of the persistence and the HWP/MOS approach, being lower for all horizons than all the other lines. However, for short horizons, it cannot do (significantly) better than persistence, whereas for long horizons, the accuracy is limited by the NWP model. Therefore, the line would rise close to the persistence results, and continue staying close to the HWP/MOS line.

The behavior shown in Figure 5.1 is quite common across all kinds of short-term forecasting models and is not specific to Prediktor, although details can vary slightly, such as the values of the RMSE or the slope of the error quality with the horizon. Typical model results today are RMSEs around 10% of the installed capacity, though with large deviations between NWP input, forecasting model, and wind farm. Improvements over the graph shown here are mostly due to improvements in NWP models. Model-specific items are to be found in the next chapter.

Typical forecast accuracies for single wind farms can vary quite dramatically. For the EU ANEMOS project, a comparison of 11 state-of-the-art tools was made for 6 sites in Europe [16], which showed that the differences between the wind farms and also between the forecasting models are quite large.

Another way to classify the error has been shown by weather and wind energy prognosis (WEPROG) [17–19]. In Figure 5.2 [20], two error sources are distinguished: the background error, which essentially is due to a suboptimal representation of the single point used for verification with the grid cell average calculated by the NWP (which is a general problem in meteorology), and a model error, where good initial data are becoming successively worse with increasing horizon due to imperfectly captured or simplified atmospheric physics, or due to the amplification of small initial errors as a result of the chaotic nature of the atmosphere.

In Figures 5.1 and 5.2, there is no obvious wind speed dependency of the error. Actually, the wind speed error of an NWP model does not seem to depend much on the level of predicted wind speed, as Lange and Heinemann [21] show in the left graph of

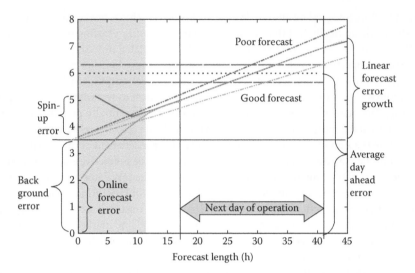

FIGURE 5.2 Typical errors introduced by the numerical weather prediction. (From Möhrlen, C., *Talk at the 3rd Workshop on Best Practice in the Use of Short-term Prediction of Wind Power*, Bremen, Germany, October 13, 2009. With permission.)

Figure 5.3 for the Deutschlandmodell of the German Weather Service (Deutscher Wetterdienst, DWD). But the nonlinear power curve (central plot [22]) skews the distribution significantly. Therefore, the distribution of errors per power bracket is nonuniformly distributed.

The complexity and importance of the problem on the wind side has resulted in many works, which have been collected and brought into the form of large reports on the state of the art. Especially the two by Giebel et al. [7] and Monteiro et al. [23] have to be mentioned here: Together, they reference over 400 papers in short-term prediction of wind power and try to be the normative references in the field. (Actually, this chapter relies heavily on text already written for Giebel et al. [7].)

Seeing such an effort being made by the institutes involved in wind power forecasting, it is satisfying to see that the hard work has borne fruit in terms of improvements of errors. The Institut für Solare Energieversorgungstechnik e.V. (ISET), Kassel, Germany (now the main part of the Fraunhofer Institut für Windenergie und Energiesystemtechnik, IWES) was the first short-term forecasting provider for transmission system operators (TSOs) in Germany. In a widely cited paper for the European Wind Energy Conference (EWEC) 2006, Lange et al. [24] presented the following plot for the accuracy of the next-day forecast in the E.On control zone. They stated that the main reasons for the improvement were: (i) taking into account the influence of atmospheric stability into the models which led to a reduction in forecast error (RMSE) by more than 20% for the example of one German TSO control zone; and (ii) a combination of different models, both for forecasting methods as well as for NWP models. The comparison of the mean RMSE of a wind power forecast for Germany obtained with the wind power management

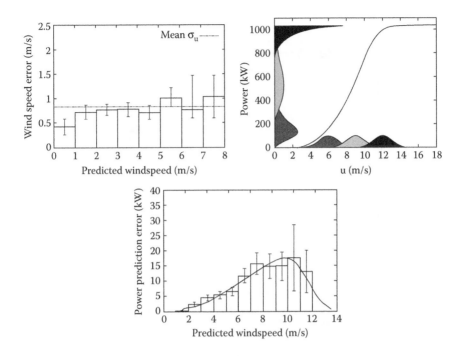

FIGURE 5.3 The error is nonlinearly distributed over the power brackets. The error is fairly linear at about 1 m/s for the shown model. However, folding this through the wind farm power curve introduces nonlinearities and increases the error in the rising part of the power curve, while decreasing it in the flatter parts. (From Lange, M., and Heinemann, D., *Poster P_GWP091 on the Global Windpower Conference and Exhibition*, Paris, France, April 2–5, 2002; Central plot: From Lange, M., and Focken, U., *Physical Approach to Short-Term Wind Power Prediction*, Springer-Verlag, Berlin, 2005. With permission.)

system (WPMS) based on artificial neural networks (ANNs) with input data from three different NWP models and with a combination of these models showed a decrease in RMSE from approximately 6% to 4.7%.

Note that their competitor, energy & meteo systems GmbH, claims a forecasting RMSE of below 5% for the day-ahead forecast for all of Germany in 2008 [25], which by now the IWES has also achieved [26].

A similar plot, though constrained to the last 2 years, was shown by Krauss et al. [27] for the EnBW (Energie Baden-Württemberg AG) TSO area (see Figure 5.4). They showed the monthly accuracy of three different forecasting systems for the aggregate error and concluded that there are significant changes in forecast accuracy from month to month, and that the ranking of the three models changes from month to month as well.

When analyzing the error of a wind power forecasting model, the usual error criterion is mean absolute error (MAE) or RMSE. Which one to choose usually depends on the cost function of the end user: if just the sum of imbalances is important, the MAE reflects this criterion well, whereas if there is a "deadband," where small errors are

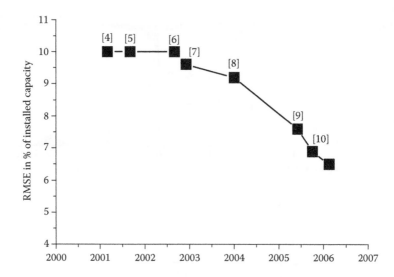

FIGURE 5.4 The development of the forecast error during the last years in the E.On Netz area. The numbers in square brackets are references from Lange et al. (From Lange, B. et al. *European Wind Energy Conference and Exhibition*, Athens (GR), 27.2.-2.3, 2006. With permission.)

essentially irrelevant, but larger errors are penalized strongly, the stronger weighting of the outliers in the RMSE better represents the system at hand. The error as such can be classified as a value error or the so-called phase error, a timing error in the forecast. In recent years, the phase error is getting a larger importance for the total error, as the overall strength of a storm can often be well predicted.

There is a wealth of different forecasting criteria, and comparability of performance values in the literature was not easy. Therefore, it was one of the tasks of the ANEMOS project to establish a common set of performance measures with which to compare forecasts across systems and locations. These common error measures are the bias, MAE, RMSE, the coefficient of determination R^2, the skill score for comparison with other models, and the error distribution as a histogram [28]. The project working group also emphasizes the need to split the dataset into separate training and validation sets, and proposes to use the normalized mean errors for a comparison across different wind farms. If there should be normalization (recommended), it should be with the installed capacity, not the mean production. The reason for this is the scalability for large regions in the case of additional wind farms: for the system operator, the installed capacity is easy to assess, whereas the mean production, especially for new wind farms, is hard to know with sufficient accuracy beforehand. An additional evaluation criterion is brought by the Spanish Wind Energy Association: the mean absolute percentage error (MAPE). This error type stems from the law that wind farm owners who want to participate in the electricity market have to predict their own power. Deviations from the declared schedule are punished according to this error measurement.

Tambke et al. [29] presented the decomposition of RMSE into the three components: bias in mean wind speed, bias in standard deviation, and dispersion. This is quite useful

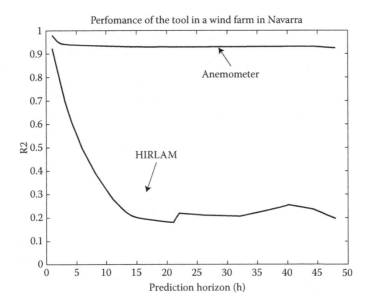

FIGURE 5.5 The error comes from the numerical weather prediction. The figure shows the difference in degree of explanation between Sipreolico run with high-resolution local area modeling (HIRLAM) input (from an older version of the Spanish HIRLAM) and Sipreolico run with on-site wind speed input. (From Sánchez, I. et al. *Proceedings of the World Wind Energy Conference*, Berlin, Germany, June 2002. With permission.)

to determine whether the main contribution to the errors of the NWP model come from level errors or biases, or rather (if the dispersion term is large) from phase errors.

During the analysis of the error, it also becomes clear that most of the error is in the NWP. The main error in the final forecast comes from the meteorological input. For example, Sánchez et al. [30] show that the Spanish statistical tool Sipreolico, run with on-site wind speed input, has a much higher degree of explanation than HIRLAM forecasts (see Figure 5.5). This means that, given a representative wind speed, Sipreolico can predict the power quite well. It is the wind speed input from the NWP model that decreases the accuracy significantly. Therefore, it is logical to try to improve the NWP input in order to come up with significant improvement in forecasting accuracy.

5.2 Time-Series Models

The easiest to use and therefore also earliest models for the short-term prediction of wind power are time-series models, usually some variants of autoregressive models, taking into account the last measured value(s) of wind power and possibly some longer-term memory of the time-series, like the mean over the last 3 weeks, months, or years. In the easiest form, this is called the persistence model, just using the last measured value as the forecast value. Although for very short horizons this model is quite good and a typical benchmark, in wind energy where the process is instationary on every time scale [31] it tends to turn into large errors already after a few hours. Examples are autoregressive with moving average (ARMA), autoregressive integrated moving average (ARIMA), or Kalman filter approaches. A variation of this are ANNs, including the variants of support vector machines, which seem to do quite well, but often deliver no higher improvements over persistence than normal time-series models. Ten years ago, we proposed the New Reference Model [15], essentially a weighting between the mean of the time-series and the last measured value with the weight depending on the autocorrelation function. This simple model seems to explain the roughly 10% improvement over persistence claimed by many other approaches. However, it always drags the values back toward the mean value, even though in some cases one would expect a current increase to continue.

A recent trend is the use of regime switching models that estimate a number of different regimes within the time-series (e.g., different weather patterns) and, for every regime, have a different set of parameters for the time-series model(s) themselves.

Bossanyi [32] used a Kalman filter with the last six values as input and got up to 10% improvement in the RMSE over persistence for 1-min averaged data for the prediction of the next time step. This improvement decreased for longer averages and disappeared completely for 1-h averages.

Fellows and Hill [33] used 2-h-ahead forecasts of 10-min wind speeds in a model of the Shetland Islands electricity grid. Their approach was to use optimized, iterative Box-Jenkins forecasting from detrended data, which then was subjected to central moving average smoothing. For 120-min look-ahead time, the RMSE reduction over persistence was 57.6%.

Nogaret et al. [34] reported that for the control system of a medium-sized island system, persistent forecasting is best with an average of the last two or three values, that is, 20–30 min.

Tantareanu [35] found that ARMA models can perform up to 30% better than persistence for 3–10 steps ahead in 4-s averages of 2.5 Hz sampled data.

Kamal and Jafri [36] found an $ARMA(p,q)$ process suitable for both wind speed simulation and forecasting. The inclusion of the diurnal variation was deemed important as the (mainly thermally driven) climate of Pakistan exhibited quite strong uniformity, especially in the summer months.

Dutton et al. [37] used a linear autoregressive model and an adaptive fuzzy logic-based model for the cases of Crete and Shetland. They found minor improvements over persistence for a forecasting horizon of 2 h, but up to 20% in RMSE improvement for an 8-h horizon. However, for longer horizons, the 95% confidence band contained most of the likely wind speed values, and therefore a meteorological-based approach was deemed more promising on this timescale.

In the same team, Kariniotakis et al. [38,39] tested various methods of forecasting for the Greek island of Crete. These included adaptive linear models, adaptive fuzzy logic models, and wavelet-based models. Adaptive fuzzy logic models were installed for online operation in the frame of the Joule II project CARE (JOR3-CT96-0119).

Torres et al. [40] used an ARMA model to forecast hourly average wind speeds for five sites in Navarra. They used site- and month-specific parameters for the ARMA model. The ARMA model usually outperformed persistence for the 1-h forecast, and always was better in RMSE and MAE for higher horizons up to 10 h ahead. The two complex sites have a slightly higher RMSE in general, but are still in the same range as the other sites. In general, 2%–5% improvements for the 1-h forecast correspond to 12%–20% improvement for the 10-h forecast.

Makarov et al. [41] described a major California ISO-led project. They developed prototype algorithms for short-term wind generation forecasting based on retrospective data (e.g., pure persistence models). The methods tested include random walk, moving average, exponential smoothing, autoregression, Kalman filtering, "seasonal" differencing, and Box–Jenkins models. The Box–Jenkins model demonstrated the best performance. They also used a bias compensation scheme to minimize the look-ahead forecast bias. For forecasts for the next hour and 1 h ahead, the total ISO-metered generation is predicted with MAE below 3% and 8% of the maximal observed generation correspondingly.

Pinson et al. [42] found that wind power, and especially wind power variability from large offshore wind farms (Horns Rev and Nysted), occurs in certain regimes, and therefore tested "regime-switching approaches relying on observable (i.e. based on recent wind power production) or non-observable (i.e. a hidden Markov chain) regime sequences" (p. 2327) for a one-step forecast of 1, 5, and 10-min power data. "It is shown that the regime-switching approach based on MSAR models significantly outperforms those based on observable regime sequences. The reduction in one-step ahead RMSE ranges from 19% to 32% depending on the wind farm and time resolution considered" (p. 2327).

Beyer et al. [43] found improvements from neural networks in RMSE for next-step forecasting of either 1 or 10-min averages to be in the range of 10% over persistence. This improvement was achieved with a rather simple topology, while more complex neural network structures did not improve the results further. A limitation was found in extreme events that were not contained in the dataset used to train the neural network.

Sfetsos [44] applied ARIMA and feed-forward neural network methods to wind speed time-series data from the United Kingdom and Greece, comparing the results of using either 10-min or hourly averaged data to make a forecast 1 h ahead. For both datasets, neither forecasting method showed a significant improvement compared to persistence using hourly averaged data, but both showed substantial (10%–20%) improvement using 10-min averages. The result is attributed to the inability of hourly averages to represent structure in the time-series on the high-frequency side of the "spectral gap," lying at a period of typically around 1 h.

Sfetsos [45,46] compared a number of methods, including a Box–Jenkins model, feed-forward neural networks, radial basis function networks, an Elman recurrent network, adaptive network-based fuzzy inference system (ANFIS) models, and a neural logic network based on their ability to forecast hourly mean wind speeds. All nonlinear models exhibited comparable RMSE, which was better than any of the linear methods. For the 1 h ahead, the best model was a neural logic network with logic rules, reducing the error of persistence by 4.9%.

EPRI, the U.S. Electric Power Research Institute, has recently [47] announced their work on the adaptation of their artificial neural network short-term load forecaster (ANNSTLF) tool to wind power forecasting. They target the range of up to 3 h with 5-min intervals.

5.3 Meteorological Modeling for Wind Power Predictions

As the pure time-series models only have a reach of 6 h or so, the use of dedicated weather predictions has received a lot of attention in the last 8–10 years. To get useful results beyond, say, 6 h, the wind field (and sometimes also other variables) from a local area model (LAM) or a global model is needed for wind power prediction of the next day. Weather systems travel across the moderate latitudes with speeds rarely exceeding a few hundred kilometers a day; hence, an LAM will usually be able to forecast the incoming weather with a domain size of 1–2.000 km. In most cases, the LAM will be initialized using the results of a global model run, and will be nested with some larger outer domains of lower horizontal resolution to a final domain size of a few hundred kilometers and a horizontal resolution of 2–5 km. This kind of modeling, also for real-time operations, is clearly within reach of a larger company's IT department, as for example a smaller Linux cluster suffices to run the weather research and forecasting (WRF) model, which is currently the most run meteorological model for wind energy. It should be noted here that the effective resolution, that is the scale at which features are actually resolved in the NWP model, is some 4–7 grid points [48–50]; that is, even for a horizontal grid resolution of 2 km, only features of the order of 10 km are really taken into account.

However, for most normal operations a forecasting client would use data from an existing model, often the model from the local or national meteorological institute. Two issues are noteworthy in this connection: the use of a second NWP model will in all cases improve the forecast (in most cases even just when averaging the two wind speed forecasts), and in recent years the use of ensemble forecasting has received a lot of attention, especially due to their ability to also yield an uncertainty forecast (see also Section 5.5).

This section gives an overview of operational NWP models having relevance for wind power prediction in Europe. Various global forecasting systems exist, designed to predict large-scale synoptic weather patterns. But the increase in computer resources during the next years will allow the global models to overtake the current role of the limited area models (LAM) down to about a 10-km horizontal resolution. The LAMs, which get their boundary conditions from the global models and operate at the moment at horizontal resolutions of 7–12 km, will be replaced by high-resolution, convection-resolving LAMs with horizontal resolutions well below 4 km.

The resolution of the most commonly used global forecast models further increase and the first models have horizontal grid resolutions well below 20 km in 2010. There is a consensus in the European SRNWP community (short-range numerical weather prediction, http://srnwp.met.hu/) that global models such as the integrated forecast system (IFS) at the European Centre for Medium-Range Weather Forecasts (ECMWF) will take over the role of LAMs in their current form. There are developments to statically nest the LAM directly into the global model (e.g., ICON at DWD, ARPEGE at Météo-France) with horizontal resolutions up to 5 km in certain target areas (e.g., Europe). In cooperation with Météo–France, a nonhydrostatic kernel of the IFS at the ECMWF will be developed.

The main European LAM model consortia are the ALADIN (http://www.cnrm.meteo.fr/aladin), COSMO (http://www.cosmo-model.org), HIRLAM (http://www.hirlam.org), and LACE (http://www.rclace.eu) projects and the U.K. MetOffice (http://www.metoffice.com/research/nwp/index.html).

At the moment most countries run their models for overlapping European areas at grid resolutions of 12–7 km. In the next few years they will move toward a grid resolution of 4–1 km and therefore will not run an intermediate nested European grid area anymore. They plan to directly nest their very-high-resolution models, which then will cover only the national area, into a global model at 25 km or less. For very-high-resolution requirements of a Europe-wide SRNWP coverage a need arises for close cooperation and exchange of NWP products.

Rife and Davis [51] compared two otherwise identical model setups with horizontal resolution of 30 and 3.3 km, respectively, for wind speed variations at and near the White Sands Missile Range in New Mexico (United States): "The authors hypothesize that the additional detail and structure provided by high resolution becomes a 'liability' when the forecasts are scored by traditional verification metrics, because such metrics sharply penalize forecasts with small temporal or spatial errors of predicted features" (p. 3368). Therefore, they use three alternative skill scores, namely (in order of tolerance of timing errors) anomaly correlation, object-based verification, and variance anomalies. "The largest improvement of the fine-grid forecasts was in the cross-mountain component" (p. 3380). In general, the higher-resolution forecasts exhibited more skill than their coarser counterparts.

A large effort aiming at meteorological forecasts for wind energy has also been made by the original ANEMOS project. A long report [52] details some work on especially downscaling techniques with microscale, mesoscale, and computational fluid dynamics models. The best parameterization for MM5 was found to be medium-range forecast, although it did not lead the competition at every forecast horizon and case study. If possible from a computational point of view, two-way nesting between domains is clearly preferred. Whereas one group using mostly physical modeling reported increased accuracy down to a grid spacing of 2 km, another one using an advanced statistical model claimed no improvement when going from a grid spacing of 9 to 3 km. This is probably due to the fact that the forecasted time-series become more "realistic" when increasing horizontal resolution, in the sense that the ups and downs of the time-series have a similar amplitude to the original series in the high-frequency domain. However, this means a higher potential for phase errors, so for the usual RMSE or MAE the error goes up. Increasing the horizontal resolution beyond the resolution of the terrain database is fairly useless. On the other hand, increasing the vertical resolution in the lowest, say, 200 m of the atmosphere improved the results in all cases. The report closed with the following recommendations: "If you have a site in complex terrain, where you even after using an advanced MOS are not happy with the forecasts, then try to use higher-resolution modelling. In many cases and with a large number of approaches, the models can improve the NWP results. When setting up a model yourself, make sure to use the best terrain DB available (e.g., SRTM data), and try to get good NWP input data. Set up the model to have good vertical resolution, and reasonable horizontal resolution. Find out for yourself what "reasonable" means in this context. Use a MOS. Use insights gleaned from high-resolution modelling to decide which parameters to employ in the MOS. In any case, setting up a model from scratch will take a long time before one is familiar with the model and its quirks, so do not plan on having a solution up and running immediately" (p. 100).

5.4 Short-Term Power Prediction Models

In the previous two sections, we described the type of models often used in short-term prediction, and the input data from the meteorological perspective. In this section, we will describe the models for the best power forecast day-ahead with the inputs given. There are two different main forecast business models: one is to provide the forecasts as a service, run from central servers at the forecasters place, and the other is to install a forecasting model at the client's site, leaving the clients to run the model themselves. There are advantages and disadvantages to both approaches. Often, it boils down to the question of whether the client is willing to send its data over the Internet to the service provider. In other cases, it is clear that the low end of the (considerable) price range is typically service providers, as the setup of a forecasting model is quite a difficult and time-consuming task. Incidentally, the two business models were already in place in the first operational installations in Eastern and Western Denmark in 1993 and 1994—Risø ran Prediktor (without online data at the time) on their site and provided the forecasts through a web server, while DTU.IMM installed the WPPT in Jutland.

The Risø DTU Prediktor has already been described earlier in this chapter. The other Danish tool, the Wind Power Prediction Tool (WPPT) has been developed by the Institute for Informatics and Mathematical Modelling (IMM) of the Technical University of Denmark. In 2006, the original developer Torben Skov Nielsen together with Henrik Madsen and Henrik Aalborg Nielsen founded the DTU spin-off company ENFOR, which now stands for all commercial activity with the model. WPPT has been running operationally in Western Denmark since 1994 and in Eastern Denmark since 1999. They use adaptive recursive least-squares estimation with exponential forgetting in a multi-step setup to predict from 0.5 up to or more than 36 h ahead. A data-cleaning module was developed, as was an upscaling model. This model has successfully operated at Elsam and other Danish utilities [53]. WPPT is a modeling system for predicting wind power production for individual wind farms, for groups of wind farms, or for a larger region. WPPT can be configured to take advantage of the following data:

- Online power production measurements for individual wind farms
- Aggregated online power production measurements for larger areas
- Offline power production measurements for individual wind farms
- Aggregated offline power production measurements for larger areas
- NWP data covering individual wind farms
- NWP data covering larger areas
- Multiple NWP forecast providers
- Scheduled availability and curtailment

The forecasts can be in the form of single-point forecasts (forecasts of the expected value) or in the form of probabilistic forecasts where the entire distribution of the expected outcome is given.

The complexity of the model structure employed by WPPT will depend on the available data. In order to illustrate the flexibility of WPPT, a complex installation for predicting the total wind power production in a larger region on the basis of a combination of online measurements of power production from selected wind farms, power measurements for all wind turbines in the region, and NWPs of wind speed and wind direction is presented here as an example.

A central part of this system is the statistical models for short-term prediction of the wind power production in wind farms or areas. The modeling system combines traditional linear models with a specific but very general class of nonlinear models—the conditional parametric models.

For online applications, it is advantageous to allow the function estimates to be modified as data become available. Furthermore, because the system may change slowly over time, observations should be down-weighted as they become older. For this reason, a time-adaptive and recursive estimation method is applied.

The time-adaptivity of the estimation is an important property as the total system consisting of a wind farm or area, its surroundings, and the NWP model itself will be subject to changes over time. This is caused by effects such as aging of the wind turbines, changes in the surrounding vegetation, and, maybe most importantly, changes in the NWP models used by the weather service as well as changes in the population of wind turbines in the wind farm or area.

Nielsen et al. [54] found a way to algorithmically optimize the tuning parameters for the time-adaptive model, like forgetting factor and bandwidth. In the same work, they also improved the robustness of WPPT against suspicious data.

Depending on the available data, the WPPT modeling system employs a highly flexible modeling hierarchy for calculating predictions of the available wind power from wind turbines in a region. For a larger region, this is typically done by separating the region into a number of subareas. Wind power predictions are then calculated for each subarea and hereafter summarized to get a prediction for the total region.

In the following an installation using online production data from a number of wind farms in a region (reference wind farms), offline production data for the remaining wind turbines in the region, and NWPs of wind speed and wind direction in the calculation of a total regional power prediction is outlined. The predictions cover a horizon corresponding to the prediction horizon of the NWPs—typical from 1 to 48 h ahead in time. The time resolution of the predictions can be chosen freely but a reasonable choice for the longer prediction horizons is to use the same time resolution as available for the NWPs.

The predictions for the total region are calculated for a number of reference wind farms using online measurements of power production as well as NWPs as input. The predictions from the reference wind farms in the region are summarized and hereafter upscaled to get the prediction of power production of all wind turbines in the region. This modeling chain takes advantage of the autocorrelation present in the power production for prediction horizons less than approximately 12–18 h, but also of the smooth properties of the total production as well as the fact that the numerical weather models perform well in predicting the weather patterns but less well in predicting the local weather at a particular wind farm.

The power prediction for the region is here calculated directly by the upscaling model, but a larger region could be separated into a number of subareas, each covered by a model chain as described above. The total power production will then be calculated as a sum of the predictions for the subareas.

A rather similar approach to Prediktor was developed at the University of Oldenburg [55]. They named it Previento [56]. A good overview of the parameters and models influencing the result of a physical short-term forecasting system has been given by Mönnich [57]. He found that the most important of the various submodels being used is the model for atmospheric stability. Mönnich also found that the submodels for orography and roughness were not always able to improve results. The use of MOS was deemed very useful. However, as the NWP model changed frequently, the use of a recursive technique was recommended. A large influence was found regarding the power curve. The theoretical power curve given by the manufacturer and the power curve found from data could be rather different. Actually, even the power curve estimated from data from different years could show strong differences. The latter might be due to a complete overhaul of the turbine. The largest influence on the error was deemed to come from the NWP model itself. In 2004, the two principal researchers behind Previento, Matthias Lange and Ulrich Focken, left the university to form energy & meteo systems GmbH, a company that had good success from the start and has now over 20 employees. Their work on the weather-dependent combination of models has also been published [22]. In essence, principal component analysis identifies between five and eight different weather types, and the model parameters are optimized according to weather type.

ARMINES and RAL have developed work on short-term wind power forecasting since 1993. Initially, short-term models for the next 6–10 h were developed on the basis of time-series analysis to predict the output of wind farms in the frame of the LEMNOS project (JOU2-CT92–0053). The developed models were integrated in the EMS software developed by AMBER S.A and installed for online operation on the island of Lemnos.

Various approaches have been tested for wind power forecasting based on ARMA, neural networks of various types (backpropagation, RHONN, etc.), fuzzy neural networks, wavelet networks, and so on. From this benchmarking procedure, models based on fuzzy neural networks were found to outperform the other approaches [39,58,59].

The wind forecasting system of ARMINES integrates:

- *Short-term models*: these are based on the statistical time-series approach able to predict efficiently wind power for horizons up to 10 h ahead.
- *Longer-term models*: these are based on fuzzy neural networks able to predict the output of a wind farm up to 72 h ahead. These models receive as input online SCADA data and NWPs [60].
- *Combined forecasts*: such forecasts are produced from intelligent weighting of short-term and long-term forecasts for an optimal performance over the whole forecast horizon.

The developed prediction system is integrated in the MORE-CARE EMS software and is installed for online operation in the power systems of Crete and Madeira [61]. A stand-alone application of the wind forecasting module is configured for online operation in Ireland [62]. An evaluation of this application is presented in Reference [63]. The average reported error is of the order of 10% of the installed power. For Ireland, they show that using a power curve derived from HIRLAM wind and measured power can improve the forecast RMSE by nearly 20% in comparison with using the manufacturer's power curve [62].

Wind power of 80 MW is installed on the island of Crete where the demand varies between 170–450 MW throughout the year. Wind penetration reaches high levels. Furthermore, the fact that the network is autonomous makes the use of wind power forecasting necessary for an economic and secure integration of wind farms in the grid. Currently, the MORE-CARE system [64] is installed and operated by PPC in Crete and provides wind power forecasts for all the wind farms for a horizon of 48 h ahead. These forecasts are based on NWPs provided by the SKIRON system, which is operated by IASA. Online data are provided by the SCADA system of the island.

In Portugal, the MORE-CARE system is operated by EEM and provides forecasts for the production of the wind farms on the island of Madeira. The prediction modules provide forecasts for the short-term up to 8 h ahead using online SCADA data as input. Moreover, MORE-CARE provides predictions for the run-of the river hydro-installations of the island.

Since 2000, the ISET has operatively worked with short-term forecasting, using the DWD model and neural networks. It came out of the German federal monitoring program—Wissenschaftliches Mess-und EvaluierungsProgramm (WMEP) [65], where the growth of wind energy in Germany was to be monitored in detail. Their first customer was E.On, who initially lacked an overview of the current wind power production and therefore wanted a good tool for nowcasting [66]. Their model was called the Advanced Wind Power Prediction Tool.

Ernst and Rohrig [67] reported in Norrköping in 2002 on the latest developments of ISET's WPMS. They then predicted for 95% of all wind power in Germany. In some areas of German TSOs, E.On Netz and Vattenfall Europe Transmission, wind power exceeded 100% coverage at times. An additional problem in Germany is that the TSOs even lack the knowledge of the currently fed in wind power. In the case of E.On Netz, the installed capacity of *ca*. 5 GW has been upscaled from 50 representative wind farms with one-third of the total installed capacity (from 16 farms totaling 425 MW). Their input model was the Lokalmodell (always the actual model) of the DWD, which they then feed into an ANN. To improve on the Lokalmodell, they tried transforming the predicted wind to the location of wind farms using the numerical mesoscale atmospheric model KLIMM (KLImaModell Mainz), but dropped it again (B. Ernst, personal communication, September 17, 2003). The Lokalmodell is run twice daily with a horizontal resolution of 7 km, forecasting up to 48 h ahead. The ANN also provides for an area power curve. The WPMS runs at E.On since 2001, at RWE since June 2003, for Vattenfall Europe since the end of 2003, and in a variety of other places as well [68]. A version for 2 h horizon has been developed for National Windpower in the United Kingdom. For the E.On total area, they claim RMSE values of 2.5% for 1 h horizon (3.3% persistence), 5.2% for 3 h (7.3% persistence), and 6% for 4 h (9% persistence), and reach the error of a purely NWP-based prognosis (7.5%) at 7 h horizon.

The strong wind energy growth in Spain led Red Eléctrica de España (the Spanish TSO) to have the Sipreólico tool developed by the University Carlos III of Madrid [69]. The tool is based on Spanish HIRLAM forecasts, taking into account hourly SCADA data from 80% of all Spanish wind turbines [30]. These inputs are then used in adaptive nonparametric statistical models, together with different power curve models. There are nine different models, depending on the availability of data: one pure time-series model, not using NWP input at all; three more include increasingly higher-order terms of the forecasted wind speed; and a further three also take the forecast wind direction into account; the last two are combinations of the other ones, plus a nonparametric prediction of the diurnal cycle. These nine models are recursively estimated with both a recursive least-squares algorithm and a Kalman filter. For the recursive least-squares algorithm, a novel approach is used to determine an adaptive forgetting factor based on the link between the influence of a new observation, using Cook's distance as a measure, and the probability that the parameters have changed. The results of these 18 models are then used in a forecast combination [70], where the error term is based on exponentially weighted mean squared prediction error with a forgetting factor corresponding to a 24-h memory. The R^2 for all of Spain is more than 0.6 for a 36 h horizon. The main problem of the Spanish case is the Spanish HIRLAM model in conjunction with the complex terrain. The resolution of HIRLAM is not enough to resolve the flow in many inland areas. The model itself works very well when driven by measured wind speeds instead of predicted ones (with R^2 over 0.9 for the whole horizon; see also Figure 5.5).

LocalPred and RegioPred [71] are a family of tools developed by Martí Perez et al. (at formerly Centro de Investigaciones Energéticas Medioambientales y Tecnológicas— CIEMAT, now Centro Nacional de Energías Renovables—CENER). Originally, it involved adaptive optimization of the NWP input based on principal component analysis, time-series modeling, mesoscale modeling with MM5, and power curve modeling. They could show, for a case of rather complex terrain near Zaragoza (Spain), that the resolution of HIRLAM was not good enough to resolve the local wind patterns [72]. The two HIRLAM models in Spain were at the time running on a $0.5° \times 0.5°$ and $0.2° \times 0.2°$ resolution. The use of WPPT as a statistical postprocessor for the physical reasoning was deemed very useful in the early stages of the development [73]. Successive research and development carried out at CENER [74] have transformed LocalPred into a multimodel wind power forecasting system. In its current form, an ensemble forecasting model takes MM5, SKIRON, and the ECMWF model as NWP inputs for learning machine techniques as cluster or support vector machines. The final prediction is offered by an adaptive model that combines all the individual inputs.

GL Garrad Hassan [75] has a forecasting model called GH Forecaster, based on NWP forecasts from the U.K. MetOffice. It uses "multi-input linear regression techniques" to convert from NWP to local wind speeds. For $T + 24$ h, they reach 35%–60% improvement over persistence.

eWind is an American model by TrueWind, Inc. (now AWS TruePower) [76]. Instead of using a once-and-for-all parameterization for the local effects, like the Risø approach does with WAsP, they run the ForeWind numerical weather model as a mesoscale model using boundary conditions from a regional weather model. This way, more physical processes are captured, and the prediction can be tailored better to the local site. In the initial configuration of the eWind system, they used the mesoscale atmospheric simulation system (MASS) model [77]. Additional mesoscale models used were: ForeWind, MM5, WRF, COAMPS, workstation-ETA, and OMEGA. To iron out the last systematic errors they use adaptive statistics, either a traditional multiple screening linear regression model, or a Bayesian neural network. Their forecast horizon is 48 h. They published a 50% improvement in RMSE over persistence in the 12–36 h range for five wind towers in Pennsylvania [78]. The current iteration of eWind uses ARPS, MASS, and WRF, fed by the global models GFS, GEM, and ECMWF, to yield an ensemble of nine different model runs [79]. For the average prediction of six wind farms in Europe, their "results reveal that the ensemble prediction outperforms the accuracy of [...] the MOS method applied to single NWP models, achieving between a 20 and 30% of improvement during the first three days of prediction" (p. 1). Zack [80] of AWS TrueWind, Inc. presented the high-resolution atmospheric model to operate in a rapid update cycle mode, called Wind Energy Forecast Rapid Update Cycle (WEFRUC). The model assimilates different types of data available in the local area environment of a wind plant such as remotely sensed data, which is the starting point for a short-term simulation of the atmosphere. So, the atmospheric simulation produced by the physics based model is incrementally corrected through the use of the measured data as it evolves. The update cycle is 2 h.

The 3Tier Environmental Forecast Group [81] works with a nested NWP and statistical techniques for very short terms in the Pacific Northwestern United States. They show performance figures in line with most other groups in the field.

This list is far from exhaustive. In the two larger overviews mentioned in the introduction, there are more models listed. It can be estimated that there are some 50 professional groups and companies providing reasonable or better forecasts or forecast models.

5.5 Uncertainty, Ramps, and Variability

The direct application of forecasting tools to NWP or time-series models leads to what could be called a deterministic forecast. The technology for this had already been developed during the 1990s, and though small improvements to the general process are still possible, the big improvements are to be found in additional services or types of forecasts. The small incremental improvements are, for example, the development of a better statistical estimation of the power curve by Pinson and Madsen [82], or the use of regime switching models by Gneiting et al. [83] or Pinson et al. [42] for a more specific conversion from wind forecast to power depending on the large-scale weather type.

5.5.1 Probabilistic Forecasting

Owing to the fact that improvements of the forecast quality as such were slow, the main focus of research went into additional issues, especially the forecast uncertainty. Already in the early days of Landberg, the plots could contain some error bars calculated from historic data. However, those error bars were static and not dependent on the actual forecast. Fully probabilistic forecasts began to appear only after 2000. Typical confidence interval methods, developed for models like neural networks, are based on the assumption that the prediction errors follow a Gaussian distribution. This, however, is often not the case for wind power prediction where error distributions may exhibit some skewness, while the confidence intervals are not symmetric around the spot prediction due to the form of the wind farm power curve. On the other hand, the level of predicted wind speed introduces some nonlinearity to the estimation of the intervals; for example, at the cut-out speed, the lower power interval may suddenly switch to zero.

A probabilistic forecast can be in different forms, most typically a quantile forecast. This means, for example, that the 20% quantile forecast is a direct forecast of the value that is exceeded 80% of the time. Another possibility is to derive the probability density function (PDF), whereas a third option is to derive uncertainty bands around the deterministic forecast. Note that the 50% quantile forecast is not the deterministic forecast—usually it is smoother than the deterministic forecast. There are two essential ways to get to the probabilistic forecast: either through some clever data manipulation of the single NWP (and SCADA) data feeds or through the use of ensembles of weather forecasts. Those ensembles can be dedicated weather ensembles either from a single source, like the 11-member NCEP (National Centers for Environmental Prediction) ensemble or the 51-member ECMWF ensemble, or from the same model with different parameterizations like the WEPROG multi-scheme ensemble prediction system (MSEPS) (see weprog. com), or it can be the deterministic runs from various institutes.

TIGGE, the THORPEX Interactive Grand Global Ensemble, is a key component of the THORPEX World Weather Research Programme to accelerate the improvements in the accuracy of 1-day to 2-week high-impact weather forecasts. For mentions of the different systems contributing to the TIGGE database, see References [84–92]. The data can be accessed via three servers in Europe (http://tigge-portal.ecmwf.int/), America (http://tigge.ucar.edu/), and Asia (http://wisportal.cma.gov.cn/tigge/). The forecasts are available for research purposes with a time delay of 48 h. In 2010, the ECMWF ensemble prediction system resolutions increased from TL399/TL255 to TL639/TL319.

The increase in available computer power led to some progressive thinking on how to make the best use of these resources. Instead of just increasing the resolution, the processing cycles might be better used in reducing the other errors. This can be done using ensembles of forecasts, either as a multimodel ensemble, using many different NWP models of different parameterizations within the same model, or by varying the input data and calculating an ensemble based on different forecast initializations. The use of this is to be able to quantify the uncertainty inherent in the forecasts. For example, if a slight variation in the initial state of the model (which is still consistent with the measured data) leads to a larger variation a few days ahead, where for example a low-pressure system takes one of two distinct tracks, then the situation is different from one where all low-pressure tracks more or less run over the same area. A number of groups in the field are currently investigating the benefits of ensemble forecasts.

Möhrlen et al. [93] used a multischeme ensemble of different parameterization schemes within HIRLAM. They make the point that, if the observational network has a spacing of 30–40 km, it might be a better use of resources not to run the NWP model in the highest possible resolution (in the study, 1.4 km), but instead to use the computer resources for calculating a large amount of forecasts and generate an ensemble. A doubling of resolution means a factor 8 in running time (as one has to double the number of points in both horizontal grid components and time). The same effort could therefore be used to generate eight ensemble members. The effects of a lower resolution would not be so bad, as effects well below the spacing of the observational grid are mainly invented by the model anyway, and could be taken care of by using direction-dependent roughness instead. Giebel et al. [94] and Waldl and Giebel [95,96] investigated the relative merits of the Danish HIRLAM model, the Deutschlandmodell of the DWD, and a combination of both for a wind farm in Germany. The RMSE of the Deutschlandmodell was slightly better than the one of the Danish model, whereas a simple arithmetic mean of both models yielded an even lower RMSE.

Giebel and Boone [97] extended this analysis to additional wind farms and used two different short-term prediction models for the analysis. The result was the same, that a combination of models is helpful. Nielsen et al. [98] showed that the combination of models can always be better than the best of the two input models and that, in most cases, even a simple average outperforms the best of the models. In their paper, they developed the theory of how to combine forecasts if bias and variance/covariance of the individual forecasts are known. They applied their approach to two wind farms in Denmark (Klim) and Spain (Alaiz) with up to four individual forecasts per wind farm, all done by WPPT with different NWP input. This "resulted in improvements as high as 15%, with an overall level of 9%, for the wind farm near Klim in Denmark. For the wind farm near Alaiz, the corresponding numbers are 9 and 4%, respectively. However, for Alaiz if one meteorological forecast and three different combinations of MOS and power-curve are used, then no improvement is obtained" (p. 481).

In the framework of the Danish PSO-funded project Intelligent Prognosis, Nielsen et al. [54] showed generic figures for the potential improvement of an additional NWP forecast depending on the correlation between the forecasts and the relative performance. The figures were verified for the two wind farms in Klim in Denmark and Alaiz in Spain. It "is recommended that two or three good meteorological forecasts are used and the forecast errors of these should have low correlation (less that approximately 0.8). This seems to be the case for meteorological forecasts originating from different global models" (p. 9).

Pinson and Kariniotakis [99,100] proposed a methodology for the estimation of confidence intervals based on the resampling approach. This method is applicable to both physical and statistical wind power forecasting models. The authors also presented an approach for assessing online the uncertainty of the predictions by appropriate prediction risk indices ["Meteo risk index" (MRI)] based on the weather stability. They used a measure of the distance (or the similarity) of subsequent predictions in a poor-man's ensemble. The approach was verified using HIRLAM forecasts and data from five wind farms in Ireland.

Lange and Waldl [21,101] classified wind speed errors as a function of look-ahead time. The errors in wind speed of the older DWD Deutschlandmodell are fairly independent of the forecast wind speed, except for significantly lower errors for the 0- and 1-m/s bins [101]. Another result was that only for some wind farms did the error depend on the Grosswetterlage (a classification system with 29 classes for the synoptic situation in Europe), as classified by the DWD. As a result of the nonlinearity of the power curve, wind speed forecasting errors are amplified in the high-slope region between the cut-in wind speed of the turbine and the plateau at rated wind speed, where errors are dampened. Landberg et al. [102] reported the same behavior. Nielsen [103] also showed the WPPT error for Western Denmark to have its peak at a forecast of half the installed capacity. This method is only applicable to models that provide intermediate forecasts of wind speed at the level of the wind park.

The WEPROG MSEPS has been operational since 2004. On the basis of WEPROG's own NWP formulation, the system is built up with three different dynamics schemes, five different condensation schemes, and five different vertical diffusion schemes, which result in an ensemble of 75 members. The characteristic of the MSEPS system is that it has the capability to develop physical uncertainties with well-defined differences among the ensemble members. This is of advantage especially for wind energy predictions, because it means that the uncertainty is not dependent on the forecast horizon as in other ensemble approaches, but instead develops in every forecast step as a result of the physically different formulations of the individual ensemble members [17] (http://www.weprog.com/publications).

In Denmark, the Zephyr collaboration had a PSO-funded 3-year project [104,105] on the use of different kinds of ensembles for utility grade forecasting. Among others, the NCEP, the National Center for Atmospheric Research (NCAR), and ECMWF ensembles were used, multimodel ensembles (with input from both DMI and DWD) were compared, and some methods for a good visual presentation of the uncertainty were researched. One main result [106] was the development of a technique to transform the quantiles of the meteorological distribution to the quantiles of the power forecast distribution. The resulting quantiles were sharp and skillful. The use of pure meteorological ensemble quantiles was shown to be insufficient, as the ensemble spread is not probabilistically correct. Even using the transformation it was not possible to get satisfactory outer quantiles (eg below 15% and above 90%), as the meteorological ensemble spread is not large enough. This is especially relevant for the first days of the ensemble runs. However, in practice, this might be less of a problem, as the ensemble runs also needed 17 h to complete, therefore making the first day impossible to use operatively. The model was used in a demo application run for two Danish test cases, the Nysted offshore wind farm and all of the former Eltra area (Denmark West). The results were quite satisfactory, with a horizon of 1 week, and were used for maintenance scheduling of conventional power plants, for the weekly coal purchase planning and for trading on the Leipzig electricity exchange, which is closed over the weekends. Besides a final project report [8], a number of more detailed reports on the model [107], the experiences with the demo application [108], the possibilities of nesting HIRLAM directly in the ECMWF ensemble members [109], and some special turbulent kinetic energy parameterizations within HIRLAM [110] came out.

Roulston et al. [111,112] evaluated the value of ECMWF forecasts for the power markets. Using a rather simple market model, they found that the best way to use the ensemble was what they called climatology conditioned on the ECMWF ensemble prediction system. The algorithm was to find 10 days in a reference set of historical forecasts for which the wind speed forecast at the site was closest to the current forecast. This set was then used to sample the probability distribution of the forecast. This was done for the 10th, 50th, and 90th percentile of the ensemble forecasts.

Taylor et al. [113] created a calibrated wind power density from the ECMWF ensemble prediction system. "The resultant point forecasts were comfortably superior to those generated by the time-series models and those based on traditional high resolution wind speed point forecasts from an atmospheric model" (p. 781).

Pinson and Madsen [82] "describe, apply and discuss a complete ensemble-based probabilistic forecasting methodology" for the example case of Horns Rev as part of the Danish PSO research project "HREnsembleHR"—High Resolution Ensemble for Horns Rev, funded by the Danish PSO Fund from 2006–2009 (see www.hrensemble.net) (p. 137). The forecasts from WEPROG's 75-member MSEPS ensemble are converted to power using the novel orthogonal fitting method. The single forecasts are then subjected to adaptive kernel dressing with Gaussian kernels, as "in theory, any probabilistic density may be approximated by a sum of Gaussian kernels," meaning that the resulting probabilistic distribution can be "a non-symmetric distribution (and possibly multimodal), thus being consistent with the known characteristics of wind power forecast uncertainty" (p. 142).

Bremnes [114] developed a probabilistic forecasting technique, estimating the different quantiles of the distribution directly. In another study [115], Bremnes described his method of local quantile regression (LQR) in more detail, and showed that, for a test case in Norway, HIRLAM forecasts have a lower interquantile range than climatology, which means that the HIRLAM forecasts actually exhibit skill. LQR HIRLAM features about 10% better in economic terms than pure HIRLAM forecasts, increasing the revenue from *ca.*75%–79% of the ideal income (without any forecast errors) to *ca.*79%–86%, depending on the horizon. However, his pure HIRLAM forecasting did not have an upscaling or MOS step, so this might have worked in favor of LQR in comparison. Bremnes proposed to use the method to reduce the large amount of information found in meteorological ensembles. The motivation for this was that he could show that the economically optimal quantile was not the central ("best") quantile, but one given by the relative prices of up- and down-regulation.

Bremnes [116] compared three different statistical models for quantile forecasts: LQR, the local Gaussian model (assuming that, around the forecasted values, the distribution can be approximated with a Gaussian), and the Nadaraya–Watson estimator. Applied to a wind farm in Norway with HIRLAM 10 forecasts, no clear preference of method was found, although the local Gaussian model produced slightly more uncertain forecasts than the other two methods. So if ease of implementation is an issue, the Nadaraya–Watson estimator might be the best.

Nielsen and Madsen [103] developed a stochastic model for Eltra, describing variance and correlation of the forecast errors of WPPT, version 2. Nielsen et al. [117,118] tried a method similar to the LQR technique for the case of the small Danish offshore wind farm Tunø Knob, using WPPT with various parameters as input, among them the MRI. They concentrated on the 25% and 75% quantiles. Also here, the predictions proved "sharp" in comparison with historic data, meaning that the interquantile range, given as the difference between the 75% and the 25% quantile, is much narrower than the historical average of the quantiles of the production distribution. There were deviations in quantiles between the training set and the test set. For the LQR approach, it did not seem important to include the MRI.

The optimal combination of forecasts is a field that has garnered attention quite recently. As Sánchez [119] points out, "It is common in the wind energy industry to have access to more than one forecaster. It is well known that the relative performances of the alternative wind power forecasts can vary with the wind farm, and also with time. In these cases, an adaptive combination of forecasts can be useful to generate an efficient single forecast" (p. 691/2). He therefore implemented for the Spanish TSO a two-step procedure involving the adaptive exponential combination (AEC): "The AEC is designed to give all of the weight to the best available forecast" (p. 691/2).

5.5.2 Ramp Forecasting

In Europe, where the high density of population and the need for a given distance to the neighbors of a wind farm led to fairly small wind farm sizes, the distribution of the installed wind power capacity over a larger area leads to smoothing of the wind power output. Therefore, sudden increases or decreases of the wind power feed were a rare occurrence, and were not usually forecast separately. This is currently changing rapidly, with wind farm sizes in the United States, Scotland, Australia, and generally offshore reaching into some hundred MWs, and even more in China, where the current plan involves development of seven large wind farm areas of 10 GW each. The ramps coming from such a large concentrated wind farm area can be quite drastic in comparison with the load at the feed-in point. Therefore, especially in the United States, in recent years talk has been on the need for a ramp forecast. For example, Bonneville Power Administration [5] held a competition dedicated to ramp forecasting. The first results [6] indicated that, for ramps, hourly predictions are not good enough, and shorter timings of the forecast lead to smaller deviations. However, as Focken [120] pointed out, in the subsequent Request for Proposals for a short-term prediction system, ramps were not mentioned at all. Focken (having been part of the ramp forecasting competition with his company energy & meteo systems GmbH) attributes this to the fact that a ramp does not have an action in the control room associated with it "the operators don't know what to do with a ramp forecast." Having said that, in the remainder of his talk Focken pointed out that the ramp forecast needs to be something separate from the usual root mean square-optimized forecast, since this tends to be too smooth.

5.5.3 Variability Forecasting

In between forecasts of the uncertainty and of single extreme events there is a third category that is slowly receiving attention: variability forecasting. Often, it is possible to forecast that the weather situation tomorrow is going to be quite variable or quite stable.

Variability forecasting refers to large amplitude, periodic changes in wind speed, and it is only recently that it has come into the sight of researchers. Davy et al. [121] defined an index of variability based on the standard deviation of a band-limited signal in a moving window, and developed methods to statistically downscale reanalysis data to predict their index. Among the important predictors of variability, they found planetary boundary layer height, vertical velocity, and U-wind speed component during June to September (southern hemisphere winter), and U-wind speed, geopotential height, and cloud water during December to February (southern hemisphere summer).

Vincent et al. [31,122] defined a variability index as the sum of all amplitudes occurring within a given frequency range based on an adaptive spectrum. They studied the climatological patterns in variability on timescales of minutes to 10 h at the Horns Rev wind farm, and showed that there were certain meteorological conditions in which the variability tended to be enhanced. For example, variability had a higher average amplitude in-flow from sea than in-flow from the land, often occurred in the presence of precipitation, and was most pronounced during the autumn and winter seasons.

Von Bremen and Saleck [123] proposed the *totalfluc*, the sum of the absolute values of gradients exceeding a certain threshold within a, say, 6-h period, as a measure of variability. The variability of wind speed data from FINO 1, converted to power with the power curve of the nearby Alpha Ventus offshore wind farm, was highest around a wind speed of 10 m/s. A clustering analysis of the principal components of the 500-hPa geopotential height showed that the largest variations occurred for north-western flow.

Vincent and Hahmann [124] showed, for a fairly frequent offshore weather pheno menon called open cellular convection, that the phase (i.e., timing) of the errors is essentially impossible to forecast, but the period of the variability and the general appearance of this weather phenomenon is quite well predicted.

5.6 Decision-Making and Value Derived from Wind Power Forecasts

If one asks the control room personnel of a large utility or TSO what they wish from a forecast, their answer usually is "just give us your best forecast," that is, they are only interested in the most probable forecast. A probabilistic forecast contains more information than can be easily digested, and at most end users there is no decision-making process attached to a quantile forecast. Therefore, the researchers and operators of forecasts try to integrate tools for the best decision-making process in the forecasting software, so as to give the user an optimal solution for a given problem with the input available. For example, the ANEMOS.plus project (see Anemos-plus.eu) integrates tools for scheduling, trading, storage management, bottleneck management, and reserve allocation in the ANEMOS platform.

Hasche et al. [125] used wind power integration in liberalized electricity markets (WILMAR model) to assess the value of improved forecasts in operations in Germany. One interesting conclusion was that "Operational costs due to forecast errors could be reduced by one third if an overall stochastic optimization were used in scheduling" (p. 21).

Also, Dobschinski et al. [126] found that the balancing cost, especially for minute reserve, could be much lower if the TSOs used the probabilistic distributions offered by modern forecasting models for the day-ahead prediction. However, this is mitigated by the fact that, as they show, a proper use of forecasts for the next few hours is even better.

Even though the necessity and advantages of wind power forecasting are generally accepted, not many analyses have examined in detail the benefits of forecasting for a utility. This lack of analyses stems partly from the fact that a lot of data input and a proper time step model are needed to be able to draw valid conclusions. In recent years, a number of wind integration studies have undertaken the effort with data backing from typically the TSO.

Nielsen et al. [127] assessed the value for Danish wind power on the NordPool electricity exchange to be 2.4 €c/kWh in a year with normal precipitation (the NordPool system is dominated by Norwegian and Swedish hydropower). This would be reduced by 0.13–0.27 €c/kWh due to insufficient predictions. The same result is expressed as the penalty due to bad prediction of wind power being 12% of the average price obtained on NordPool by Sørensen and Meibom [128].

Gilman et al. [129] found that AWS TrueWind Inc.'s forecasting saved Southern California Edison US$2 million in imbalance cost for December 2000 alone, compared to a system based on pure climatology.

Parkes et al. [130] did an analysis using the GH Forecaster service for the United Kingdom and Spain. Although the two markets are different, both work under the assumption that it should pay to have better forecasts. In the United Kingdom, the best forecast was the centered one, meaning that the technically best forecast was also the economically best for the wind farm owner. A 50 MW wind farm with 30% capacity factor could gain £660.000 from forecasting. Owing to the 5% lower MAE for a total portfolio of three wind farms, another £3/MWh could be gained. In Spain, the exercise yielded about 7 €/MWh for the single wind farm and another 3 €/MWh for a portfolio. Using a better power model, their group estimated [131] for a 100 MW wind farm in the United Kingdom an added income of €177.000 per year for a 1.2% MAE improvement.

The importance and impact of good forecasts were also stated by Operations Manager Carl Hilger from Eltra (the antecessor of Energinet.dk) [132]: "If only we improved the quality of wind forecasts with one percentage point, we would have a profit of two million Danish crowns." Similar orders of magnitude are quoted infrequently by other utilities or traders, but usually not for publication. For the Xcel energy forecasting project, arguably the largest and most ambitious privately funded forecasting project to date, Parks [133] reported savings of US$6 million for 1 year alone for three different regions. This significantly exceeds their investment (which is not a public figure; Keith Parks, personal communication, May 9, 2011).

In a widely quoted paper, Pinson et al. [134] "formulate a general methodology for deriving optimal bidding strategies based on probabilistic forecasts of wind generation" (p. 1148). By taking into account the uncertainty structure of the forecast, the bidding strategy based on probabilistic choice can lead to a reduction of more than half the regulation cost for the wind power producer, in their example of a multi-megawatt wind farm participating in the Dutch electricity market in 2002.

For users of short-term predictions, there is a series of workshops, probably the closest thing to an actual forecast user group, run by Giebel (see powwow.risoe.dk/ BestPracticeWorkshop.htm). The slides of the participant talks are available from the website. Most notably, it is interesting to see the different challenges that the different utilities or TSOs have, and how they use the wind power forecasts to address those challenges. During the POW'WOW project, a report was written on the first two workshops [135]. "Some major results of the workshops were:

- Competition improves accuracy.
- The value of accurate wind power predictions is appreciated.
- The market for wind power prediction models is mature, with many service providers.

The Best Practice in the use of short-term forecasting of wind power can be summarised as:

- Get a model.
- Get another model (NWP and/or short-term forecasting model).
- Get a good nationwide model instead of many simple and cheap models.
- Balance all errors together, not just wind.
- Use the uncertainty/PDF.
- Use intraday trading.
- Use longer forecasts for maintenance planning.
- Meteorological training for the operators.
- Meteorological hotline for special cases.

Additionally, if you are setting up a system for dealing with wind power in your country," (p. 6) there are essentially two ways to deal with forecasting: a demand on every wind power producer, as for example in Spain or the United Kingdom, or with a centralized system as in Germany or Denmark. As "the system operator needs to have a good quality forecasting tool anyway, so all the other producers of wind power might as well forego the need to get forecasts themselves" (p. 6).

5.7 Conclusions

During the last 25 years, wind power forecasting has developed greatly, from the first approaches using just time-series for the forecast, to the use and power conversion of NWP products, to dedicated probabilistic forecasts and decision support tools relying on them. At the same time, a professional market place has opened, where a handful of companies dedicated to wind power forecasting compete for customers, and sometimes collaborate on common issues. For example, ENFOR and energy & meteo systems GmbH, two of the players on the market with some of the longest experience, collaborate in the ANEMOS consortium. Forecast accuracy has increased significantly during the last 25 years, and additional forecast types, such as ramp, variability, or icing forecasts, have started to appear.

Short-term prediction consists of many steps. For a forecasting horizon of more than 4 h ahead, it starts with an NWP model. Further steps are the downscaling of the NWP model results to the site, the conversion of the local wind speed to power (these two can be done in one step with the right statistical model of the power curve), and upscaling from the single wind farms power to a whole region. On all these fronts, improvements have been made since the first models. Typical numbers in accuracy are an RMSE of about 10%–15% of the installed wind power capacity for a 36 h horizon.

The main error in a short-term forecasting model stems from the NWP model. One current strategy to overcome this error source, and to give an estimate of the uncertainty of one particular forecast, is to use ensembles of models, either by using multiple NWP models or by using different initial conditions within those.

Acknowledgments

The writing of this chapter and the underlying report was largely funded by the Marie-Curie Fellowship JOR3-CT97-5004, the ANEMOS project ENK5-CT-2002-00665, ANEMOS.plus, EU FP6 contract number 038692, and SafeWind, EU FP7 grant number 213740.

References

1. Pestana, R. Dealing with limited connections and large installation rates in Portugal. *Presentation at REN (Portuguese TSO): 2nd Workshop on Best Practice in the Use of Short-term Forecasting*, Madrid, May 28, 2008. Available at: http://powwow.risoe. dk/publ/RPestana_(REN)-DealingWLimitedConnALargeInstallRatesInPT_ BestPracticeSTP-2_2008.pdf

2. Courtney, M. *UpWind WP6—Remote Sensing. Final Report—Deliverable D6.17.* Roskilde, Denmark, 2011.

3. Ulianov, Y., G. Martynenko, V. Misaylov, and I. Soliannikova. Wind turbines adaptation to the variability of the wind field. Poster on the EGU General Assembly, Vienna (AT), May 2–7, 2010; see also *Geophysical Research Abstracts* 12, 13650-2, 2010. Available at: http://meetingorganizer.copernicus.org/ EGU2010/EGU2010-13650-2.pdf

4. Brown, B.G., R.W. Katz, and A.H. Murphy. Time-series models to simulate and forecast wind speed and wind power. *Journal of Climate and Applied Meteorology* 23(8), 1184–1195, 1984.

5. Pease, J. Wind power forecasts in the US context. *Presentation at the 2nd Workshop on Best Practice in the Use of Short-term Forecasting*, Madrid, Spain, May 28, 2008. Available at: http://powwow.risoe.dk/publ/JPease_(BPA)-ChallengesWIntOLargeS caleWindByRegionalUtility_BestPracticeSTP-2_2008.pdf

6. Pease, J. Critical short-term forecasting needs for large and unscheduled wind energy on the BPA system. Presentation at the *3rd Workshop on Best Practice in the Use of Short-term Forecasting*, Bremen, Germany, October 13, 2009. Available at: http:// powwow.risoe.dk/publ/JPease_(BPA)-BPAWindRampEventTrackingSystem_ BestPracticeSTP-3_2009.pdf

7. Giebel, G., R. Brownsword, G. Karioniotakis, M. Denhard, and C. Draxl. *The State of the Art in Short-term Prediction of Wind Power—A Literature Overview*, 2nd ed. Project report for the ANEMOS.plus and SafeWind projects. Roskilde, Denmark: Risø DTU, 2011, 110pp.

8. Giebel, G. (ed.), J. Badger, L. Landberg, H.Aa. Nielsen, T.S. Nielsen, H. Madsen, K. Sattler et al. Wind power prediction using ensembles. Risø-R-1527, September 2005.

9. McCarthy, E. Wind speed forecasting in the Central California wind resource area. *Paper presented at the EPRI-DOE-NREL Wind Energy Forecasting Meeting*, Burlingame, California, March 23, 1998.

10. Troen, I., and L. Landberg. Short-term prediction of local wind conditions. *Proceedings of the European Community Wind Energy Conference*, Madrid, Spain, September 10–14, 1990, pp. 76–78, ISBN 0-9510271-8-2.

11. Landberg, L. *Short-term Prediction of Local Wind Conditions*. PhD Thesis, Risø-R-702(EN). Roskilde, Denmark: Risø National Laboratory, 1994, ISBN 87-550-1916-1.

12. Troen, I., and E.L. Petersen. *European Wind Atlas*. Published for the EU Commission DGXII by Risø National Laboratory, Denmark, 1998, ISBN 87-550-1482-8.

13. Landberg, L. Short-term prediction of the power production from wind farms. *Journal of Wind Engineering and Industrial Aerodynamics* 80, 207–220, 1999.

14. Giebel, G., L. Landberg, C. Bjerge, M.H. Donovan, A. Juhl, K. Gram-Hansen, H.-P. Waldl et al. CleverFarm—First results from an intelligent wind farm. *Paper presented at the European Wind Energy Conference and Exhibition*, Madrid, Spain, June 16–19, 2003.

15. Nielsen, T.S., A. Joensen, H. Madsen, L. Landberg, and G. Giebel. A new reference for predicting wind power. *Wind Energy* 1, 29–34, 1998.

16. Martí, I., G. Kariniotakis, P. Pinson, I. Sánchez, T. S. Nielsen, H. Madsen, G. Giebel et al. Evaluation of advanced wind power forecasting models—Results of the Anemos project. *Proceedings of the European Wind Energy Conference and Exhibition*, Athens, Greece, 2006.

17. Möhrlen, C., and Jørgensen, J.U. Forecasting wind power in high wind penetration markets using multi-scheme ensemble prediction methods. *Proceedings of the German Wind Energy Conference*, Bremen, Germany, November 2006.

18. Pahlow, M., C. Möhrlen, and J. Jørgensen. Application of cost functions for large-scale integration of wind power using a multi-scheme ensemble prediction technique. In *Optimization Advances in Electric Power Systems*, Edgardo D. Castronuovo, ed. New York: NOVA Publisher, 2008, ISBN: 978-1-60692-613-0.

19. Pahlow, M., L. Langhans, C. Möhrlen, and J. Jørgensen. On the potential of coupling renewables into energy pools. *Zeitschrift für Energiewirtschaft* 31(1), 35–46, 2007.

20. Möhrlen, C. First experiences of the new EEG trading rules in Germany and introduction of an ensemble based short-term forecasting methodology for intra-day trading. *Presentation at the 3rd Workshop on Best Practice in the Use of Short-term Prediction of Wind Power*, Bremen, Germany, October 13, 2009. Available at: http://powwow.risoe.dk/publ/CM%C3%B6hrlen_(Weprog)-EnsembleKalmanFilter_BestPracticeSTP-3_2009.pdf

21. Lange, M., and D. Heinemann. Accuracy of short term wind power predictions depending on meteorological conditions. *Poster P_GWP091 on the Global Windpower Conference and Exhibition*, Paris, France, April 2–5, 2002.

22. Lange, M., and U. Focken. *Physical Approach to Short-Term Wind Power Prediction*. Berlin: Springer-Verlag, 2005.

23. Monteiro, C., R. Bessa, V. Miranda, A. Botterud, J. Wang, and G. Conzelmann. _Wind Power Forecasting: State-of-the-Art 2009_. Argonne National Laboratory ANL/DIS-10-1, November 2009 (see also http://www.dis.anl.gov/projects/windpower-forecasting.html).

24. Lange, B., K. Rohrig, B. Ernst, F. Schlögl, Ü. Cali, R. Jursa, and J. Moradi. Wind power prediction in Germany—Recent advances and future challenges. _European Wind Energy Conference and Exhibition_, Athens, Greece, 27.2.-2.3, 2006.

25. Lange, M., U. Focken, R. Meyer, M. Denhardt, B. Ernst, and F. Berster. Optimal combination of different numerical weather models for improved wind power predictions. _Proceedings of the 6th International Workshop on Large-Scale Integration of Wind Power and Transmission Networks for Offshore Wind Farms_, Delft, The Netherlands, 2006.

26. Wessel, A., R. Mackensen, and B. Lange. Development of a shortest-term wind power forecast for Germany including online wind data and implementation at three German TSOs. _Presentation at the 3rd Workshop for Best Practice in the Use of Short-term Forecasting_, Bremen, Germany, October 13, 2009. Available at: http://powwow.risoe.dk/publ/RMackensen_(IWES)-IntradayForecastIncludingOffsiteMe asurements_BestPractice-3_2009.pdf

27. Krauss, C., B. Graeber, M. Lange, and U. Focken. Integration of 18GW wind energy into the energy market—Practical experiences in Germany. _Workshop on the Best Practice in Short-term Forecasting_, Delft, The Netherlands, October 25, 2006.

28. Madsen, H., P. Pinson, G. Kariniotakis, H.Aa. Nielsen, and T.S. Nielsen. Standardizing the performance evaluation of short-term wind power Prediction models. _Wind Engineering_ 29(6), 475–489, 2005.

29. Tambke, J., M. Lange, U. Focken, J.-O. Wolff, and J.A.T. Bye. Forecasting offshore wind speeds above the North Sea. _Wind Energy_ 8(1), 3–16, 2005.

30. Sánchez, I., J. Usaola, O. Ravelo, C. Velasco, J. Domínguez, and M.G. Lobo. Sipréolico—A wind power prediction system based on flexible combination of dynamic models. Application to the Spanish power system. _Proceedings and Poster of the World Wind Energy Conference_, Berlin, Germany, June 2002.

31. Vincent, C., G. Giebel, P. Pinson, and H. Madsen. Resolving non-stationary spectral information in wind speed time-series using the Hilbert-Huang transform. _Journal of Applied Meteorology and Climatology_ 49(2), 253–267, 2010.

32. Bossanyi, E.A. Short-term wind prediction using Kalman filters. _Wind Engineering_ 9(1), 1–8, 1985.

33. Fellows, A., and D. Hill. Wind and load forecasting for integration of wind power into a meso-scale electrical grid. _Proceedings of the European Community Wind Energy Conference_, Madrid, Spain, pp. 636–640, September 10–14, 1990.

34. Nogaret, E., G. Stavrakakis, J.C. Bonin, G. Kariniotakis, B. Papadias, G. Contaxis,M. Papadopoulos et al. Development and implementation of an advanced control system for medium size wind-diesel systems. _Proceedings of the EWEC '94 in Thessaloniki_, October 10–14, 1994, pp. 599–604.

35. Tantareanu, C. _Wind Prediction in Short Term: A First Step for a Better Wind Turbine Control_. Denmark: Nordvestjysk Folkecenter for Vedvarende Energi, October 1992, ISBN: 87-7778-005-1.

36. Kamal, L., and Y.Z. Jafri. Time-series models to simulate and forecast hourly averaged wind speed in Quetta, Pakistan. *Solar Energy* 61(1), 23–32, 1997.
37. Dutton, A.G., G. Kariniotakis, J.A. Halliday, and E. Nogaret. Load and wind power
   forecasting methods for the optimal management of isolated power systems with
   high wind penetration. *Wind Engineering* 23(2), 69–87, 1999.
38. Kariniotakis, G., E. Nogaret, and G. Stavrakis. Advanced short-term forecasting of
   wind power production. *Proceedings of the European Wind Energy Conference*,
   Dublin, Ireland, October 1997, pp. 751–754, ISBN: 0 9533922 0 1.
39. Kariniotakis, G.N., E. Nogaret, A.G. Dutton, J.A. Halliday, and A. Androutsos.
   Evaluation of advanced wind power and load forecasting methods for the optimal
   management of isolated power systems. *Proceedings of the European Wind Energy
   Conference*, Nice, France, March 1–5, 1999, pp. 1082–1085, ISBN: 1 902916 00 X.
40. Torres, J.L., A. Garcia, M. De Blas, and A. De Francisco. Forecast of hourly average
   wind speed with ARMA models in Navarre (Spain). *Solar Energy* 79(1), 65–77, 2005.
41. Makarov, Y., D. Hawkins, E. Leuze, and J. Vidov. *California ISO Wind Generation
   Forecasting Service Design and Experience*. California: California Independent
   System Operator Corporation.
42. Pinson, P., L.E.A. Christensen, H. Madsen, P.E. Sørensen, M.H. Donovan, and L.E.
   Jensen. Regime-switching modelling of the fluctuations of offshore wind generation. *Journal of Wind Engineering and Industrial Aerodynamics* 96(12), 2327–2008.
43. Beyer, H.G., T. Degner, J. Hausmann, M. Hoffmann, and P. Ruján. Short term prediction of wind speed and power output of a wind turbine with neural networks.
   Proceedings of the EWEC '94 in Thessaloniki, October 10–14, 1994, pp. 349–352.
44. Sfetsos, A. A novel approach for the forecasting of mean hourly wind speed time-
   series. *Renewable Energy* 27, 163–174, 2001.
45. Sfetsos, A. Time-series forecasting of wind speed and solar radiation for renewable
   energy sources. PhD thesis Imperial College, UK, 1999.
46. Sfetsos, A. A comparison of various forecasting techniques applied to mean hourly
   wind speed time-series. *Renewable Energy* 21, 23–35, 2000.
47. Annendyck H., and D. Becker. Advanced artificial neural network short term load
   forecaster (ANNSTLF). *Paper presented at the Northern African Power Industry
   Convention (NAPIC)*, Algiers, Algeria, November 22–24, 2004.
48. Pielke, R.A., and E. Kennedy. Mesoscale terrain features, January 1980. Report #
   UVA-ENV SCI-MESO-1980-1. Charlottesville, VA: Department of Environmental
   Sciences, University of Virginia, 29pp.
49. Young, G. S., and R.A. Pielke. Application of terrain height variance spectra to meso
   scale modelling. *Journal Of the Atmospheric Sciences* 40, 255–2560, 1983.
50. Skamarock, W.C. Evaluating mesoscale NWP models using kinetic energy spectra.
   Monthly Weather Review 132, 3019–3032, 2004.
51. Rife, D.L., and C.A. Davis. Verification of temporal variations in mesoscale numerical wind forecasts. *Monthly Weather Review* 133(11), 3368–3381, 2005.
52. Giebel, G. (ed.), J. Badger, P. Louka, G. Kallos, I. Martí Perez, C. Lac, A.-M.
   Palomares, G. Descombes. Results from mesoscale, microscale and CFD modelling.
   Deliverable D4.1b of the Anemos project, December 2006, 101pp. Available online
   from Anemos.cma.fr.

53. Nielsen, T.S., H. Madsen, and J. Tøfting. Experiences with statistical methods for wind power prediction. *Proceedings of the European Wind Energy Conference*, Nice, France, March 1–5, 1999, pp. 1066–1069, ISBN: 1 902916 00 X.

54. Nielsen, H.Aa., P. Pinson, L.E. Christiansen, T.S. Nielsen, H. Madsen, J. Badger, G. Giebel, and H.F. Ravn. Improvement and automation of tools for short term wind power forecasting. *Scientific Proceedings of the European Wind Energy Conference and Exhibition*, Milan, Italy, May 7–10, 2007.

55. Beyer, H.G., D. Heinemann, H. Mellinghoff, K. Mönnich, and H.-P. Waldl. Forecast of regional power output of wind turbines. *Proceedings of the European Wind Energy Conference*, Nice, France, March 1–5, 1999, pp. 1070–1073, ISBN: 1 902916 00 X.

56. Focken, U., M. Lange, and H.-P. Waldl. Previento—A wind power prediction system with an innovative upscaling algorithm. *Proceedings of the European Wind Energy Conference*, Copenhagen, Denmark, June 2–6, 2001, pp. 826–829, ISBN: 3-936338-09-4.

57. Mönnich, K. Vorhersage der Leistungsabgabe netzeinspeisender Windkraftanlagen zur Unterstuetzung der Kraftwerkseinsatzplanung. PhD thesis, Carl von Ossietzky Universität Oldenburg, 2000.

58. Kariniotakis, G.N., G.S. Stavrakakis, and E.F. Nogaret. Wind power forecasting using advanced neural network models. *IEEE Transactions on Energy Conversion* 11(4), 762–767; *Presented at the 1996 IEEE/PES Summer Meeting*, July 28–August 1, 1996, Denver, Colorado.

59. Kariniotakis, G.N. *Contribution au développement d'un système de contrôle avancé pour les systèmes éolien-diesel autonomes* [in French]. Thèse de doctorat européen, Ecole des Mines de Paris, Spécialité énergétique, December 1996.

60. Kariniotakis, G.N., and D. Mayer. An advanced on-line wind resource prediction system for the optimal management of wind parks. *Paper presented at the 3rd MED POWER Conference 2002*, Athens, Greece, November 4–6, 2002.

61. Kariniotakis, G.N., D. Mayer, J.A. Halliday, A.G. Dutton, A.D. Irving, R.A. Brownsword, and P.S. Dokopoulos, M.C. Alexiadis. Load, wind and hydro power forecasting functions of the More-Care EMS system. *Paper presented at the 3rd MED POWER Conference 2002*, Athens, Greece, November 4–6, 2002.

62. Costello, R., D. McCoy, P. O'Donnell, A.G. Dutton, and G.N. Kariniotakis. Potential benefits of wind forecasting and the application of More-Care in Ireland. *Paper presented at the 3rd MED POWER Conference 2002*, Athens, Greece, November 4–6, 2002.

63. Kariniotakis, G., and P. Pinson. Evaluation of the More-Care wind power prediction platform. Performance of the fuzzy logic based models. *Paper presented at the European Wind Energy Conference and Exhibition*, Madrid, Spain, June 16–19, 2003.

64. Hatziargyriou, G.C., M. Matos, J.A. Pecas Lopes, G. Kariniotakis, D. Mayer, J. Halliday, G. Dutton et al. "MORE CARE" advice for secure operation of isolated power systems with increased renewable energy penetration & storage. *Proceedings of the European Wind Energy Conference*, Copenhagen, Denmark, June 2–6, 2001, pp. 1142–1145, ISBN: 3-936338-09-4.

65. Durstewitz, M., C. Ensslin, B. Hahn, and M. Hoppe-Kilpper. *Annual Evaluation of the Scientific Measurement and Evaluation Programme (WMEP)*. Kassel, Germany, 2001.

66. Ernst, B., K. Rohrig, H. Regber, and Dr. P. Schorn, Managing 3000 MW wind power in a transmission system operation center. *Proceedings of the European Wind Energy Conference*, Copenhagen, Denmark, June 2–6, 2001, pp. 890–893, ISBN: 3-936338-09-4.

67. Ernst, B., and K. Rohrig. Online-monitoring and prediction of wind power in German transmission system operation centres. *Proceedings of the First IEA Joint Action Symposium on Wind Forecasting Techniques*. Norrköping, Sweden: FOI— Swedish Defence Research Agency, December 2002, pp. 125–145.

68. Lange, B., K. Rohrig, F. Schlögl, Ü. Cali, R. Mackensen, and L. Adzic. Lessons learnt from the development of wind power forecast systems for six European transmission system operators. *Proceedings of the EWEC*, Brussels, Belgium, March 30–April 3, 2008 (only abstract available).

69. González Morales, G. Sipreólico. Wind power prediction experience. Talk slides accompanied by the paper in Reference [30] *Proceedings of the First IEA Joint Action Symposium on Wind Forecasting Techniques*. Norrköping, Sweden: FOI—Swedish Defence Research Agency, December 2002, pp. 197–214.

70. Sánchez, I. Short-term prediction of wind energy production. *International Journal of Forecasting* 22(1), 43–56, 2006.

71. Martí Perez, I. Wind forecasting activities. *Proceedings of the First IEA Joint Action Symposium on Wind Forecasting Techniques*. Norrköping, Sweden: FOI—Swedish Defence Research Agency, December 2002, pp. 11–20.

72. Martí Perez, I., T.S. Nielsen, H. Madsen, J. Navarro, A. Roldán, D. Cabezón, and C.G. Barquero:.Prediction models in complex terrain. *Proceedings of the European Wind Energy Conference*, Copenhagen, Denmark, June 2–6, 2001, pp. 875–878, ISBN: 3-936338-09-4.

73. Martí Perez, I., T.S. Nielsen, H. Madsen, A. Roldán, S. Pérez. Improving prediction models in complex terrain. *Poster P_GWP185 on the Global Windpower Conference and Exhibition*, Paris, France, April 2–5, 2002, 4pp. on the Proceedings CD-ROM.

74. Martí, I., M.J. San Isidro, D. Cabezón, Y. Loureiro, J. Villanueva, E. Cantaro, and I. Pérez. Wind power prediction in complex terrain: From the synoptic scale to the local scale. *Proceedings of the Science of Making Torque from Wind*, Delft, The Netherlands, 19–21 April 2004, pp. 316–327.

75. Gow, G. Short term wind forecasting in the UK. *Proceedings of the First IEA Joint Action Symposium on Wind Forecasting Techniques*. Norrköping, Sweden: FOI— Swedish Defence Research Agency, December 2002, pp. 3–10.

76. Bailey, B., M. C. Brower, and J. Zack. Short-term wind forecasting. *Proceedings of the European Wind Energy Conference*, Nice, France, March 1–5, 1999, pp. 1062–1065, ISBN: 1 902916 00 X; see also http://www.truewind.com/.

77. Electric Power Research Institute (EPRI). *California Wind Energy Forecasting System Development and Testing, Phase 1: Initial Testing*. EPRI Final Report 1007338, January 2003.

78. Zack, J.W., M.C. Brower, and B.H. Bailey. Validating of the forewind model in wind forecasting applications. *Talk at the EUWEC Special Topic Conference on Wind Power for the 21st Century*, Kassel, Germany, September 25–27, 2000.

79. Vidal, J., A. Tortosa, O. Lacave, J. Aymamí, J. Zack, and D. Meade. Validation of a wind power forecast system - The multi-model NWP ensemble strategy. *Proceedings of the European Wind Energy Conference*, Warsaw, Poland, April 20–23, 2010.

80. Zack, J. W. A new technique for short-term wind energy forecasting: A rapid update cycle with a physics-based atmospheric model. *Proceedings of the Global Wind Power Conference and Exhibition*, Chicago, United States, March 28–31, 2004.

81. Westrick, K. Wind energy forecasting in the Pacific Northwestern U.S. *Proceedings of the First IEA Joint Action Symposium on Wind Forecasting Techniques*. Norrköping, Sweden: FOI—Swedish Defence Research Agency, December 2002, pp. 65–74.

82. Pinson, P., and H. Madsen. Ensemble-based probabilistic forecasting at Horns Rev. *Wind Energy* 12(2), 137–155, 2009.

83. Gneiting, T., K. Larson, K. Westrick, M.G. Genton, and E. Aldrich. *Calibrated Probabilistic Forecasting at the Stateline Wind Energy Center: The Regime-Switching Space-Time (RST) Method*. Technical Report No. 464. Seattle, Washington: Department of Statistics, University of Washington, September 2004.

84. Bourke, W., R. Buizza, and M. Naughton. Performance of the ECMWF and the BoM ensemble prediction systems in the Southern Hemisphere. *Monthly Weather Review* 132(10), 2338–2357, 2004.

85. Seaman, R., W. Bourke, P. Steinle, T. Hart, G. Embery, M. Naughton, and L. Rikus. Evolution of the Bureau of Meteorology's global assimilation and prediction system. Part 1: Analysis and initialisation. *Australian Meteorological Magazine* 44(1), 1–18, 1995.

86. Buizza, R., and T.N. Palmer. The singular-vector structure of the atmospheric general circulation. *Journal of the Atmospheric Sciences* 52(9), 1434–1456, 1995.

87. Molteni, F., R. Buizza, T.N. Palmer, and T. Petroliagis. The ECMWF ensemble prediction system: Methodology and validation. *Quarterly Journal of the Royal Meteorological Society* 122(529), 73–119, 1996.

88. Buizza, R., M. Miller, and T.N. Palmer. Stochastic representation of model uncertainties in the ECMWF ensemble prediction system. *Quarterly Journal of the Royal Meteorological Soceity* 125(560), 2887–2908, 1999.

89. Buizza, R., J.-R. Bidlot, N. Wedi, M. Fuentes, M. Hamrud, G. Holt, and F. Vitart. The new ECMWF VAREPS (variable resolution ensemble prediction system). *Quarterly Journal of the Royal Meteorological Society* 133(624), 681–695, 2007.

90. Courtier, P., C. Freydier, J.-F. Geleyn, F. Rabier, and M. Rochas. The ARPEGE project at Météo-France. *Workshop on Numerical Methods in Atmospheric Models*, Vol. 2. Reading, UK: ECMWF, 1991, pp. 193–231.

91. Nicolau, J. Short-range ensemble forecasting. *WMO/CSB Technical Conference Meeting*, Cairns, Australia, December 2002.

92. Bowler, N.E., A. Arribas, K.R. Mylne, K.B. Robertson, and S.E. Beare. The MOGREPS short-range ensemble prediction system. *Quarterly Journal Royal Meteorological Society* 134(632), 703–722, 2008.

93. Möhrlen, C., J. Jørgensen, K. Sattler, and E. McKeogh. Power predictions in complex terrain with an operational numerical weather prediction model in Ireland including ensemble forecasting. *Poster on the World Wind Energy Conference*, Berlin, Germany, June 2002.

94. Giebel, G., L. Landberg, K. Mönnich, and H.-P. Waldl. Relative performance of different numerical weather prediction models for short term prediction of wind energy. *Proceedings of the European Wind Energy Conference*, Nice, France, March 1–5, 1999, pp. 1078–1081, ISBN: 1 902916 00 X.

95. Waldl, H.-P., and G. Giebel. The quality of a 48-hours wind power forecast using the German and Danish weather prediction model. *Talk at the EUWEC Special Topic Conference on Wind Power for the 21st Century*, Kassel, Germany, September 25–27, 2000.

96. Waldl, H.-P., and G. Giebel. *Einfluss des dänischen und des deutschen Wettervorhersagemodells auf die Qualität einer 48-Stunden-Windleistungsprognose.* 5. Deutsche Windenergiekonferenz DEWEK 2000, Wilhelmshaven (DE), June 7–8, 2000, pp. 145–148.

97. Giebel, G., A. Boone. A comparison of DMI-Hirlam and DWD-Lokalmodell for short-term forecasting. *Poster on the European Wind Power Conference and Exhibition*, London, United Kingdom, November 22–25, 2004.

98. Nielsen, H.Aa., T.S. Nielsen, H. Madsen, M.J. San Isidro Pindado, I. Marti. Optimal combination of wind power forecasts, *Wind Energy* 10(5), 471–482, 2007.

99. Pinson, P., and G. Kariniotakis. On-line assessment of prediction risk for wind power production forecasts. *Proceedings of the European Wind Energy Conference and Exhibition*, Madrid, Spain, June 16–19, 2003.

100. Pinson, P., and G. Kariniotakis. On-line adaptation of confidence intervals based on weather stability for wind power forecasting. *Proceedings of the Global Wind Power Conference and Exhibition*, Chicago, United States, March 28–31, 2004.

101. Lange, M., and H.-P. Waldl. Assessing the uncertainty of wind power predictions with regard to specific weather situations. *Proceedings of the European Wind Energy Conference,* Copenhagen, Denmark, June 2–6, 2001, pp. 695–698, ISBN: 3-936338-09-4. (Note: This paper is misprinted in the proceedings; please access this from physik.uni-oldenburg.de/ehf.)

102. Landberg, L., G. Giebel, L. Myllerup, J. Badger, T.S. Nielsen, and H. Madsen. *Poor Man's Ensemble Forecasting for Error Estimation.* Portland/Oregon: AWEA, June 2–5, 2002.

103. Nielsen, H.Aa. *Analyse og simulering af prædiktionsfejl for vindenergiproduktion ved indmelding til NordPool.* Report, IMM, DTU, 20. February 2002.

104. Giebel, G., L. Landberg, J. Badger, K. Sattler, H. Feddersen, T.S. Nielsen, H.Aa. Nielsen, and H. Madsen. Using ensemble forecasting for wind power. *Paper presented at the European Wind Energy Conference and Exhibition*, Madrid, Spain, June 16–20, 2003.

105. Giebel, G., J. Badger, L. Landberg, H.Aa. Nielsen, H. Madsen, K. Sattler, and H. Feddersen. Wind power forecasting using ensembles. *Paper presented at the Global Wind Power Conference and Exhibition*, United States, March 28–31, 2004.

106. Nielsen, H.Aa., T.S. Nielsen, H. Madsen, J. Badger, G. Giebel, L. Landberg, H. Feddersen, and K. Sattler. Wind power ensemble forecasting. *Paper presented at the Global Wind Power Conference and Exhibition*, Chicago, United States, March 28–31, 2004.

107. Nielsen, H.Aa., H. Madsen, T.S. Nielsen, J. Badger, G. Giebel, L. Landberg, K. Sattler, and H. Feddersen. Wind power ensemble forecasting using wind speed and direction ensembles from ECMWF or NCEP. *Project Report*, Lyngby, 2005.

108. Nielsen, H.Aa., D. Yates, H. Madsen, T.S. Nielsen, J. Badger, G. Giebel, L. Landberg, K. Sattler, and H. Feddersen. Analysis of the results of an on-line wind power quantile forecasting system. *Project Report*, Lyngby, 2005.

109. Feddersen, H., and K. Sattler. Verification of wind forecasts for a set of experimental DMI-HIRLAM ensemble experiments. *DMI Scientific Report 05-01*, 2005.

110. Boone, A., G. Giebel, K. Sattler, H. Feddersen. Analysis of HIRLAM including turbulent kinetic energy. *Project Report*, 2005.

111. Roulston, M.S., D.T. Kaplan, J. Hardenberg, and L.A. Smith:.Value of the ECMWF ensemble prediction system for forecasting wind energy production. *Proceedings of the European Wind Energy Conference*, Copenhagen, Denmark, June 2–6, 2001, pp. 699–702, ISBN: 3-936338-09-4.

112. Roulston, M.S., D.T. Kaplan, J. Hardenberg, and L.A. Smith. Using medium-range weather forecasts to improve the value of wind energy production. *Renewable Energy* 28, 585–602, 2003.

113. Taylor, J.W., P.E. McSharry, and R. Buizza. Wind power density forecasting using ensemble predictions and time-series models. *IEEE Transactions on Energy Conversion* 24, 775–782, 2009.

114. Bremnes, J.B. Probabilistic wind power forecasts by means of a statistical model. *Proceedings of the First IEA Joint Action Symposium on Wind Forecasting Techniques.* Norrköping, Sweden: FOI—Swedish Defence Research Agency, December 2002, pp. 103–114.

115. Bremnes, J.B. Probabilistic wind power forecasts using local quantile regression. *Wind Energy* 7(1), 47–54, 2004.

116. Bremnes, J.B. A comparison of a few statistical models for making quantile wind power forecasts. *Wind Energy* 9(1–2), 3–11, 2006.

117. Nielsen, H.Aa., H. Madsen, and T.S. Nielsen. Using quantile regression to extend an existing wind power forecasting system with probabilistic forecasts. *Paper in the Scientific Track of the European Wind Energy Conference*, London, United Kingdom, November 22–25, 2004.

118. Nielsen, H.Aa., H. Madsen, and T.S. Nielsen. Using quantile regression to extend an existing wind power forecasting system with probabilistic forecasts. *Wind Energy*, 9(1–2), 95–108, 2006.

119. Sánchez, I. Adaptive combination of forecasts with application to wind energy. *International Journal of Forecasting* 24(4), 679–693, 2008.

120. Focken, U. Experiences with extreme event warning and ramp forecasting for US wind farms. *Talk at the 4th Workshop on Best Practice in the Use of Short-term Prediction of Wind Power*, Quebec City, California, October 16, 2010.

121. Davy, R.J., M.J. Woods, C.J. Russell, and P.A. Coppin. Statisical downscaling of wind variability from meteorological fields. *Boundary Layer Meteorology* 135(1), 161–175, 2010.

122. Vincent, C.L., P. Pinson, and G. Giebel. Wind fluctuation over the North Sea. *International Journal of Climatology* 2011; doi: 10.1002/joc.2175.

123. Von Bremen, L., and N. Saleck. Minimizing the risk in offshore wind power integration induced by severe wind power fluctuations. *Proceedings of the EAWE Special Topic Conference: The Science of Making Torque from Wind*, Crete, Greece, 2010.

124. Vincent, C.L., and A. Hahmann. Hour-scale wind fluctuations over the North Sea. *European Wind Energy Conference*, Brussels, Belgium, March 14–17, 2011.

125. Hasche, B., R. Barth, and D.J. Swider. Effects of improved wind forecasts on operational costs in the German electricity system. *Paper presented at the Conference on Energy and Environmental Modelling*, Moscow, Russia. Available at EcoMod: http://www.ecomod.org/files/papers/366.pdf.

126. Dobschinski, J., E. De Pascalis, A. Wessel, L.v. Bremen, B. Lange, K. Rohrig, and Y-M. Saint Drenan. The potential of advanced shortest-term forecasts and dynamic prediction intervals for reducing the wind power induced reserve requirements. *Scientific Proceedings of the European Wind Power Conference*, Warsaw, Poland, April 20–23, 2010, pp. 177–182.

127. Nielsen, L.H., P.E. Morthorst, K. Skytte, P.H. Jensen, P. Jørgensen, P.B. Eriksen, A.G. Sørensen et al. Wind power and a liberalised North European electricity exchange. *Proceedings of the European Wind Energy Conference*, Nice, France, March 1–5, 1999, pp. 379–382, ISBN: 1 902916 00 X.

128. Sørensen, B., and P. Meibom. Can wind power be sold in a deregulated electricity market? *Proceedings of the European Wind Energy Conference*, Nice, France, March 1–5, 1999, pp. 375–378, ISBN: 1 902916 00 X.

129. Gilman, B., M. Cheng, J. Isaac, J. Zack, B. Bailey, and M. Brower. The value of wind forecasting to Southern California Edison. *Paper 021 presented at the Conference Proceedings CD-ROM of the AWEA Windpower 2001 Conference*, Washington, United States, June 3–7, 2001.

130. Parkes, J., J. Wasey, A. Tindal, and L. Munoz. Wind energy trading benefits through short term forecasting. *European Wind Energy Conference and Exhibition*, Athens, Greece, 27.2.-2.3, 2006.

131. Collins, J., J. Parkes, and A. Tindal. Short term forecasting for utility-scale wind farms—The power model challenge. *Wind Engineering* 33(3), 247–258, 2009.

132. Hilger, C. Oral statement made at the *Fuel and Energy Technical Association Conference on "Challenges from the Rapid Expansion of Wind Power,"* April 3, 2005.

133. Parks, K. Xcel energy/NCAR wind energy forecasting system. *Talk at the UWIG Forecasting Workshop*, Albany, United States, February 23–24, 2011.

134. Pinson, P., C. Chevallier, and G. Kariniotakis. Trading wind generation with short-term probabilistic forecasts of wind power. *IEEE Transactions on Power Systems* 22(3), 1148–1156, 2007.

135. Giebel, G., and G. Kariniotakis. *Best Practice in the Use of Short-Term Forecasting. A Users Guide.* Project Report for the POW'WOW project, 2008, 6pp. Available at: http://powwow.risoe.dk/publ/GiebelKariniotakis-BestPracticeInSTF_156_Ewec2007fullpaper.pdf

6

Price-Based Scheduling for Gencos

Govinda B. Shrestha

Songbo Qiao

6.1 Introduction

Optimal economic operation of power systems has always been a very important subject in the planning and operation of power systems. While the minimization of the overall cost considering both the investment cost as well as the operating cost has been the foundation of most planning approaches, the minimization of operation cost by operating the power system at the minimum marginal cost is the most common basis of optimal power system operation. These principles have played an invaluable role in the optimal operation of the traditional power systems so far. However, the basic tenet of these principles has been challenged by two fundamental changes in the power industry around the world: the development of competitive power markets, and the emergence of renewable energy sources.

Incorporation of renewable energy posed some difficulty in the purely cost-based approaches, as these sources of energy incurred only capital costs and exhibited very little operating cost so that the marginal cost of energy production is virtually zero. Therefore, coordinating hydropower generation in the traditional power systems required slightly more creative approaches, one common method being to assign the cost of the energy replaced by the hydroelectricity as its own cost. In general, hydro-power has always been included in such a way that the total system cost is minimized. Such approaches were adequate as the proportion of renewable resources in the power systems remained very low. With the recent emphasis on the development of renewable resources, particularly wind power and solar photovoltaic (PV) power to meet the bulk of the energy need in the future, the substitution methods described above are not expected to remain very effective.

Deregulation of power markets leading to independent generating companies (Gencos) participating in the grid operated by the independent system operator (ISO) has brought the price as the basis for market operation. The system-wide operation is carried out by ISO through market clearing decided on the basis of the bids submitted by Gencos. Individual Gencos are responsible for their own unit commitment (UC) and generation scheduling according to the market clearing by the ISO.

This chapter describes some studies carried out to investigate price-based (profit-based) unit commitment (PBUC) and price-based scheduling for Gencos utilizing a variety of approaches. The symbols and acronyms used in this chapter are listed below for easy reference.

6.1.1 List of Symbols

Symbol	Definition
δ_i	Cold startup cost
σ_i	Hot startup cost
σ_j	Standard deviation of the forecast market price at time period j ($/MWh)
τ_i	Unit cooling time constant
$\Pi_{i,s,\text{spot}}$	Profit from unit i in the spot market for scenario s
$a_i, b_i,$ and c_i	Cost coefficients of unit i
B	Bidding blocks ($B \in 1, 2 \dots NB$)
$B(\text{HL})$	Block of contract with length of HL hours
BI	Block index of contract $B(\text{HL})$
$C_i(G_i)$	Cost of producing G_i units of power
FC_i	Fixed annual cost for the generator unit i
$G_{i,t}$	Generation in period t by unit i (MWh)
$G_i^{\max}$	Generation upper limit of unit i
$G_i^{\min}$	Generation lower limit of unit i

h_t	Hedging ratio in period t such that $0 \le h_t \le 1$
HL	Length in hours of a block of contract
i	Index of thermal units ($i \in 1, 2 \ldots$ NT)
j	Index for electricity trading time period in one day ($j \in 1, 2 \ldots 48$)
J	Total system cost
k	Index for days in the year ($k \in 1, 2 \ldots 365$)
MCP	Market clearing price ($/MWh)
MUT_i	Minimum up time of unit i
MDT_i	Minimum down time of unit i
n	Stage number in dynamic programming
N	Index for bid steps or blocks ($N \in 1, 2, 3$ in this study)
NT	Number of thermal generating units
P_{G_i}	Power rating of generator i (MW)
$PB_{B,t}$	Bid price for block B
$PB_{N,j}$	Bid price for step N at time period j ($/MWh)
PD_t	Power demand at time t
Pen_t	Penalty function at time t
PF_t	Forward price at time t
PN	Number of periods in each stage n
PR_j	Market clearing price at time period j ($/MWh)
$PR_{t,s}$	Price corresponding to scenario s
$prob_s$	Probability associated with scenario s
$\overline{PS_j}$	Mean value of forecast market price at time period j
PS_t	Expected average spot price forecast in period t ($/MWh)
$PS_{s,t}$	Spot price in time t corresponding to scenario s
PS_t^{rand}	Random variable in time t for the next day price
$QB_{B,t}$	Bid quantity for block B
$QB_{N,j}$	Bid quantity of bid step N at time period j (MW)
QF_t	Quantity forward in time t
R_t	Reserve requirement in each period ($)
s	Index of a scenario ($s = 1, 2, \ldots S$)
S	Total number of price scenarios
$SD_{i,t}$	Shutdown cost of unit i in time t
$SDev_t$	Standard deviation of PS_t^{rand} in time t
$ST_{i,t}$	Startup cost of unit i in time t
t	Time period ($t \in 1, 2, \ldots N$)
T	Total number of periods
$T_{t,i}^{off}$	Time unit i in "off" state at time t
$T_{t,i}^{on}$	Time unit i in "on" state at time t
U	Utility function
$UB_{N,j}$	Bid status of bid step N at time period j (1 = successful, 0 = unsuccessful)
$USD_{i,t}$	Unit shutdown decision of unit i in time t (1 = shutdown, 0 = no shutdown)

USEP	Uniform Singapore energy price
$UST_{i,t}$	Startup decision of unit i in time t (1 = startup, 0 = no startup)
$W_{i,t}$	Unit status (1 = running, 0 = off) of unit i at time t
Wpen	Weight of the penalty function
Y_t	Sales/purchase in the spot market (MWh)

6.2 Spot Price Modeling

The power industry has been under the process of deregulation around the world although the restructuring occurred under different circumstances for different countries. However, the main objective of this exercise has been to establish separate entities (companies) for generation (Gencos), transmission (Transcos), and distribution (Discos) and to create a competitive open market where different market participants could freely compete. Then the electricity price—the spot price—is determined on the basis of the bids by the market participants, which varies every trading period. The spot price has become the *de facto* basis for the success of bids by Gencos and hence their UC and the generation scheduling. A clear understanding of the price behavior and a suitable representation of the price become essential for scheduling of Gencos.

6.2.1 Price Uncertainty

Electricity is a special commodity in that supply and demand must be matched instantaneously. Its special characteristics in storing, generation, and transmission make its prices more volatile than those of many other ordinary commodities. In the short term, the nonstorability of electricity often leads to situations where any imbalance of power in a system leads to wide fluctuations in the price. As it is, electricity price at peak hours is usually much higher. If some unforeseen generator outages occur, the price can rise drastically. At the same time, many usual factors such as political uncertainties, fuel price fluctuation, physical characteristics of the power system, and so on, also play important roles in long-term price variation. Price uncertainty may also be seen as arising from many well-known factors such as seasonality, day type (weekdays, weekends, and holidays), and so on.

Variations of the daily and monthly mean price values in the Singapore market in 2006 are shown in Figure 6.1. It is clearly seen that the monthly average value varies in a much narrower range compared to the daily average price, which may even be considered irregular and drastic.

Figure 6.2 shows the variation of half-hourly prices in the Singapore market which exhibit even higher volatility and spikes. For the year 2006, there are 12 periods when the price exceeded $1000/MWh, 8 of them larger than $2000/MWh.

The USEP in 2006 reached its ceiling price of $4500/MWh on 21 December. This peak price was caused by the forced outage of two combined cycle gas turbine (CCGT) units and an unplanned disruption of gas supply resulting in a power deficiency of 77.8 MW.

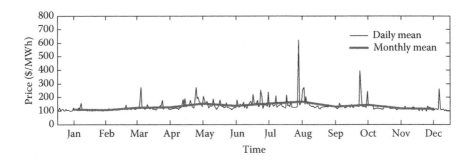

FIGURE 6.1 Monthly and daily average electricity price (the uniform Singapore electricity price in 2006).

Although the price fluctuated less in 2007 there was still a high percentage of unusual prices, most of which occurred in the first and third quarters of the year. Year 2007 also saw 12 periods when the price exceeded $1000/MWh, 7 of them higher than $2000/MWh. A forced outage of a CCGT and offer changes caused the USEP peak price of $4330/MWh on 6 January. On 28 July, the USEP climbed to its ceiling price ($4500/MWh) after registering a deficit caused by a series of supply crunches—forced outage of one CCGT unit, scheduled maintenance of one CCGT unit, with two CCGT units remaining out at the same time.

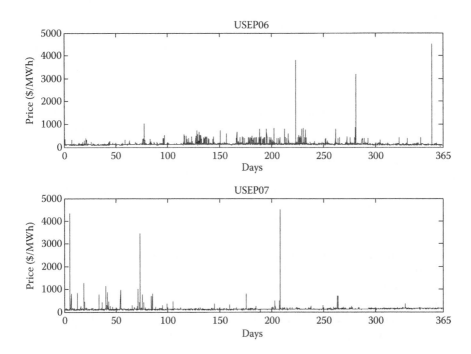

FIGURE 6.2 Half-hourly price variations for 2006 and 2007.

6.2.2 Statistical Distribution of Price

Market price forecasting techniques are required to support planning and operations activities of Gencos. Although it may be possible to qualitatively identify the reasons behind certain specific variations in electricity price, it would be extremely difficult to model such cause–effect relationships that can be used in price forecasting necessary for regular operations planning. Although the historical data shows that the mean values of the prices are quite stable and may be forecast with some ease, it is the individual price at each interval that becomes important for planning purposes. Therefore, it is found preferable to (i) represent the price at any one period as a random variable with a reasonably stable mean value, and (ii) find ways to represent the deviations from the mean value in a statistical way [1].

Historical USEPs from the Singapore market are used to study the behavior of the market price in order to identify possible statistical distributions, and then hypothesis tests are formulated and carried out to determine the theoretical distributions that are best supported by the historical data.

6.2.2.1 Graphical Analysis

As observed above, the mean value of electricity price is less volatile and less difficult to forecast than individual values. However, it is the individual spot price at a particular period that plays an important part in optimizing system operation, and hence planning studies, of market participants. Thus, it may be worthwhile to explore whether the spot price could be represented as a stochastic variable with a mean value so that the individual price values could be treated probabilistically. Further, it would be very useful to investigate whether the price values follow any particular statistical distribution. Then it may be possible to represent the spot price at any one given period probabilistically using a few parameters such as the mean and the standard deviation that could be estimated from sample price data.

In the Singapore market, market clearing is done every half an hour, giving 48 spot price values in one day. The following analysis attempts to study the stochastic behavior of price at individual periods, that is, USEP.

The histogram depicting the spot price values for 2006 and 2007 is shown in Figure 6.3. Although the histogram shows a smooth distribution it is quite widespread at higher values. In 2006 [2], about 9% of the USEP was lower than $110/MWh, 87.41% of the USEP was between $110 and $200/MWh and about 3.58% of the USEP was greater than $200/MWh. In addition, the USEP was less volatile in 2007 [3] than in 2006, for only 1.09% of the USEP in 2007 (which is one-third of the USEP in 2006) was higher than $200 and a total of 48 trading periods in 2007 (compared to 85 trading periods in 2006) were observed in which the USEP closed at or above $500/MWh.

Figure 6.4 shows the histogram of the USEP for one particular period. Similar to the annual histogram, it also exhibits some extreme values although the majority seems to conform to a distinct pattern. It would be very useful if the pattern could be modeled statistically. The background theory of statistics used for this purpose is outlined in the following section.

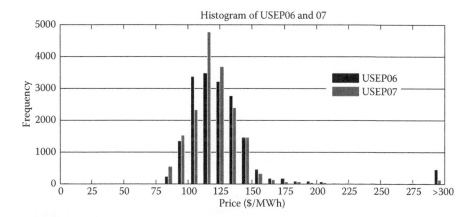

FIGURE 6.3 Frequency distribution of monthly and daily electricity price (USEP 2006 and USEP 2007).

6.2.2.2 Theoretical Distribution Functions

On the basis of the shape of the histogram, three common statistical distributions are identified for investigation: normal, lognormal, and Weibull distributions. The latter two are indicated by the presence of a definite minimum value and rather extended larger values of the price data. These distributions have been used in studies as reported earlier and are described by the following probability density functions (PDFs).

Lognormal:

$$f(x;\mu,\sigma) = \frac{1}{x\sigma\sqrt{2\pi}} e^{-\frac{(\ln x - \mu)^2}{2\sigma^2}}, \quad x \in (0,+\infty) \tag{6.1}$$

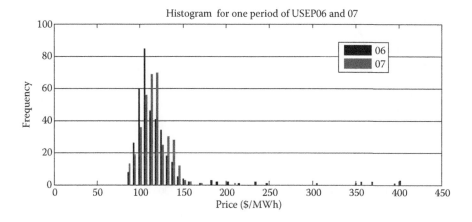

FIGURE 6.4 Price histograms for one period for 2006 and 2007.

Normal:

$$f(x;\mu,\sigma) = \frac{1}{\sigma\sqrt{2\pi}}e^{-\frac{(x-\mu)^2}{2\sigma^2}}, \quad x \in (-\infty,+\infty) \tag{6.2}$$

Weibull:

$$f(x;\gamma,\theta) = \frac{\gamma}{\theta}x^{\gamma-1}e^{-x^\gamma/\theta}, \quad x \in (0,+\infty) \tag{6.3}$$

Hypothesis testing will be utilized to investigate whether the historical USEP data will reasonably fit any of these statistical distributions. For continuous functions, the testing is carried out by checking whether the theoretical cumulative distribution function agrees well with the sample data. For example, Figure 6.5 shows the cumulative distribution of the USEP data shown in Figure 6.3. Three theoretical cumulative distribution functions formed from the parameters computed from the sample data are also shown in Figure 6.5. The procedure of hypothesis can indicate whether the sample data conforms to any of these theoretical distributions.

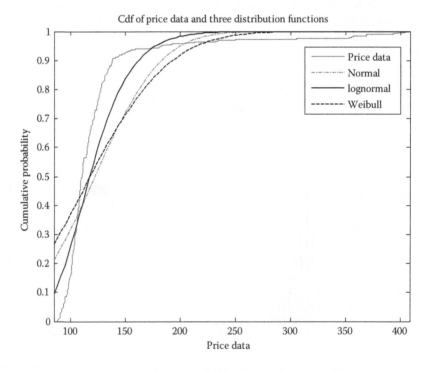

FIGURE 6.5 Cumulative distribution of uniform Singapore electricity price and three distribution functions.

6.2.2.3 Hypothesis Testing

Hypothesis testing is carried out to investigate:

1. The null hypothesis H_0: The price data come from a specific statistical distribution.
2. The alternative H_1: The prices do not come from that distribution.

The three candidate distributions under consideration are normal, lognormal, and Weibull distributions. The tests are carried out for all three distributions at two different levels of significance ($\alpha = 0.05$ and $\alpha = 0.01$). The statistical tool box under MATLAB® is utilized for this purpose, which uses the Lilliefors test and the Kolmogorov–Smirnov test for hypothesis testing. The Lilliefors test is a two-sided goodness-of-fit test suitable when a fully specified null distribution is unknown and its parameters must be estimated. It is specific for the normal family distribution, in this case for the normal and lognormal distributions. Similarly, the Kolmogorov–Smirnov test is used for the Weibull distribution. Investigations are carried out to determine whether the price data can be characterized by any of the candidate distribution functions.

6.2.2.4 Data Analysis

The first attempt is to fit the raw data without any adjustments to the theoretical functions. A graphical analysis of the match between the raw data and the theoretical distributions were studied. A graphical depiction of the fit is shown in Figure 6.6 for periods 16, 24, and 40, covering the prices at various load levels during the day. The sample for each period is 365. It was observed that certain price values (e.g., \$3320.7/MWh at period 24) are clearly outliers that can distort the fit, because of the large standard deviation caused by such extreme values.

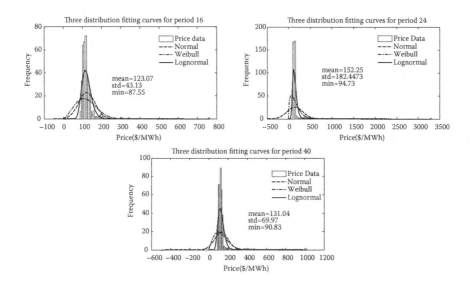

FIGURE 6.6 Three distributions fitting the raw sample data price.

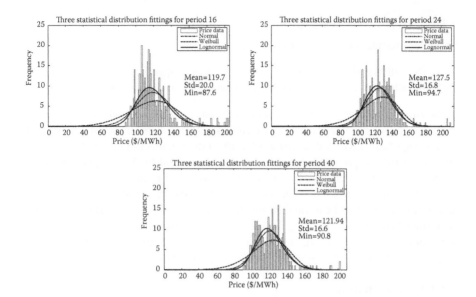

FIGURE 6.7   Three distributions fitting the sample data after filtering extreme data.

Therefore, it is felt necessary that such extreme data be filtered out so that the general nature of the data may be suitably characterized. In this process, extreme data exceeding the average by five standard deviations (USEP $\geq \mu \pm 5\sigma$) were filtered.

Figure 6.7 depicts the resulting fit between the sample data and the theoretical functions, which show better match between the corresponding curves. It can be observed from these figures that the lognormal distribution fits the data best, followed by the normal and Weibull curves. The normal curve seems to fit better during peak periods whereas the Weibull distribution seems better during off-peak periods. If these observations could be generalized, it would make them useful for modeling the price behavior in planning studies and operations scheduling.

It can be observed from Figure 6.7 that the lognormal and Weibull distributions do not fit well in the lower tail of the figures. This discrepancy occurs because the USEP has a definite minimum value, but the range of theoretical distribution is $(0, +\infty)$. The range of USEP data can be easily converted to $(0, +\infty)$ by shifting the USEP data to the left by a value close to the minimum value ($c \approx \text{USEP}_{min}$) observed in the sample. Such a shift should improve the performance of the lognormal and Weibull distributions. The resulting fit between the USEP data shifted by a value ($c \approx 0.8 \times \text{USEP}_{min}$) and the theoretical distributions is shown in Figure 6.8. It is observed that the lognormal distribution fits the data better than the original sample without the shift.

6.2.2.5 Hypothesis Testing on Annual Data

Hypothesis testing was conducted taking into account USEP data conditioned as discussed above. Table 6.1 shows the result of the test that lists the frequency of acceptance of the null hypothesis for different distributions. The lognormal distribution is accepted the most to fit the data in both 2006 and 2007. At 5% significance level, the test failed to

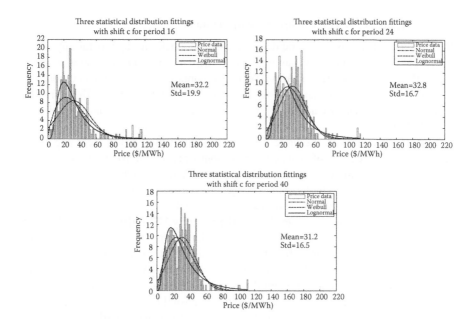

FIGURE 6.8 Three distributions fitting the filtered data after a shift.

reject the null hypothesis in 30 out of 48 (62.5%) periods in 2006 and 28 out of 48 (58.3%) periods in 2007. At 1% significance level, the acceptance becomes larger but has less statistical power.

The results of the test on the data after the shift c show improvements in terms of the acceptance of all the distributions. The lognormal distribution is still found to be the best fit with acceptance of the null hypothesis in 36 (75%) periods at 5% significance level in 2006. These results, however, do not provide sufficient support for any of the distributions.

TABLE 6.1 Acceptance of Distribution Functions for Annual Uniform Singapore Electricity Price (USEP)

Distribution Function	α	USEP 2006	USEP 2007
Lognormal	0.05 (5%)	30 (62.5%)	28 (58.3%)
Lognormal with shift		36 (75%)	17 (35.4%)
Weibull		3 (6.25%)	0
Weibull with shift		10 (20.8%)	11 (22.9%)
Normal		15 (31.3%)	11 (22.9%)
Lognormal	0.01 (1%)	39 (81.3%)	37 (77.1%)
Lognormal with shift		42 (87.5%)	25 (52.1%)
Weibull		6 (12.5%)	2 (4.2%)
Weibull with shift		18 (37.5%)	19 (39.6%)
Normal		22 (45.8%)	27 (56.3%)

6.2.2.6 Hypothesis Testing on Monthly Data

The outcomes of the above testing do not provide any definite support for any of the theoretical distributions. This may be because the data in different seasons may not exhibit the same characteristics. Annual data may contain distinctly different groups of data that exhibit different characteristics because of seasonal variations. Therefore, the price data are grouped on a monthly basis so that each monthly sample has a sample size of 29–31 points.

Seasonal differences in the price data are confirmed by plotting the comparison between the average prices of each period on a monthly basis. The average for two different months and the annual average are illustrated in Figure 6.9, which shows a distinct difference between the various sample averages. Therefore, the monthly samples should provide better results if they are tested as separate samples.

The results of the hypothesis testing on the monthly USEP data for four different months, conducted exactly in the same fashion as in Section 6.2.2.5, are listed in Table 6.2. The acceptance of all the distributions improves with the shift. The highest acceptance is 47 out of 48 (97.9%) for both the lognormal and Weibull distributions in February. The acceptance for different distributions drops in August compared to the other three months because of causes beyond the scope of this analysis. (The market suffered from serious generator outages in that month disturbing the market price behavior.) Thus, it is found that various ways of refining the data can influence the acceptance of different distributions to various extents; the lognormal distribution stands out as the best fit for the data obtained from the Singapore market.

Graphical and statistical analyses have been conducted to determine the possible theoretical distributions for electricity price. It was found that the raw price data are not amenable to direct statistical representation. However, suitable processing of the historical data, such as (i) filtering out the extreme data, (ii) adopting suitable shifts to match

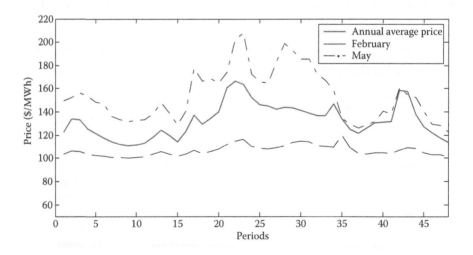

FIGURE 6.9   Comparison of average price on an annual basis and a monthly basis.

TABLE 6.2 Acceptance of Distribution Functions for Monthly Uniform Singapore
Electricity Price

Distribution Function	α	February 2006	May 2006	August 2006	November 2006
Lognormal	0.05 (5%)	37 (77.1%)	25 (52.1%)	13 (27.1%)	36 (75%)
Lognormal with shift		43 (89.6%)	37 (77.1%)	35 (72.9%)	45 (93.8%)
Weibull		26 (54.2%)	18 (37.5%)	8 (16.7%)	23 (47.9%)
Weibull with shift		33 (68.8%)	26 (54.2%)	31 (64.6%)	31 (64.6%)
Normal		33 (68.8%)	19 (39.6%)	12 (25%)	32 (66.7%)
Lognormal	0.01 (1%)	43 (89.6%)	34 (70.8%)	22 (45.8%)	44 (91.7%)
Lognormal with shift		47 (97.9%)	45 (93.8%)	39 (81.3%)	48 (100%)
Weibull		36 (75%)	25 (52.1%)	12 (25%)	37 (77.1%)
Weibull with shift		47 (97.9%)	38 (79.2%)	35 (72.9%)	43 (89.6%)
Normal		39 (81.3%)	23 (47.9%)	13 (27.1%)	43 (89.6%)

the range of the theoretical distribution, and (iii) regrouping the data in monthly samples instead of yearly samples to separate the seasonality, significantly enhances the cohesive statistical characteristics of the sample data.

Hypothesis tests are conducted at two different levels of significance and it shows that the lognormal distribution is best supported by the historical data, although the Weibull and normal distributions also were accepted reasonably well.

6.3 Generation Characteristics

Generation cost characteristics and their representations have been thoroughly investigated for optimizing UC and generation scheduling problems in the operation of traditional power system. In the competitive electricity markets, although cost characteristics are no longer the basis for UC and scheduling exercises, they remain very important for successful operation and maximization of profits of Gencos. However, the diverse types of generators available these days, and the increasing presence of renewable generation, particularly wind and solar generation, have made the treatment of generation characteristics more complicated. Traditional and renewable generator cost characteristics will be discussed in some detail in this section.

6.3.1 Traditional Generator Cost Characteristics

UC problems have usually attempted to minimize the cost of generating units in an electric power system, especially for traditional fuel cost units considering the following.

6.3.1.1 Operation Cost

The use of operation costs, including running cost, startup cost, shutdown cost, is explained in the following equations.

Running Cost:

$$C_i(G_i) = a_i + b_iG_i + c_iG_i^2 \tag{6.4}$$

Here, $C_i(G_i)$ is the unit cost characteristic of the generating units, which is a function of G_i and is derived from heat-rate characteristics.

Startup Cost:

Either an exponential function or a two-step function is used to represent the startup cost of the generator.

Exponential Startup Function:

$$ST_i = \sigma_i + \delta_i \left[1 - \exp\left(\frac{-T_{t,i}^{off}}{\tau_i} \right) \right] \tag{6.5}$$

Two-step Startup Function:

$$ST_i = \begin{cases} \delta_i & \text{if } T_{t,i}^{off} \geq \tau_i \\ \sigma_i & \text{otherwise} \end{cases} \tag{6.6}$$

Here, τ_i is the number of hours that it takes for the boiler to cool down.

Shutdown Cost:

Shutdown cost is usually given a constant value for each unit.

6.3.1.2 Marginal Cost Analysis

Occasionally, small Gencos in the competitive deregulated market are assumed to be price-takers, that is, their generation bid cannot influence the market price. The marginal cost of the generating units therefore is involved in the scheduling problem.

The scheduling task for a price-taker Genco would be straightforward if the future market prices were known. The optimal generation at a particular moment is simply where the incremental generation cost (marginal cost) equals the market price. Nevertheless, the electricity market is more likely to be an oligopoly market and the market price cannot be precisely predicted. The marginal cost bidding in this environment for the small Genco who plays as a price-taker may not be the optimal scheme. The bid at marginal cost may be driven out of competition by bids from bigger generators with market power. Scheduling based on marginal cost bidding is investigated in Section 6.5.2, as a comparison.

6.3.1.3 Screening Curve Analysis

The screening curve analysis has been used to establish the relative economic merits of different generator types that can be utilized to determine the optimal generation combination [4]. Such analyses not only consider the running cost of a generation unit but also include the capital cost that plays an important role in the long-term system planning. The screening curve will enable the Genco to bid in the most economical way such

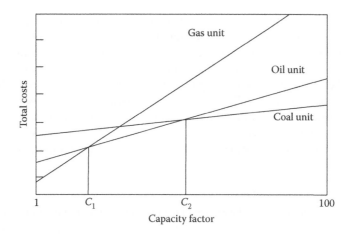

FIGURE 6.10 Screening curves of generation units.

that the generating units will run with capacity factors in the range where they are the most efficient units.

Figure 6.10 shows a typical screening curve traditionally used to illustrate generation mix concepts. This figure shows that the coal unit with a very high capital cost and a low operation cost would be the cheapest generator when operated with a capacity factor larger than C_2. Similarly, the gas unit with a low capital cost but a high fuel cost would be the most economical unit only if it has to be operated with a capacity factor less than C_1. In the same way, if the capacity factor for a planning plant stays in the interval between C_1 and C_2, then the oil unit will be the most economical unit. This observation indicates an effective way to specify the bid for the generation unit which is to maintain the running time of each unit within its economical capacity factor range, which can be expected to keep generation cost at its minimum.

6.3.2 Renewable Generation Characteristics

The growth of renewable generation, particularly wind and solar generation, has been significant in recent years. The participation of renewable generation in the deregulated market is also a challenge due to the nondispatchable and unpredictable nature of generation. Therefore, it is very important to investigate the characteristics of renewable generation, such as wind and solar generation, in order to become competitive and survive in the deregulated market.

6.3.2.1 Wind Generation Behavior

Wind power generators cannot be dispatched like those conventional generators. This poses problems for generation schedules. High penetration of wind generation in electric power industries poses new challenges in the short-term market and for system operators, mainly due to the variability of wind. Thus, it is quite important to efficiently and economically integrate wind generation into the competitive electricity market.

The Genco participating in an electricity market needs to submit supply bids and to commit to the delivery of an agreed amount of energy in a given period. If the actual delivered energy differs from the agreed amount, the system operator has to change the schedules of other generators in order to maintain the balance between supply and demand. The cost of this rescheduling must be paid by those that cause it; therefore, penalty or imbalance prices will be charged to the producers of wind power with short fall. As the power produced by a wind farm depends heavily on the fluctuation of wind, unplanned deviations of a wind farm output may increase operating costs for the electricity system, because of increased requirements of primary reserves, and add potential risks to the reliability level.

6.3.2.1.1 Imbalance Cost Characteristics

The major difference between the bids for a traditional generator and wind energy is the real dispatchable power. Failure at the bidding auction level does not entail penalties, whereas failure at wind power availability will incur penalty. The imbalance cost that occurs in the process of scheduling wind power generation will be introduced in this section.

For any trading period, the price-taker needs to propose a level of contracted energy QB_j. The profit of this Genco with the submission of the amount of energy QB_j but actually generating Q_j can be formulated as

$$\Pi_j = QB_j PR_j - C(Q_{im_j}) \tag{6.7}$$

where PR_j is the market clearing price and $C(Q_{im_j})$ is the imbalance cost on the regulation market. The imbalance Q_{im_j} can be defined as

$$Q_{im_j} = QB_j - Q_{W_j} \tag{6.8}$$

and, consequently, $C(Q_{im_j})$ can be given by

$$C(Q_{im_j}) = \begin{cases} PP_j Q_{im_j}^+ & \text{if } Q_{im_j}^+ \geq 0 \\ Pc_j Q_{im_j}^- & \text{if } Q_{im_j}^- < 0 \end{cases} \tag{6.9}$$

where Q_j is the wind power actually generated by the wind farm at period j. PP_j and Pc_j indicate positive and negative imbalance prices, respectively. It should be noted that these two types of imbalance prices will be determined by the particular regulation mechanism. In some specific electricity market, they simply equal a certain proportion of the MCP: for instance, in Spain, these two categories of regulation prices could be $PP_j = (1 + \pi)PR_j$ and $Pc_j = (1 - \pi)PR_j$. More generally, the imbalance prices are asymmetric and come from a more complex function of the spot price. In the regulation mechanism of NordPool, they may even depend on the sign of the imbalance as a whole. Therefore, the market participants will not be penalized if they offset the system imbalance. The imbalance cost here will be taken as a parameter so as to get the sensitivities of the revenues with respect to the imbalance deficiency.

6.3.2.1.2 Probabilistic Representation of Wind Generation

To analyze the characteristics of wind power generation, it is important to model the behavior of wind speed. The Weibull distribution function has been widely used to characterize the uncertain nature of wind speed. The PDF of wind speed is given by

$$f(v) = \left(\frac{k}{c}\right)\left(\frac{v}{c}\right)^{k-1} e^{-\left(\frac{v}{c}\right)^k}, \quad 0 < v < \infty \tag{6.10}$$

where v is the wind speed, k is the shape parameter, and c is the scale parameter. The Weibull distribution function with $k = 2$ is called the Rayleigh distribution.

If the wind speed is assumed to come from a Weibull distribution, it is then important to transform the wind speed distribution to a wind power distribution. As the wind speed can be viewed as a random variable, the output wind power generation may also be regarded as a random variable through the transformation. The output of wind turbines can be stated as follows:

$$Q = \begin{cases} 0 & v < v_{\text{cut-in}} \quad \text{or} \quad v > v_{\text{cut-out}} \\ \dfrac{1}{2} C_p \rho A v^3 & v_{\text{cut-in}} \le v < v_{\text{satur}} \\ Q_{\max} & v_{\text{satur}} \le v \le v_{\text{cut-out}} \end{cases} \tag{6.11}$$

where Q is the output power.

Equation 6.11 shows the relationship between the hourly wind speed and the output power. The power output is zero for speed below the cut-in wind speed. In the interval of wind speed between the cut-in and saturation speed, the maximum coefficient power can be obtained using the wind power equation. The output power remains at the maximum value for the wind speed in the interval between v_{satur} and $v_{\text{cut-out}}$, where the upper limit is the cut-out point. For wind speed larger than the cut-out point, for safety, no power is generated and therefore remains zero.

Considering the statistical nature of wind speed, the transformation can be accomplished in the following manner:

$$f_Q(q) = f_V[g^{-1}(q)]\left[\frac{d(g^{-1}(q))}{dq}\right] \tag{6.12}$$

The probability of the wind speed being smaller than the cut-in speed or larger than the cut-out speed can be expressed as follows:

$$\text{Prob}(v_{\text{cut-out}} \le v) = 1 - F(v_{\text{cut-out}}) = e^{-(v_{\text{cut-out}}/C)^k} \tag{6.13}$$

$$\text{Prob}(v \le v_{\text{cut-in}}) = F(v_{\text{cut-in}}) = 1 - e^{-(v_{\text{cut-in}}/C)^k} \tag{6.14}$$

The probability of the wind speed, v, being greater than v_{satur} but lower than $v_{cut-out}$, can be represented as

$$\text{Prob}(v_{satur} \leq v \leq v_{cut-out}) = F(v_{cut-out}) - F(v_{satur})$$
$$= e^{-(v_{satur}/C)^k} - e^{-(v_{cut-out}/C)^k} \tag{6.15}$$

Thus, the PDF of the wind speed can be utilized to derive the PDF of the corresponding output power. It is noted that the probability of the power being zero coincides with the probability of the wind speed being lower than the cut-in point and higher than the cut-out point, which can be calculated as follows:

$$\text{Prob}(P = 0) = 1 + e^{-(v_{cut-out}/C)^k} - e^{-(v_{cut-in}/C)^k} \tag{6.16}$$

Accordingly, the probability of the wind power generation being equal to P_{max} equals the probability of the wind speed being higher than v_{satur} but lower than $v_{cut-out}$, which has been expressed in Equation 6.15.

Therefore, the probability of the output power can be expressed in the following mixed PDF equation that is derived from the PDF of wind speed:

$$f_q(Q) = \begin{cases} (1 + e^{-(v_{cut-out}/C)^k} - e^{-(v_{cut-in}/C)^k})\delta(Q) & Q = 0 \\ 0 & 0 < Q \leq Q_{min} \\ (k/3)c_t C^3 (Q/c_t C^3)^{(k/3)-1} e^{-(Q/c_t C^3)^{k/3}} & Q_{min} < Q < Q_{max} \\ e^{-(v_{satur}/C)^k} - e^{-(v_{cut-out}/C)^k}\delta(Q - Q_{max}) & Q = Q_{max} \\ 0 & Q_{max} < Q \end{cases} \tag{6.17}$$

where $c_t = 1/2AC_{v_{satur}}$, $\delta(Q)$ is the impulse function, and Q is the wind power output.

The forecast of wind speed as well as wind power within a short time lag is more accurate and the variance of the distribution is smaller than those forecasts conducted for a longer time lag. In this chapter, the wind power production is modeled using the PDF expressed in Equation 6.17.

6.3.2.2 PV Power Behavior

The growth of PV electricity generation has been significant in recent years. However, the cost of electricity generated by PV power remains high and a subsidy is usually provided to PV suppliers. Currently, PV electricity is commonly directly sold back to the energy supplier at a fixed market price and subsidy. Another disadvantage of PV power is that it depends on irradiation which is unpredictable. The probability of PV power not being able to deliver adequately as scheduled is obvious, and, consequently, related penalty costs are incurred when trading PV electricity in power markets.

A simplified model could be applied for PV power to participate in the energy and imbalance system that involves penalty cost as shown:

$$\Pi_j = Q_j PR_j + k_j Pc_j Q_{im_j} - C(Q) - C(Q_{im_j}) \tag{6.18}$$

where $C(Q)$ is the cost function of PV power to produce Q and $C(Q_{im,j})$ is the penalty cost in the imbalance market.

As PV irradiation cannot be exactly predicted, the penalty cost will be applied to the nondelivered energy, which is their obligation. The statistical distributions could be assumed to model the deployment of the imbalance capacity. Joining the market may be more beneficial for PV power generation than to sell at fixed prices if the produced electricity is sufficient.

6.4 Price-Based (Profit-Based) Unit Commitment

The UC problem in traditional power systems used to be solved for cost minimization. Zhai et al. [5] have proposed a new method that combines the concepts of augmented Lagrangian relaxation and surrogate subgradient to solve the UC problem with identical units. Senjyu et al. [6] proposed a fast, extended priority list method for the UC problem where a technique for getting a better initial solution rapidly has been developed.

However, in deregulated markets, Gencos are usually entities owning generation resources and competing with other participants with the sole objective of maximizing the profits without concerns about the system cost [7]. In the deregulated environment, Gencos have no more obligations to serve the full demand which allows the UC schedule to be more flexible. Under softer demand constraints, a Genco may prefer PBUC generating units to maximize profit. In this line, Richter and Sheble [8] have proposed a genetic algorithm (GA) for the PBUC problem, which considers softer demand constraints and allocates fixed and transitional costs to scheduled hours. Arroyo and Conejo [9] propose a 0/1 mixed integer linear programming to maximize the unit profit by selling both energy and spinning reserve in the spot market.

6.4.1 PBUC Problem Formulation

In general, by modifying the UC problem through changing the demand constraints and altering objective function from cost minimization to profit maximization [10], the PBUC problem based on forecast market price can be represented as follows:

Objective Function:

$$\max \Pi = \sum_{t=1}^{T} \sum_{i=1}^{NT} \left\{ \begin{array}{c} (G_{i,t} \cdot PS_t)W_{i,t} - C_{i,t}(G_{i,t}) \cdot W_{i,t} \\ -ST_{i,t} \cdot UST_{i,t} - SD_{i,t} \cdot USD_{i,t} \end{array} \right\} \tag{6.19}$$

And then the objective function is solved with the following constraints:

Rated Maximum and Minimum Capacities:

$$G_i^{\min} W_{i,t} \le G_{i,t} \le G_i^{\max} W_{i,t} \tag{6.20}$$

Minimum Up/Down Time:

$$\left(T_{i,t-1}^{on} - MUT_i \right)\left(W_{i,t-1} - W_{i,t} \right) \ge 0$$
$$\left(T_{i,t-1}^{off} - MDT_i \right)\left(W_{i,t} - W_{i,t-1} \right) \ge 0 \tag{6.21}$$

Demand Constraint:

$$\sum_{i=1}^{NT} G_{i,t} \leq PD_t \qquad (6.22)$$

It is worth noting that the demand in Equation 6.22 is not the system demand but the quantity a producer is assigned to produce on the basis of the bids submitted and the market cleared. The decision to commit a generating unit is price-based and the schedules are allowed to be more flexible.

6.4.2 Solution Methodology

In the traditional economic dispatch, all marginal units have the same incremental energy cost (system lambda) at the solution, but, in PBUC, ideally the marginal units incremental cost should be equal to that of the marginal price. As it is profitable to produce as long as the cost of production is lower than the revenue obtained by selling the quantity produced, the average cost is utilized as the basis of enumerating the best possible combination for the operation of the generating units. In this study, the proposed methods are applied to four different sets of data. The comparison studies and analysis are then conducted on the basis of these study results.

6.4.2.1 Dynamic Programming

The solution is obtained using dynamic programming (Figure 6.11) in coordination with a selective enumeration method (described in Section 6.4.2.2) and nonlinear programming.

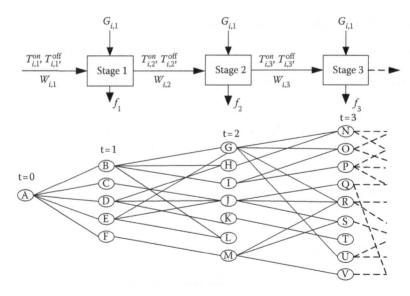

FIGURE 6.11 Dynamic programming.

The stages of dynamic programming are the nodes and the lines joining the nodes represent time periods. The state variables associated with the individual stages are $T_{i,t}^{on}$, $T_{i,t}^{off}$ and $W_{i,t}$, which are derived from the selective enumeration process described in Section 6.4.2.2.

The decision variable, G_n, is obtained by treating the input state, W_{n-1}, as a parameter. Hence, the cost at each stage is obtained by solving subproblem 6.24 for constraints 6.20 and system demand constraints 6.23:

Demand Balance:

$$\sum_{i=1}^{NT} G_{i,t} = PD_t \tag{6.23}$$

$$\min J_t = \sum_{i=1}^{NT} \left[C_{i,t}(G_{i,t}) \cdot W_{i,t} + ST_{i,t} \cdot UST_{i,t} + SD_{i,t} \cdot USD_{i,t} \right] \tag{6.24}$$

where $t \in 1, 2, \ldots N$.

With $W_{i,t}$ and $W_{i,t-1}$ known, $ST_{i,t}$ and $SD_{i,t}$ can be treated as constant and hence the problem reduces to

$$\min J_t = \sum_{i=1}^{NT} \left[C_{i,t}(G_{i,t}) \cdot W_{i,t} \right] \tag{6.25}$$

This simple quadratic problem can be solved by quadratic/nonlinear programming methods.

The decision functions for each stage, 1, 2, ... N, can be represented by the forward recursion formula

$$f_1(G_{i,1}) = \min J_1 \tag{6.26}$$

$$f_t(G_{i,t}) = \min(J_t + F_{t-1}(G_{i,t-1})) \quad \text{for } t = 2,3...N \tag{6.27}$$

where $F_{t-1} = f_{t-1}^{min}$.

However, at each stage, two or more least-cost options obtained from decision functions corresponding to each node are chosen. For example, for node G at stage 2, as shown in Figure 6.11, both cost options representing paths A–B–G and A–E–G are stored. This is required to account for $T_{i,t}^{on}, T_{i,t}^{off}$, MUT_i, and MDT_i of units i, which will have an effect on the cost function at future stages. Finally, to decrease the dimensionality of the problem, not all but best few least-cost nodes are selected at the end of each stage. The overall process is outlined in the flow chart in Figure 6.12.

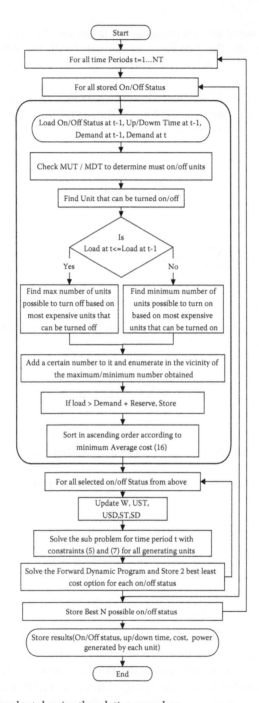

FIGURE 6.12   Flow chart showing the solution procedure.

6.4.2.2 Selective Enumeration Method

A selective enumeration method is proposed to solve this UC problem. The decision to commit or not to commit a unit is made on the basis of the cost characteristics of the units available and the heuristics involved is explained in the following sections with the help of Figure 6.13.

6.4.2.2.1 Average Cost at Maximum Output

The orders in which the units can be turned on/off are stacked on the basis of increasing/decreasing average cost. The minimum average cost is calculated considering both the average and startup costs for each unit within the limits of rated maximum and minimum capacities as shown:

$$\min AC = \frac{C_i(G_i)}{G_i} + \frac{ST_i}{G_i MUT_i} \tag{6.28}$$

The importance of MUT_i is to reduce the effect of startup cost on the basis of the minimum number of hours it should operate if the unit is turned on. The treatment of cost in this way makes sure that a unit with low running cost but high starting cost will not be left out from the selective enumeration process.

6.4.2.2.2 Selective Enumeration

Selective enumeration is adopted following an increase or decrease in demand in each period. Figure 6.13 shows an example of where the demand has increased from period $t-1$ to t. The arrow pointing to unit C suggests that, on the basis of unit minimum

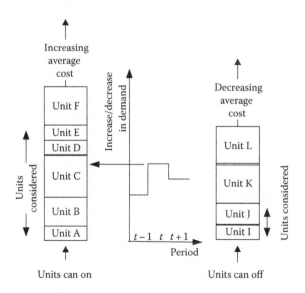

FIGURE 6.13 Selective enumeration process.

average costs, at least three units, that is, A, B, and C, should be turned on to meet the increased demand.

Consider the case where more units are selected, say D and E. This allows for the difference in the startup costs and marginal cost characteristics of the different units, which may affect overall costs The number of extra units considered are usually problem-dependent and can be chosen depending on the cost characteristics and the maximum generation capacity of the units under consideration. The units are then enumerated in different combinations.

Even fewer than the minimum required number of units (e.g., two units in this case) can be considered particularly if the next unit being considered is a very large unit. On the other hand, all five units (A, B, C, D, and E) may be considered. Moreover, to allow for expensive units that could not have been turned off in the previous periods because of MUT constraints, a combination of a few units being turned off, say unit I and J, may also be taken into account.

6.4.2.2.3 *Filtering Process*

From all the enumerated combinations, the ones not satisfying the demand constraint are first discarded. For the remaining, the minimum average cost is calculated and sorted out in ascending order of cost. As the minimum average costs considered represent a very close approximation of the actual costs, the best few combinations are chosen for the remaining calculations. Thus, the best among the combinations chosen are obtained. In the case under study, results for the 10 unit systems were obtained by considering only the best 5 combinations, whereas for the 40 and 110 unit systems the best 20 combinations were considered.

The method described above is detailed in the bold rectangular box in Figure 6.12. For the profit–maximizing objective, the same method described above is used with different objective functions.

6.4.2.3 Genetic Algorithms

GAs are adaptive search techniques based on the principles and mechanisms of natural selection and the "survival of the fittest" from natural evolution [8,11]. Gas belongs to the class of probabilistic algorithms as they combine elements of directed search and stochastic search and hence are more robust than existing directed search methods.

The increasing use of GAs to solve mixed-integer nonlinear problems can be seen from its application in solving UC problems [12,13]. The use of GA is common when the problem is complex and the solution is time-consuming. The choice of GA is based on the performance and flexibility with respect to solution procedure, solution time, and the need for Gencos to adapt to the changing market conditions. Moreover, the use of UC decisions from PBUC for the expected mean price obtained from the authors' previous work [14,15] provided highly fit schemas in the initialization stage of GA. This improved the performance of GA to a great extent thereby providing good solutions in a reasonable time frame. In addition, the advantage of this method is that a group of good UC schedules for the objective function can be obtained.

For a large Monte Carlo sample, S, Equation 6.19 can be rewritten as

$$
\max \Pi_{i,\text{spot}} = \frac{1}{S} \sum_{s=1}^{S} \sum_{t=1}^{T} \left\{ G_{i,t} \cdot PS_{s,t}^{\text{rand}} - C_{i,t}(G_{i,t}) \cdot W_{i,t} \right\}
$$
$$
- \sum_{t=1}^{T} (ST_{i,t} \cdot UST_{i,t} + SD_{i,t} \cdot USD_{i,t}) \tag{6.29}
$$

Equation 6.29 along with constraints 6.20 through 6.22 gives the optimal quantity to generate in the spot market for a risk-neutral Genco.

In GA terminology, the maximized profit, as shown in Equation 6.29, is the fitness function. To deal with the MUT/MDT constraints, a penalty function is introduced to penalize for any violations in constraints. This function helps to produce feasible solutions by making sure that nonfeasible solutions are assigned lower fitness than that of feasible solutions and is derived in the following way:

if $UST_{i,t} = 1$ and $T_{i,t}^{\text{off}} < MDT_i$
       $\text{Pen}_{i,t} = \left(MDT_i - T_{i,t}^{\text{off}} \right) \text{Wpen}$
else if $USD_{i,t} = 1$ and $T_{i,t}^{\text{on}} < MUT_i$
       $\text{Pen}_{i,t} = \left(MUT_i - T_{i,t}^{\text{on}} \right) \text{Wpen}$
else
       $\text{Pen}_{i,t} = 0$
End

In the above function, the violations of constraints, which are far from the feasibility boundary, are assigned higher penalties. Moreover, weight of the penalty function, Wpen, is set to a low value in the initial generations and increases steadily with the number of generations. This ensures that the large search space is covered in the initial generations to reach the optimal region and the higher penalty function in later generations helps in fine-tuning the optimal solution while satisfying all the constraints. Now the objective function for GAs can be written as

$$
\max \Pi_i = \frac{1}{S} \sum_{s=1}^{S} \sum_{t=1}^{T} \left\{ G_{i,t} \cdot PS_{s,t}^{\text{rand}} - C_{i,t}(G_{i,t}) \cdot W_{i,t} \right\}
$$
$$
- \sum_{t=1}^{T} (ST_{i,t} \cdot UST_{i,t} + SD_{i,t} \cdot USD_{i,t} + \text{Pen}_{i,t}) \tag{6.30}
$$

The GA consists of the following stages:

Initialization: The sequences of stored UC decisions (best 10) are obtained from the method described in Reference [15] for the forecast price. As these are the sequences fulfilling all the constraints associated with the unit, it can be said that these form highly fit schemas. These building blocks for the GA problem will perform computations by recombining among each other to form better strings in each generation.

Reproduction, Crossover, and Mutation: Stochastic universal sampling, two-point crossover, and nondynamic mutation are used for reproduction, crossover, and mutation, respectively. The probability of crossover and mutation is set at 0.7 and 0.1, respectively. Moreover, to prevent premature convergence, linear fitness scaling with a scaling parameter of 2 is used.

Replacement: An elitist strategy is used as the replacement method in each generation.

Objective Function: Here each chromosome in the population represents a candidate solution, that is, a possible generation schedule. This means that W, UST, and USD in Equation 6.30 are known for each time period for each candidate solution and hence ST, SD, and Pen can be treated as constants. Therefore, the last three terms in Equation 6.30 can be calculated separately for all the chromosomes in the population.

6.4.3 Price-Based Self-Commitment Under PBUC

In markets such as the New Zealand energy market, UC is the sole responsibility of individual Gencos. For this type of market, PBUC gives optimal dispatch for the forecast price if it can be achieved through bidding. Hence, the information on optimal production obtained is still valuable when making bidding strategies. These strategies may, however, include uncertainty in price, the behavior of other participants, and risk averseness of the Genco. Moreover, a cumulative bid for all the units owned by a Genco may also be submitted to the pool. However, only after the market is cleared will each Genco know their individual demand in the spot market. Now, based on these demands, the Gencos can again carry out self-commitment to obtain optimal decisions when the demand constraints become relevant for competitive Gencos. This makes the UC similar to the traditional power systems where the objective is to minimize system cost to meet system demand.

6.4.3.1 Self-Commitment for Individual Generating Units

Solving the objective function of PBUC will provide an optimal power scheduling decision corresponding to each decision period for the forecast price, as shown in Figure 6.14. The area within the dotted lines in the figure corresponds to the range of marginal cost of the unit between its capacity limits. It can be seen that price being equal to marginal cost (period 7–14) does not necessarily mean that turning the units on in those periods are profitable because other factors like startup/shutdown costs and up/down time constraints also have effect on the decisions being made. The UC decisions obtained for each unit in this way is the optimal solution for the unit in view of the forecast spot price. The power schedule, as shown in Figure 6.14, can be considered the target schedule for a unit of the Genco and hence provides information for the Genco to make bidding strategies accordingly to obtain the targets.

PBUC for individual units described in Section 6.4.2.2 gives the best commitment decision for a particular price for individual units. Now, when PBUC is solved for the same set of prices for all the units owned by a Genco and summed up optimal quantities cleared, a total quantity for the Genco at that price is obtained. This means that for that

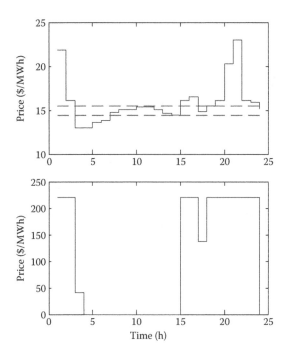

FIGURE 6.14 Power dispatch for single unit for forecast price.

cumulative quantity, the UC decision is optimal. However, the demand of Gencos is determined by the pool and can be quite different from the target depending on their bidding strategies and various other reasons like system security, reserve requirements, and so on, to be met by the ISO. In this case, PBUC solution considering individual generating units would not be optimal and hence needs adjustment to meet the actual demand. Here, the problem is resolved by adjusting (increasing or decreasing) the price fictitiously in each period until the demand condition is satisfied. This result of using this adjustment in price on the optimal dispatch is seen in Figure 6.15. This is the same example as in Figure 6.14, but for the increased price (shown in bold) for period 9. The effect on dispatch can be seen in the changes in periods 9–11.

6.4.3.2 Self-Commitment for Groups of Generating Units

The same technique can be applied for committing a group of units to meet the demand target. An iterative method is proposed where price in different periods is subsequently increased or decreased so that the units committed also increase or decrease, respectively, until the demand constraint is satisfied. This method, which uses price as a guideline for solving the self-commitment problem with demand constraint, is explained with the help of the following steps:

BEGIN

Step 1: Initialize the price for $t \in 1, \ldots N$ to the expected average spot price forecast PS_t.

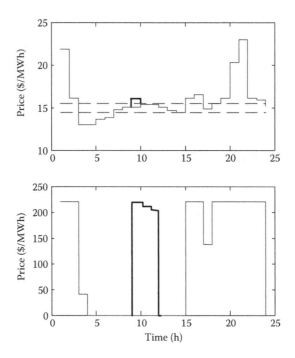

FIGURE 6.15 Power dispatch for change in price.

Step 2: Begin iteration.

Step 3: For the price PS$_t$, solve PBUC described in Section 6.5 to obtain $W_{i,t}$ and $G_{i,t}$.

Step 4: For all $t, t \in 1, \ldots N$

If $\sum_{i=1}^{NT} G_{i,t} < PD_t$, then PS$_t$ = PS$_t$ + ΔPS.

Else if $\sum_{i=1}^{NT} G_{i,t}^{\max} > PD_t$, then PS$_t$ = PS$_t$ − ΔPS.

Else PS$_t$ = PS$_t$.

Here, the change in price for the given conditions prevents over-scheduling and under-scheduling, which increases the cost of production and violates the demand constraint, respectively.

Step 5: For the $W_{i,t}$ obtained, solve the objective function, which is now a simple quadratic equation and can be solved using the "quadprog" function for the following demand constraint:

$$\sum_{i=1}^{NT} G_{i,t} = PD_t \qquad (6.31)$$

Step 6: Store the optimal solution obtained so far.

Step 7: After every two iterations, set

PS$_t$ = max(PS$_t$−1, PS$_t$−2) and ΔPS = ΔPS/2.

Here, PS_t is set so that the UC decision obtained using this price guarantees the fulfillment of demand constraints after every iteration. In addition, reducing the amount by which the price is changed (ΔPS) helps in fine-tuning the scheduling decisions in subsequent iterations to reach the optimal solution.

Step 8: Repeat steps 3–7 until the maximum number of iteration is reached.

END

The proposed procedure was implemented in different systems reported in the literature, and Figure 6.16 shows the effect of this self-commitment method on the target dispatch obtained through PBUC.

The solid lines in the graphs indicate the target commitment plan for the forecast price obtained from PBUC for a Genco owning a 10-unit system given for a 12-h period. Now, let us assume that the total power cleared by ISO for the Genco is shown by the dotted lines in the topmost graph in Figure 6.16. Now, self-commitment of the units as described above suggests an optimal power dispatch schedule for the 10 units as shown by dotted lines in the figure.

6.4.3.3 Consideration of Price Uncertainty in Bidding Under PBUC

Stochastic models for constructing bids have been developed by Baillo et al. [16] and Plazas et al.[17]. The models developed in view of the day-ahead market and adjustment markets are solved using bender decomposition in a two-stage stochastic decision process in the study by Baillo et al. [16] and mixed integer linear programming in the study by Plazas et al. [17]. However, these models cannot be optimal in hour-ahead markets like the National Electricity Market of Singapore, where the gate closure time is shorter and Genco commitment decisions can respond quickly to market conditions.

Bidding for power producers with low generation cost units can be done to ensure that the units are dispatched each hour. Similarly, for units with very high cost, the producers prefer not to dispatch them. However, for marginal or near-marginal units, the producers should consider, among options of committing the unit to run, keeping it in banking mode or shutting down the unit completely. To make decisions on bid curves that maximize profits, two factors must be considered: the uncertainty in price, which has direct impact on profit, and PBUC, which takes into consideration time-dependent unit constraints such as MUT/MDT, and so on, while maximizing profits [18].

Some reasonable assumptions are made in the development of bidding strategies.

i. While making a price quantity bid, the Gencos choice of quantity is fixed in blocks, that is, if Gencos bid in 5 MWh blocks of quantity, then the quantity bids can be, for example 50, 55, 60 . . . 85, and so on, in intervals of 5.

ii. Similarly, price bids are also limited to intervals of 0.25, for example, 10.25, 10.50, 10.75, . . . and so on.

6.4.3.3.1 Price Quantity Relationship

Using the method for solving PBUC as described in Section 6.4.3, optimal profit and quantity corresponding to each price scenario can be obtained. The price, quantity, and profit for each of the 2500 scenarios obtained from PBUC are sorted out in ascending order of optimal quantity dispatch as required for a stepwise nondecreasing bid curve.

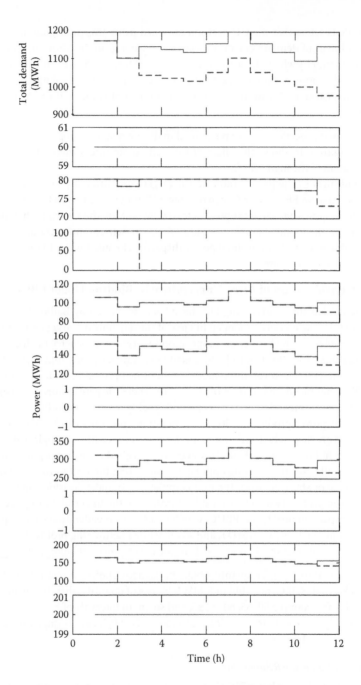

FIGURE 6.16 Effect of self-commitment on target dispatch of profit-based unit commitment.

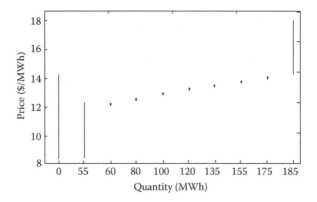

FIGURE 6.17 Price distribution for discrete quantities.

Figures 6.17 and 6.18 and Table 6.3 are outcomes from a sample implementation carried out to observe the price–quantity relationship for a marginal generation unit. The following observations were made:

i. The solution of generation quantity from PBUC results in a number of discrete quantities. For example, in the sample considered, 10 discrete quantities are obtained for all price scenarios, as can be seen from Figure 6.17 and Table 6.3.

ii. With the increase in quantity dispatched it can be seen that the price range for which the quantity is dispatched has, in general, an upward slope, that is, more quantities are dispatched for higher prices.

Figure 6.17 is a typical graph for a period, showing the relationship between price and quantity. Quantities shown in the *x*-axis are the discrete values of quantities obtained for a period while solving PBUC for 2500 price scenarios. The figure shows the range of price at which those quantities give optimal solutions. For instance, PBUC gives optimal

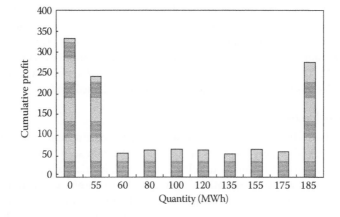

FIGURE 6.18 Cumulative profit distribution for discrete quantities.

TABLE 6.3   Price Distribution

Quantity (MWh)	Price ($/MWh) From	To
0	8.50	14.50
55	8.50	12.25
60	12.50	12.50
80	12.75	12.75
100	13.00	13.00
120	13.25	13.25
135	13.50	13.50
155	13.75	13.75
175	14.00	14.00
185	14.25	18.00

profit with a quantity of 0 MWh in several scenarios when the price in the period is between $8.5 and $14.5. In the same manner, for a quantity of 55 MWh, optimal profit is obtained when the price in the period is between $8.5 and $12.25. From the characteristics of the unit and optimal solution from PBUC indicated in Figure 6.17, it can be deduced that:

  i. Above a certain maximum price ($14.25 in this case), it is in the interest of Gencos to bid maximum capacity, that is, 185 MWh.
 ii. The main difficulty here is, however, to choose the price below which to turn the units "off," that is, the decision between 0 and 55 MWh. This is very important because this is where the intertemporal constraints (MUT/MDT) become binding and the decision in one period is interlinked with the decision in other periods.
iii. Once the above decisions are made the other price–quantity relationships are mostly straightforward as they follow the cost characteristics of the unit.

Figure 6.18 shows cumulative profit corresponding to various quantities. To obtain the cumulative profit, the ratio of profit (obtained from PBUC) in one scenario to the maximum profit in all the price scenarios is first calculated. These profit ratios are aggregated for each set of discrete quantities to obtain the cumulative profit. For example, cumulative profit obtained for the optimal quantity as zero for prices between $8.5 and $14.5 is 333.42. This may seem to suggest that the decision to bid zero is a favorable decision for a price less than $14.5, but is not so as will be explained next.

6.4.3.3.2 Building Bid Curve

After identifying the price–quantity relationship, a stepwise nondecreasing price–quantity curve will be developed. From Figure 6.17 it can be seen that, when the price scenarios are treated individually to solve PBUC, similar prices at different scenarios tend to give different optimal quantities. Hence, to obtain the bid curve, profit information corresponding to each price and quantity, similar to Figure 6.18, is used. Table 6.3 shows the price distribution obtained based on 2500 price scenarios. These price intervals are first categorized in an ascending order of intervals (e.g., <8.50, 8.50 to

<10.50, 12.25 to <12.50, 12.5 to <12.75, etc.) so that the best quantity corresponding to each of those periods can be obtained as demanded by the bidding curve.

However, before making decisions on all bid quantities, the price at which to start the unit or, in other words, the price below which to put the unit "off" should be established. The process for this adopted here is shown in Figure 6.19. The optimal power corresponding to each scenario is considered to be either "on" ($\geq$55 MWh, i.e., greater than generation lower limit G_i^{min}) or "off" (=0 MWh). Initially, small price intervals between maximum and minimum price are considered and the status ("on" or "off") that gives maximum cumulative profit for each interval is calculated. As the bid curves are required to be stepwise nondecreasing price–quantity curves, it can be seen that iteration 1 in Figure 6.19 for a small price interval does not fulfill this condition; that is, condition $QB_{B-1} \leq QB_B$ is not satisfied for all bidding blocks B and B − 1($\forall$B $\in$ 1,2, ... NB). Hence, price intervals are increased (at the rate in accordance with the second assumption at the beginning of Section 6.4, i.e., \$0.25) iteratively until the condition is satisfied. In the sample considered, it can be seen that the condition is satisfied in the third iteration, and hence for the price range shown by the arrow in iteration 3, the unit is turned "off." In this way, PB_1 is derived.

To derive $QB_{B,t}$ corresponding to each price interval, profit weight, $\Pi_{s,B}^{wt}$, for each price scenario is calculated as the ratio of the profit corresponding to a scenario in the price range to the sum of profits corresponding to all the scenarios in the range:

$$\Pi_{s,B}^{wt} = \frac{\Pi_{s,B}}{\displaystyle\sum_{s,B} \Pi_{s,B}} \qquad (6.32)$$

Subsequently, $QB_{B,t}$ is calculated as

$$QB_{B,t} = \sum_{s,B} \Pi_{s,B}^{wt} \cdot QB_{s,t} \qquad (6.33)$$

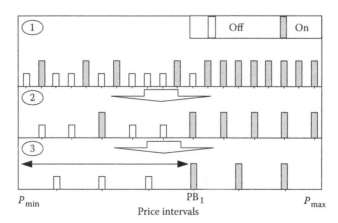

FIGURE 6.19 Determining "on"/"off" price.

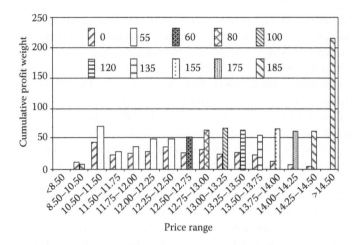

FIGURE 6.20 Cumulative profit distribution for small price intervals.

The effect of the quantity chosen for bidding in this way by considering profit weights as the basis of making decision on bid quantity is expected to increase the overall profit under uncertainty.

Figure 6.20 shows the cumulative profit when the price range is divided into different price intervals. It can be seen that the decision to turn the units "off" (0 MWh) has more cumulative profit for a price less than \$10.5/MWh. Similarly, from 10.50 to less than \$12.50/MWh, a quantity of 55 MWh has a higher cumulative profit. In this example, if the market allows a 10-step bid function, then 10 price bids corresponding to 10 discrete quantities, as shown in Figure 6.20, are expected to give the optimal bid. These price intervals correspond to $PB_{B,t}$ and can be interpreted as $PB_{1,t} = \$10.25$ for 0 MWh, $PB_{2,t} = \$12.25$ for 55 MWh, and so on.

However, if the market allows less bid steps, then the decisions on the minimum price to start (i.e., \$10.5 for 55 MWh) and the price above which maximum quantity is dispatched (i.e., \$14.25 and above for 185 MWh) are first made. From the remaining price range, for instance, \$10.50–\$14.25 in the example, the quantities having higher profit weight are chosen.

In this way, a bid curve, as shown in Figure 6.21, which is a stepwise nondecreasing curve, can be obtained which considers multiple price scenarios based on the profit weight corresponding to each scenario.

6.5 Price-Based Scheduling for Gencos

Generation scheduling and dispatch are determined by individual power producers' bids in a deregulated power market. In addition to the status of the unit, a bid includes information on how much power, at which price, at what time, and at which stage the Genco is willing to sell. Thus, the bid curves need to be determined with the objective of maximizing profit in the competing process with other participants. This calls for added emphasis

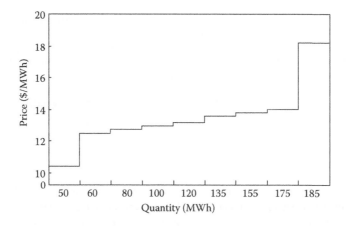

FIGURE 6.21 Price and quantity bid based on profit weight.

on the influencing factors such as price uncertainty, financial instruments, and so on. Different generation scheduling methodologies will be described in this section.

6.5.1 Stochastic Self-Scheduling Technique

The benefits obtained by a power producer will depend largely on how effectively it can incorporate the variation of the market price in its generation scheduling. A stochastic scheduling technique is presented in this section for maximizing a producer's benefit considering the stochastic nature of power price. Two approaches to the solution, namely, two-stage and multistage stochastic methods, are presented that accommodate the features of the day-ahead and hour-ahead power markets, respectively.

Implementation of the technique is illustrated by applying the technique to scheduling by a power producer with 11 generators in two different seasons. Price data from a real system have been used for price forecasting and the formation of a price scenario tree. The result indicates that this stochastic self-scheduling technique is particularly effective when the uncertainty in the price is high and the price forecast is not very accurate [19].

6.5.1.1 Price Forecast and Formation of Price Scenarios

A price forecast is used to estimate prices (expected values) for the entire scheduling period of 24 h along with their standard deviation. The forecast is used to identify a number of likely price scenarios at each hour along with the likelihoods of their occurrences. Three scenarios (normal, high, and low price) are used in this study. A price scenario tree is then formed for the 24-h period, as shown in Figure 6.22. The task of stochastic scheduling is to determine a path in this tree from hour 1 to hour 24 in order to maximize the expected value of the benefits over the entire 24-h period. The optimization problem will be solved as a problem coupled for 24 h. This will yield the optimal UC for each hour on the basis of the price forecast for 24 h. Then, the optimal output at each hour is determined from the more precise estimate price for the next hour.

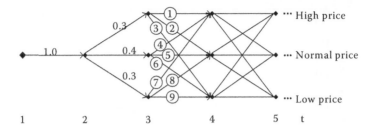

FIGURE 6.22   Price scenario tree.

The autoregressive integrated moving average (ARIMA) model is utilized to forecast the price on the basis of the recent actual historical data. The variances can also be obtained in addition to the mean values of the price for every forecast interval. This method is very suitable to form different price scenarios based on the forecast expected prices and standard deviations obtained from the corresponding variances. In the ARIMA model, the seasonal part of 1 week (168 h) is first considered as the electricity prices are inherently seasoned by weeks. In addition, the time difference is considered based on the feature of the historical data.

Using the ARIMA model, the variances σ_i^2 along with expected values y_i of the forecast prices are obtained for every forecast interval i. Generally, a larger number of discrete price scenarios would yield more accurate results, but with additional complexity of the problem and heavier computational tasks. In this section, for each hour, three discrete price states, high, normal, and low prices, with probabilities of w_{hl}, $(1-2w_{hl})$, and w_{hl}, respectively, are used. The volatility of the market price is represented by the value of w_{hl}. The values of discrete prices for a given w_{hl} can be easily computed and Table 6.4 shows these values for $w_{hl} = 0.3$.

It is noted that, as only three discrete prices are obtained for each hour, the infeasible task of exponential computation of the 24th power is avoided. Other ways for discretization and scenario generation are possible.

Detailed stochastic scheduling studies for a period of 30 days in the summer and the winter are presented in the next section. Price forecast is therefore performed for these two periods. The data used for forecasting are the historical hourly prices. Historical data of the previous five weeks are utilized to forecast the price of a particular day. Figure 6.22 shows the forecast prices for a typical day in winter (February 5, 2000) and a typical day in summer (August 30, 1999). The actual historical prices and the high and low prices computed as $y_i + 1.1616\sigma_i$ and $y_i - 1.1616\sigma_i$, respectively, are also shown in the figure. When the prices $y_i \pm 1.1616\sigma_i$ do not exceed the price limits $y_{max,i}$ and $y_{min,i}$

TABLE 6.4   Discrete Price Values with Probabilities at an Hour i

Node No.	1 (High Price)	2 (Normal Price)	3 (Low Price)
Discrete price	$y_i + 1.1616\sigma_i$	y_i	$y_i - 1.1616\sigma_i$
Probability	0.3	0.4	0.3
Price range	$>y_i + 0.52\sigma_i$	$y_i \pm 0.52\sigma_i$	$<y_i - 0.52$

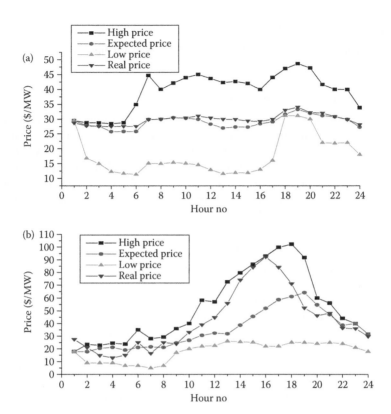

FIGURE 6.23 Forecast price for forming price scenarios. (a) Forecasted price for a winter day: Feb. 5, 2000. (b) Forecasted price for a summer day: Aug. 30, 1999.

(maximum and minimum price values observed in historical data), the three values are distinct. When the discretized low or high prices exceed the maximum or minimum observed value, the price is capped at this value.

It may be noted from Figure 6.23 that the forecast expected prices are quite close to the real prices in the winter. But the forecast expected prices at hours 13–17 in the summer are significantly lower than the real prices. The implication of these forecast accuracies will be discussed in the next section. The price scenario tree shown in Figure 6.22 is formed with these three price scenarios at each stage, except at hour 1 where the price is considered to be known.

The probabilities of the three price scenarios shown in Figure 6.22 are for $w_{hl} = 0.3$ for the first period. However, the probabilities of reaching a scenario at the successive hour $(i + 1)$ from a scenario at the previous hour (i) for any w_{hl} can be easily calculated and are shown in Table 6.5 for $w_{hl} = 0.3$.

TABLE 6.5 Arc Probabilities

Arc no.	1	2	3	4	5	6	7	8	9
Probability	0.09	0.12	0.09	0.12	0.16	0.12	0.09	0.12	0.09

6.5.1.2 Solution Procedure

Two approaches to the solution of the problem have been attempted to represent different levels of flexibility in the scheduling.

6.5.1.2.1 Two-stage Stochastic Procedure

The spot market is usually a day-ahead market. The supplier submits bids 24 h ahead, and then they have limited adjustments in their operation in different price scenarios in the same time period, allowing only the outputs of the units to change. A two-stage stochastic procedure is appropriate to represent this feature of bidding by power producers. In the first stage, only the UC schedule is determined, which is kept unchanged for all the scenarios at that period. The unit outputs are specified only 1 h ahead on the basis of a more accurate estimate of the price for the next hour.

6.5.1.2.2 Multistage Stochastic Procedure

As the bids are committed only 1 h ahead it may be desirable to adjust generation significantly, beyond the capacity of a fixed UC for all scenarios adopted in the above approach if and when there is sudden or fast change in the price. Multistage stochastic procedure accommodates this feature as both UC and unit outputs can be adjusted in this procedure for different scenarios at the same time interval. As different UC decisions can be made for different scenarios for any one time interval, this approach can respond to the change in power price more effectively than the two-stage approach. In the two-stage procedure, only the output levels of the units can be adjusted, whereas in the multistage procedure, the outputs as well as the UC decisions of all the units can be adjusted while subject to the "on"/"off" time-delay constraints. Therefore, the multistage procedure is expected to produce more expected benefits.

6.5.1.3 Illustrative Case Studies

The stochastic self-scheduling technique is applied to the operation of a power producer with 11 generating units to optimize the generation schedule (bids) in day-ahead and hour-ahead power markets. The generator data used in these studies are listed in Table A6.1.1 in Appendix A6.1. Historical power prices were obtained from the website of a real power market for the periods July 1–September 30, 1999 and January 1–March 31, 2000 to represent two different seasons, summer and winter, respectively. A bilateral contact of $d_b = 1750$ MW at a fixed price p_d has been incorporated in the studies. The issues of how to determine p_d and d_b are not explored in this study. The average price of the known historical data is taken as the contract price p_d. The benefits are calculated for a relatively long time period of 30 days for comparison because the techniques are stochastic in nature. The benefits derived by the power producer using the following four methods were computed.

    i. First, the benefits were calculated by applying deterministic optimal generation scheduling methods using the actual market prices and are denoted as R. Obviously, the unit schedule and the power dispatch so determined yield the maximum possible benefit. It should be noted, however, that the real prices can only be known after the actual event and cannot be used in practice. The benefits computed here are purely for comparison.

ii. A convenient way of generation scheduling is to use the expected values of the market prices. It will not be computationally demanding as it requires only the expected values of prices for the next 24 h at the beginning of the day and then schedules its generating units deterministically on the basis of the forecast prices to maximize its benefits. The unit outputs are adjusted in response to the real market prices hour by hour while the UC is unchanged. The benefits obtained by this method are denoted as *D*.

iii. The benefits enjoyed by the power producer using the two-stage stochastic scheduling described earlier are denoted as *T*.

iv. The benefits obtained by applying the multistage stochastic scheduling described earlier are denoted as *M*.

The benefits calculated using these four methods for a period of 30 days in summer and winter are listed in Table 6.6. The high/low price probability w_{hl} is taken as 0.3, which means that the probability of high, normal, and low price scenarios are 0.3, 0.4, and 0.3, respectively.

As expected, the benefits using the actual price, *R*, are largest in all cases. This is because the values of *R* are the maximum possible benefits as described above. It should also be noted that the benefits obtained using the multistage stochastic scheduling (*M*) are always larger than the benefits obtained using the two-stage stochastic scheduling (*T*). The benefits derived by the power producer in an hour-ahead market are $51,712 and $1932 more in the summer and in the winter, respectively, compared to the benefits derived in a day-ahead market.

The benefits achieved using the proposed stochastic methods *M* and *T* exceed the benefit achieved applying the deterministic method using the expected values of prices *D* in the summer by $80,350 and $28,603, respectively. However, despite the additional complexities involved in the stochastic methods, the benefits obtained using these methods are found to be less than those obtained by the deterministic method in winter by $7317 and $9250. Thus, it appears that specific consideration of the variability of the power prices in the optimization process may not necessarily enhance the total expected benefits.

A careful look at the forecast price in Figure 6.23a reveals that the expected value of the forecast values is very close to the actual prices in the summer and the fluctuation in price is very little. Therefore, *ad hoc* assignment of high probabilities to the high or low prices (w_{hl}) may not be quite appropriate. The forecast price in summer shown in Figure 6.23b exhibits a much larger variation from the actual prices. The proposed stochastic

TABLE 6.6 Benefits Obtained Using Different Optimization Procedures for $w_{hl} = 0.3$

Time Period	Total Benefit ($)				Benefit Difference ($)		
	R	T	M	D	T − D	M − D	M − T
Summer	28,156,205.33	27,942,355.85	27,994,068.2	27,913,752.53	28,603.31	80,315.65	51,712.34
Winter	24,112,467.90	23,964,416.56	23,966,349.08	23,973,666.66	−9250.09	−7317.58	1932.51

techniques seem to enhance the benefits under these conditions. Thus, it appears that it is important to assign proper values to w_{hl}, which reflects the variability of the market prices.

This hypothesis was tested by computing the benefits using the stochastic methods by varying the high/low price probability w_{hl} from 0.02 to 0.46 in steps of 0.04. The variation of the difference between the benefits obtained using the two-stage stochastic method (T) and the deterministic method (D) with different values of w_{hl} is shown in Figure 6.24.

The benefit difference declines with higher values of w_{hl} in the winter when the variation of the forecast price is very small. The benefit difference becomes negative for $w_{hl} \geq 0.14$. On the other hand, the benefit difference increases with w_{hl} in the summer when the price variation is high.

The variation of the difference between the benefits obtained using the multistage stochastic method (M) and the deterministic method (D) with different values of w_{hl} is

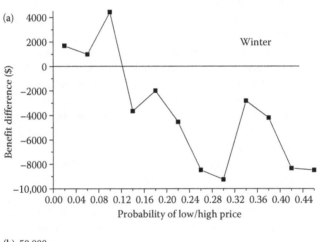

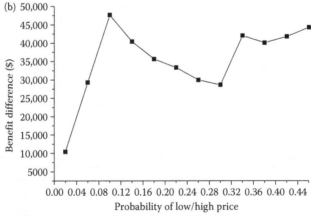

FIGURE 6.24   Variation of benefit difference ($T - D$) with w_{hl}.

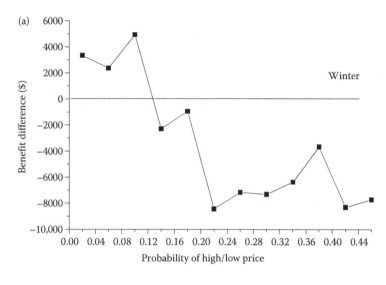

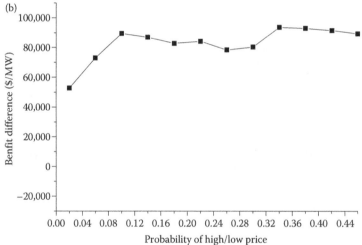

FIGURE 6.25 Variation of benefit difference ($M - D$) with w_{hl}.

shown in Figure 6.25. These curves also exhibit the same pattern as those of Figure 6.24. The benefit difference decreases with an increase in w_{hl} in winter when the variation in actual price is low and increases with an increase in w_{hl} in summer when the variation of the actual price is high.

Thus, it is evident that the proposed stochastic methods can enhance the benefits of a power producer if the high/low price probability w_{hl} is properly assigned.

The benefit of using the multistage over the two-stage stochastic procedure is highlighted in Figure 6.26, which shows the variation of the benefit difference ($M - T$) for different values of w_{hl}. It is clearly seen that the multistage method consistently gives

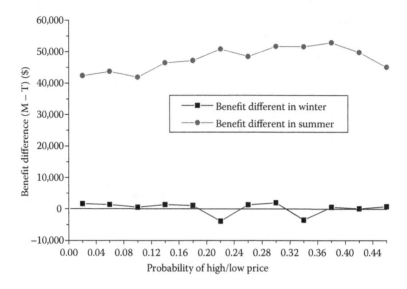

FIGURE 6.26   Variation of benefit difference $(M - T)$ with w_{hl}.

more benefit in summer when the price variation is high. However, there is hardly any difference between the benefits produced by the two stochastic methods in winter when the price variation is very small. This is explained by the fact that the ability to adjust the UC and unit outputs in a spot market will not produce additional benefits when the market prices do not vary significantly.

6.5.2 Scheduling Based on Generator Cost Characteristics and Market Price Dynamics

In this section, a new generation scheduling approach is presented that takes into account both price uncertainty and cost characteristics of the generating units in an effort to make a reasonable profit from the operation [20,21].

6.5.2.1 Bidding Strategy

The nature of the generator cost may affect the viability of a price-taker Genco in the market. In the traditional power system, the scheduling of generators considers only the marginal cost as the fixed costs are taken care of in the process of energy rate-making. Following the same logic, the fixed cost may not seem relevant in the daily bidding process in the competitive market, because for an existing Genco the fixed cost is sunk cost and it should not influence operation of the unit in any period. However, consideration of the average costs becomes necessary to ensure the viability of a generating unit. A reasonable equilibrium is such that total costs of efficient units are just covered, yielding zero profit for all units in the mix.

The proposed strategy is meant for a price-taker Genco that owns different types of generating units, which are economically competitive in the market. As these Gencos have no influence on the market price, their best strategy might be to operate their generators in the most economical way for the expected market price behavior. However, the economical operation proposed in this approach does not focus on individual bidding period but on the entire year of operation. Therefore, the bidding strategy does not attempt to follow the marginal cost or the average cost of the generating units. Rather, it tries to adopt a bidding procedure that allows the generating units to operate economically on an annual basis. The screening curves described in Section 6.3.1.3 can be utilized to this effect. In this context, the most economical operation is such that the generating units will run with capacity factors in the range where they are the most efficient units. Thus, the bid for a generator should be developed such that the probability of it being successful should be in the range of the capacity factor where the generator is the most economical unit.

As this approach emphasizes the bid success over a year rather than any individual bidding period, the detailed and precise models of the market or the Genco are not very crucial. Rather, the following broad assumptions suffice for the development of this strategy:

- i. The market is a spot price market, and the Genco is a price-taker to the extent that its bids do not affect the behavior of the spot market price. Effects of contracts have not been considered in this formulation.
- ii. The generating units of the Genco are competitive in the market such that, if the generating units are operated by devising suitable bid curves to make them successful in bidding to run them economically, they can remain profitable.

6.5.2.2 Formation of Bids

Following the bidding strategy presented in Section 6.5.2.1, a three-step bid for a Genco with three generators can be developed in the following manner. For this preliminary study, the three bid quantities are taken to be the generator capacities of the coal, oil, and gas units, respectively. Thus, for the three-step bid:

$$QB_1 = P_{G1}, \quad QB_2 = P_{G2} + P_{G1}, \text{ and } QB_3 = P_{G3} + P_{G2} + P_{G1}.$$

In case the screening curves do not yield such distinct bidding steps or if it is desired to further divide the generator capacities into more than one bid level, multilevel bidding for a generator capacity can be pursued by preparing screening curves corresponding to different marginal costs at various output levels of the generators. These implementations may be considered in more detailed studies.

These price levels can be obtained from the distribution functions of the market price for the purchasing period of time. In this study, the bid curve is set as follows:

$$QB_{1,j} = P_{G_1} \quad \text{and} \quad PB_{1,j} = PS_{j,\alpha_1}, \text{ where } (\alpha_1 \geq C_2)$$

$$QB_{2,j} = P_{G_2} + P_{G_1} \quad \text{and} \quad PB_{2,j} = PS_{j,\alpha_2}, \text{ where } (C_1 \leq \alpha_2 \leq C_2) \tag{6.34}$$

$$QB_{3,j} = P_{G_3} + P_{G_2} + P_{G_1} \quad \text{and} \quad PB_{3,j} = PS_{j,\alpha_3}, \text{ where } (0 \leq \alpha_3 \leq C_1)$$

This procedure allows a range of bid prices for the three-bid steps as indicated by the range of α_1, α_2, and α_3. The initial study is carried out with $\alpha_1 = C_2$, $\alpha_2 = C_1$, and $\alpha_3 = 0.05$, which are randomly chosen from the most economical capacity region. Further studies are also conducted to investigate whether the choice of different values of α_1, α_2, and α_3 in the above ranges affect the outcome significantly. The optimal setting of these bid prices within the optimal range needs further investigation.

6.5.2.3 Implementation of the Bidding Strategy

The profit for the Gencos with three generators by adopting the proposed bidding strategy can be expressed as

$$\Pi = \sum_{k=1}^{365}\sum_{j=1}^{48}\sum_{N=1}^{3} PR_{j,k}QB_{N,j,k}UB_{N,j,k} - \sum_{k=1}^{365}\sum_{j=1}^{48}\sum_{N=1}^{3} C(QB_{N,j,k})UB_{N,j,k} - \sum_{i=1}^{3} FC_i \quad (6.35)$$

The bid quantities (QBs) are set as discussed above. The bid status (UBs) depend on the actual MCP PRs at each operating period. Actual historical price data from the Singapore electricity market is used as the MCP in this study. It is worth noting that Equation 6.35 is not optimized any further. Rather, the specifying QBs and the corresponding PBs as formulated earlier are expected to produce appropriate levels of bidding success and reasonable amount of profit, considering both the behavior of the market price and the generator cost characteristics.

6.5.2.4 Illustrative Case Study

6.5.2.4.1 Generating Unit Characteristics

This study will consider a Genco having three generating units that are competitive in the market having the following characteristics:

i. A base unit (e.g., a coal unit) whose cost will be less than the weekly mean price 99.9% of the time.
ii. An intermediate unit (e.g., an oil unit) whose cost will be less than the weekly mean price 50% of the time.
iii. A peaking unit (e.g., a gas unit) whose cost will be less than the weekly mean price 15% of the time.

The basic data of the generators in Table A6.1.2 are used to compute the cost components of the generators that are shown in Table A6.1.3 in Appendix A6.1. The screening curves derived from these data are shown in Figure 6.27, which determines the most economical capacity factor of the generators shown in Table 6.7.

6.5.2.4.2 Bid Formulation

According to the strategy explained earlier, the three bid quantities and the corresponding bid prices for the bid curve of each period will be determined in the following way:

$$QB_{1,j} = P_{G_1}(= 300 \text{ MW}) \quad \text{and} \quad PB_{1,j} = PS_{j,\alpha_1}, \text{ where } (\alpha_1 = 0.70)$$
$$QB_{2,j} = P_{G_1} + P_{G_2}(= 500 \text{ MW}) \quad \text{and} \quad PB_{2,j} = PS_{j,\alpha_2}, \text{ where } (\alpha_2 = 0.15) \quad (6.36)$$
$$QB_{3,j} = P_{G_1} + P_{G_2} + P_{G_3}(= 650 \text{ MW}) \quad \text{and} \quad PB_{3,j} = PS_{j,\alpha_3}, \text{ where } (\alpha_3 = 0.05)$$

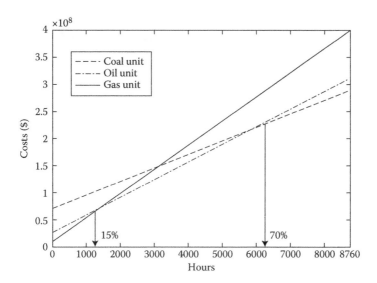

FIGURE 6.27 Screening curves of Genco generator units.

The values of PSs will depend on the parameters of the statistical distribution of the forecast market price for the respective time period. The parameters of the statistical distribution for different periods of the day are listed in Table A6.1.3 in Appendix A6.1. The bid prices are computed from the price distribution curve with probabilities of 70%, 15%, and 5% corresponding to bid quantities of 300, 500, and 650 MW, respectively. The determination of bid prices at each period from the cumulative density function is illustrated in Figure 6.28. These bid prices derived for different periods of the day are listed in Table A6.1.4 in Appendix A6.1.

The operation of the Genco is simulated using the following algorithm according to the proposed bidding strategy outlined earlier.

1. Set $k = 1$ (for day 1).
2. Set $j = 1$ (for period 1).
3. Determine the bid prices ($PB_{1,j}$, $PB_{2,j}$, and $PB_{3,j}$) from Equation 6.9.
4. Get the MCP from USEP 2006.
5. Determine the bid status by comparing the bid prices and the MCP.
6. Calculate the operating cost and the revenue for G_1, G_2, and G_3 on the basis of the bid status.

TABLE 6.7 Screening Curve Analysis Results for Genco Units

Unit Type	Size (MW)	Most Economical Capacity Factor (%)
G1: Base unit (coal)	300	70–100
G2: Intermediate unit (oil)	200	15–70
G3: Peaking unit (gas)	150	0–15

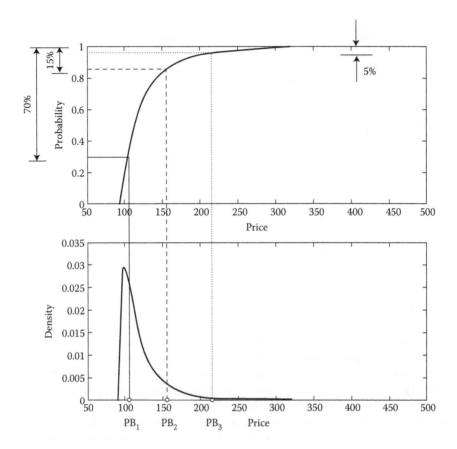

FIGURE 6.28   Use of probability and cumulative density functions of price distribution.

7. If $j < 48$, set $j = j + 1$ and go back to step 3.
8. If $k < 365$, set $k = k + 1$ and go back to step 2.
9. Summarize the operating cost and the revenue for G_1, G_2, and G_3.
10. Calculate the fixed cost for G_1, G_2, and G_3.
11. Calculate the profit by using revenue minus the fixed cost and the operating cost.

This operation is simulated for a full year, and the three generator units are committed to run according to the bid strategy developed above.

It should be noted that many details of UC constraints such as minimum up/down time have been ignored in this preliminary study to keep the analysis simple and to keep the focus on the bidding strategy.

The investment and operation and maintenance cost components contribute to a significant portion of average generation cost and therefore become important in the calculation of profits. These cost components are computed and shown in Table A6.1.3. MCP, which is in Singapore dollar, has been converted to U.S. dollar using an exchange rate of S$1. 4/US$.

TABLE 6.8 Results of Genco Operation with the Proposed Bidding Strategy

Unit Type	Revenue ($ \times 10^6$)	Cost ($ \times 10^6$)	Profit ($ \times 10^6$)
G1: Coal	417.76	243.73	174.04
G2: Oil	44.83	37.99	6.84
G3: Gas	24.27	18.15	6.12
Total	486.86	299.87	187.00

Profits generated by the three generating units by adopting the proposed bidding strategy are listed in Table 6.8. All three units make profits and the total profit for the year is found to be 187 million dollars.

6.5.2.4.3 Important Extensions

In the study concluded above, the bid prices for the three blocks were fixed at arbitrary values within the feasible range by selecting arbitrary values of α_is. It would be worthwhile to check whether the performance can be improved by varying the values of α_i within the acceptable range. Also, the statistical distribution of the market price was taken as lognormal ad hoc, because this distribution fitted the historical data. It would be desirable to investigate the impact on the performance of the bidding strategy if the distribution function did not fit the data so well.

Impact of α_is—The Desired Bid Success Rate Settings: The effect of setting the bid success rates at different values within the acceptable range was investigated by repeating the above simulation with different values of α_is. In order to keep the number of combinations at a reasonable value, six different values of α_is within their respective ranges were adopted by varying the probability in Equation 6.36 as follows:

    i. α_1 varied between C_2 and 1.00 in six equal steps, that is, between 0.70 and 1.00 with a step size of 0.06.
    ii. α_2 varied between C_1 and C_2 in six equal steps, that is, between 0.15 and 0.70 with a step size of 0.11.
    iii. α_3 varied between 0 and C_1 in six equal steps, that is, between 0 and 0.15 with a step size of 0.03.

The simulation results where the maximum profit was achieved are shown in Table 6.9. The values of the α_i settings and the actual bid success rates are also shown in the table.

TABLE 6.9 Results of Genco Operation Using the Lognormal Distribution Model

Unit Type	α_i (%)	Using Lognormal Distribution for Market Price				Using Actual Price	
		Bid Success Rate (%)	Cost ($ \times 10^6$)	Revenue ($ \times 10^6$)	Profit ($ \times 10^6$)	Profit ($ \times 10^6$)	Bid Success Rate (%)
G1: Coal	100	99.1	287.263	494.205	206.942	208.036	100
G2: Oil	59	66.9	214.461	244.651	30.190	31.609	66
G3: Gas	9	3.5	21.477	28.136	6.659	10.835	3.23
Total	—	—			243.792	250.480	—

It is seen that the maximum profit achieved in this case is considerably higher and it occurs when the bid success rate α_i settings are set at the higher end of the respective ranges. The table also lists the maximum profit achieved when the actual MCP was assumed to be known rather than estimated from the forecast distribution. This is the absolute maximum profit achievable. The comparison of the two profits shows that the performance of the bidding strategy has been quite good.

Impact of Market Price Model Accuracy: To investigate the possible effect of accuracy of the adopted statistical distribution of the market price, the simulation was repeated by adopting two more distribution models, namely, Weibull and normal distributions. The results are tabulated in Tables 6.10 and 6.11, respectively. The observations made regarding the results using the lognormal distribution are also valid for these two distributions, except that the maximum profit has decreased to some extent. This is rather expected as these two distributions are not as good a fit for the MCP as the lognormal distribution. This clearly indicates that an accurate model of the market price behavior is important for the strategy to work well.

Comparison with Marginal Cost Bidding: For comparison, the bidding exercise has been conducted for the Genco with the comparable marginal cost bidding. Although accurate replication of the bidding exercise is difficult, the three-step bids and the corresponding marginal costs have been specified as follows:

i. The bid price for the coal unit (a base unit) capacity is kept at a marginal cost at full load output power.
ii. The bid price for the oil unit (an intermediate unit) capacity is kept at a marginal cost at a medium output power of 60%.
iii. The bid price for the gas unit (peaking unit) capacity is kept at a marginal cost at a low output power of 25%.

The marginal costs are computed using the generator cost data tabulated in Appendix A6.1 and are found to be

$$PB_{1,j} = MC_1|_{P_{G1}=300MW} = 44.184 \ (\$/MWh)$$

$$PB_{2,j} = MC_2|_{P_{G2}=120MW} = 97.9071 \ (\$/MWh)$$

$$PB_{3,j} = MC_3|_{P_{G3}=37.5MW} = 245.3632 \ (\$/MWh)$$

for the coal, oil, and gas unit capacities, respectively.

TABLE 6.10 Results of Genco Operation Using the Weibull Distribution Model

		Using Weibull Distribution for Market Price			Using Actual Price		
Unit Type	α_i (%)	Bid Success Rate (%)	Cost ($\times 10^6$)	Revenue ($\times 10^6$)	Profit ($\times 10^6$)	Profit ($\times 10^6$)	Bid Success Rate (%)
G1: Coal	100	98.3	285.351	490.571	205.220	208.036	100
G2: Oil	59	64.3	207.019	237.474	30.455	31.609	66
G3: Gas	3	3.09	19.871	25.746	5.876	10.835	3.23
Total	—	—			241.550	250.480	—

TABLE 6.11 Results of Genco Operation using the Normal Distribution Model

		Using Normal Distribution for Market Price				Using Actual Price	
Unit Type	α_i (%)	Bid Success Rate (%)	Cost ($\$ \times 10^6$)	Revenue ($\$ \times 10^6$)	Profit ($\$ \times 10^6$)	Profit ($\$ \times 10^6$)	Bid Success Rate (%)
G1: Coal	100	100	289.137	497.136	207.999	208.036	100
G2: Oil	59	61.6	199.365	229.291	29.926	31.609	66
G3: Gas	9	4.27	24.467	28.234	3.767	10.835	3.23
Total	—	—			241.692	250.480	—

The total profit obtained from the simulation of the Genco operation using these marginal cost biddings is found to be 235.069M$, the profits obtained from the individual units being 208.036M$, 20.658M$, and 6.375M$ for the coal, oil, and gas units, respectively. This is slightly lower than the total profit of 243.792M$ achieved by the method proposed in this chapter using lognormal distribution of electricity price. This further indicates the ability of the proposed method to yield reasonable profits.

6.5.3 Management of Price Uncertainty in Short-Term Generation Scheduling

The way to deal with the uncertainty can be extended by taking into account the risk preference of the Genco. Two important means to manage the price uncertainty are (i) suitable flexible bids and (ii) the use of hedging tools such as forward contracts. The influence of these factors in the Genco's short-term generation planning and the corresponding profit performances is studied in this section.

6.5.3.1 Price Uncertainty

The random variable PS_t^{rand} corresponding to the next-day price is represented by the lognormal distribution [22]:

$$PS_t^{rand} = \text{Lognormal}(PS_t, SDev_t) \tag{6.37}$$

As price forecasting is not the main emphasis of this work, random price behavior conforming to this distribution is generated using the Monte Carlo technique for the given values of mean and variance.

6.5.3.2 Risk Behavior

To include risk behavior, each scenario in Equation 6.29 is first segregated as

$$\Pi_{i,s,spot} = \sum_{t=1}^{T} \left\{ \begin{array}{l} G_{i,t} \cdot PS_{s,t}^{rand} - C_{i,t}(G_{i,t}) \cdot W_{i,t} \\ -ST_{i,t} \cdot UST_{i,t} - SD_{i,t} \cdot USD_{i,t} \end{array} \right\} \tag{6.38}$$

To account for risk behavior at different generation levels, the objective is written as maximization of the expected utility for each scenario as follows:

$$\max E(U \mid \text{prob}_s) = \sum_{s=1}^{S} \text{prob}_s U(\Pi_{i,s,\text{spot}})$$

$$= \frac{1}{S} \sum_{s=1}^{S} U(\Pi_{i,s,\text{spot}}) \tag{6.39}$$

The determination of the utility function is subjective as it depends on the attitude of an individual toward accepting risk. The von Neumann–Morgenstern concept based on identifying the most favorable and least favorable outcome can be applied to find the individual utility function. These limiting outcomes are taken as the reference for identifying other possible outcomes. In this study, the exponential expected utility function is adopted as follows:

For risk-averse Gencos:

$$U(\Pi_{i,s,\text{spot}}) = \lambda\left(1 - e^{-\alpha(\Pi_{i,\text{spot}})-\beta}\right) \tag{6.40}$$

For risk-seeking Gencos:

$$U(\Pi_{i,s,\text{spot}}) = \lambda\left(e^{\alpha(\Pi_{i,\text{spot}})+\beta} - 1\right) \tag{6.41}$$

where, α, β, and λ are constants that define the risk behavior of the decision-maker, that is, the Genco.

6.5.3.3 Incorporation of Contracts in the Objective Function

Gencos may further utilize forward contracts to manage risk. The hedging method used in this study is based on Daniel Bernoulli's (1738) expected utility hypothesis on risk behavior, which was axiomized in 1944 by John von Neumann and Oskar Morgenstern [23]. Hedging with contract changes the profit in each price scenario s by $\Delta\Pi_{i,s}$.

$$\Delta\Pi_{i,s} = \sum_{t=1}^{T} QF_{i,t}(PF_t - PS_{s,t}) \tag{6.42}$$

where

$$QF_{i,t} = h_{i,t}G_i^{\max} \tag{6.43}$$

and h_t is the hedging ratio in period t such that $0 \leq h_t \leq 1$.

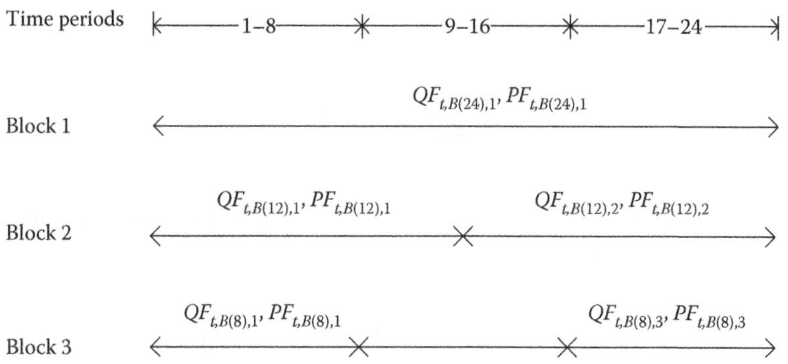

FIGURE 6.29 Blocks of contracts.

Figure 6.29 shows a possible way in which the contracts may be available in blocks of hours. Block 1, for instance, represents a 24-h (day) contract block with forward price $PF_{t,B(24),1}$. To represent these blocks, Equation 6.42 is modified to

$$\Delta\Pi_{i,s} = \sum_{t=1}^{T} \sum_{\text{all } B} QF_{i,t,B(HL),BI}(PF_{t,B(HL),BI} - PS_{s,t}) \tag{6.44}$$

Explaining further, $QF_{t,B(8),3}$ in Figure 6.29 represents quantity forward in time t for 8-h block contracts (HL = 8) between time period 17–24 (BI = 3). It should be noted that the time span of the blocks are market-specific and Figure 6.29 is given just for illustrative purposes.

A set of constraints is imposed to limit the volume of contract. Firstly, the volume of contract with same block hour length, HL, and the same block index, BI, must be equal in the time frame represented by the block index.

Hence,

$$QF_{t,B(HL),BI} = QF_{t+1,B(HL),BI} \tag{6.45}$$

In terms of the hedging ratio,

$$h_{t,B(HL),BI} = h_{t+1,B(HL),BI} \tag{6.46}$$

Secondly, the sum of all block contracts in each time period should be less than the maximum capacity of the unit.

Hence,

$$\sum_{\text{all } B} QF_{i,t,B(HL)} \leq G_i^{\max} \tag{6.47}$$

In terms of the hedging ratio,

$$\sum_{\text{all } B} h_{i,t,B(HL)} \leq 1 \tag{6.48}$$

The attitude toward risk under price uncertainty can be expressed in terms of the expected utility function as

$$E(U \mid \text{prob}_s, \Delta\Pi_{i,s}) = \frac{1}{S}\sum_{s=1}^{S} U(\Pi_{i,s,\text{spot}} + \Delta\Pi_{i,s}) \tag{6.49}$$

6.5.3.4 Bidding Methodology

Simple two-step bids, though shown to give good results for price-takers [24,25], may never be optimal as those bids cannot capture price volatility. Even price-takers cannot neglect the impact of fluctuating prices, although their individual decisions may not have any impact on the market price. To realize the benefits beyond two-step bids, we have adopted a generation schedule where:

  i. Gencos' decisions to turn "on" or "off" a unit (UC) are still determined by the two-step bid function considered as in the study by Conejo et al. [25].
  ii. Gencos can vary output between $G_i^{\min}$ and $G_i^{\max}$ at any price level.

These characteristics of bids would be similar to those where the UC is decided on the basis of day-ahead bidding using a two-step bid function and the actual generator output is decided on the basis of hour-ahead market, when the price forecast would be much more accurate. The hour-ahead price forecasting is assumed to be accurate enough to predict the MCP and the scheduling is based on this price. (In other words, it is equivalent to saying that the distribution of hour-ahead price has a standard deviation of zero.) It is further assumed that through a proper multistep bidding, it would be possible to attain a scheduling where the MCP will match the marginal cost of the generation units. The difference in the profit obtained using these hour-ahead schedules over that obtained using only the day-ahead two-step bids would indicate the additional profits possible from an ideal multistep bidding in the hour-ahead market.

It should be noted that the hour-ahead scheduling would be particularly useful for smaller and lighter generating units used by Gencos to manage the power output in short-term scheduling. The ramp rates of these units are high enough to enable them to achieve the necessary output excursions. However, many slower and larger generating units are limited by their ramping rates [26,27]. It has been shown, however, that a simple ramp rate representation is not adequate to evaluate the short-term scheduling abilities of these units [28]. Such elaborate treatment of the ramping process is not in the scope of this study. Therefore, ramp rate constraints have not been included in the model so that possible distortions from inadequate representation of ramping constraints do not affect the evaluation of possible benefits from the scheduling introduced in the hour-ahead market.

A method based on GAs, as mentioned in Section 6.4.2.3, has been developed to obtain UC decisions according to the above assumptions and the objective function can be solved using the following procedure, written in pseudo-code [29]:

BEGIN
For all population
    For all scenarios
        Calculate profit for each scenario

$$\Pi_{p,i,s,\text{spot}} = \sum_{t=1}^{T} \left\{ \begin{matrix} G_{p,i,t} \cdot PS_{s,t} - C_{i,t}(G_{p,i,t}) \cdot W_{p,i,t} \\ + ST_{p,i,t} + SD_{p,i,t} + \text{Pen}_{p,t} \end{matrix} \right\} \tag{6.50}$$

where p is the chromosome index
End
Calculate fitness function

$$\text{Fitness}_p = \frac{1}{S} \sum_{s=1}^{S} (\Pi_{p,i,s,\text{spot}}) \tag{6.51a}$$

or

$$\text{Fitness}_p = \frac{1}{S} \sum_{s=1}^{S} U(\Pi_{p,i,s,\text{spot}} + \Delta\Pi_{i,s}) \tag{6.51b}$$

End
Store fitness for the population size
END

The maximum profit for risk-neutral Gencos represented by Equation 6.29 is obtained for all the chromosomes in the population using the fitness function in Equation 6.51a. Equation 6.50 is first solved using quadratic programming (MATLAB [30]) for all scenarios and fitness is calculated through Equation 6.51a. This is not similar to the solution obtained using expected mean price in the study by Conejo et al. [25] because multistep bids provide the opportunity to vary dispatch quantity depending on the realized prices in the spot market. In case of risk-averse and risk-seeking Gencos, the maximum utility in Equation 6.49 is derived through the fitness function in Equation 6.51b. The 'Fmincon' function in MATLAB is used to solve Equation 6.51b to obtain optimum fitness.

6.5.3.5 Numerical Example

The following examples are based on generator units provided in the study by Orero and Irving [12]. The characteristics are given in Table A6.1.4 in Appendix A6.1. The forecast price [31] and standard deviation considered for analysis are given in Table A6.1.5 in

TABLE 6.12   Best Three Unit Commitment (UC) Status for the Highest Fitness Values for Unit 2

S. No.	Unit Status (1–24 h)	Mean ($) [24]
UC1	111000011111111111111111	3650.6
UC2	111000111111111111111111	3650.1
UC3	111111111111111111111111	3577.7

Appendix A6.1. To consider the effect of an increase in the uncertainty of price on the profit objective, two sets of standard deviations in price are taken for the same expected mean price. On the basis of the lognormal price distribution, Monte Carlo simulation is performed and 10,000 sets of MCPs are randomly generated to investigate the effects of price uncertainty on the expected profit.

Table 6.12 shows the best three UC status obtained from GA that gives the highest fitness values (not in order of fitness) for unit 2 in Table A6.1.4. Each of the UC status is referred to as UC1, UC2, and UC3 hereafter. The mean value in Table 6.12 is the expected profit obtained for the unit status considering the price forecast and two-step bid function in a similar way as in the study by Conejo et al. [25]. UC1 is the optimal solution obtained from PBUC whose target generation level is given in Table A6.1.6 (see Appendix A6.1) for risk-neutral Gencos.

The fitness values obtained from GA for the unit status in Table 6.12 for high and low levels of uncertainty in next-day price are shown in Table 6.13. From Table 6.13, it can be said that:

- The flexibility in scheduling provided by assumptions described in Section 6.5 shows quite an improvement in the expected profits (i.e., $4054.3 as compared to $3650.6 for UC1). This indicates that multiple bid curves should be able to represent price uncertainty better and hence can be expected to improve the expected profits.
- Considering that higher prices would make higher profits if more generation is scheduled, it is seen that UC2 gives higher expected profits than UC1 as the unit is turned on for period 7.
- Higher uncertainty in price means even higher profits if prices reach higher levels. Hence, for a higher price variance, both UC2 and UC3 give higher expected profits than UC1.

TABLE 6.13   Fitness Values for Low and High Levels of Price Uncertainty

Price Variance	UC Cases	Mean ($)	Maximum ($)	Minimum ($)	Standard Deviation
Low	UC1	4054	6264	2195	625
	UC2	4081	6425	2155	632
	UC3	4033	6583	2124	638
High	UC1	4272	7316	1486	806
	UC2	4317	7422	1501	814
	UC3	4301	7612	1512	828

This shows that a simple two-step bid function limits the possibility of higher profits as it fixes the quantity produced at either zero or a fixed value in each time period. Hence, even for price-takers, a multiple bid function should be considered that could take advantage of the possibility of higher prices by bidding at multiple quantity levels for different prices. It can be said that proper representation of uncertainty can be the key to improving profits. However, the method of obtaining proper bid curves under price uncertainty is not the main focus of this study. Hence, for analysis and comparisons, the bid function is still considered a two-step function as has been described in Section 6.4.2.3. It should be noted that for a two-step bid function, a solution using multiple price scenarios or an expected mean price will result in the same solution as in the study by Conejo et al. [25]. Hence, for risk-neutral Gencos, Equation 6.51a is directly solved using quadratic programming [30].

The risk behavior is expressed by utility functions 6 and 7, with the constants:

$$\alpha = 0.0003, \beta = 0.001, \lambda = 1, \text{ for risk-averse Gencos}$$
$$\alpha = 0.0003, \beta = 0.001, \lambda = 1/8.5, \text{ for risk-seeking Gencos.}$$

For contract decisions, four cases are considered with two types of block contracts available for hedging, as shown in Table 6.14. Forward price for each block is considered to be the mean of the expected prices in the time periods that represent the blocks. For instance, the forward price of the 24-h contract is taken as the average of the expected prices of the 24-h time periods the contract covers.

Table A6.1.6 in Appendix A6.1 shows the effect of hedging by controlling the generation level for Gencos differing in their willingness to take risk. In addition to the generation level, Table A6.1.7 shows the effect of using contracts for the four cases of Table 6.14 considering risk attitudes of Gencos.

Changes in the generation level in Table A6.1.6 in Appendix A6.1 are shown in Figure 6.30. It can be seen that risk-averse Gencos opt for lower generations to maximize their utility values whereas risk-seeking Gencos decide on producing more to maximize their utility. Considering that Gencos use two-step bids, it can be seen from Table A6.1.7 that risk-neutral Gencos looking for maximum expected profit select a UC1 schedule status, which gives an expected profit of $3660. Similarly, risk-averse Gencos choose a UC1 schedule as it gives the highest utility of 0.6556 (i.e., case 4 in Table A6.1.7). However, risk-seeking Gencos bid as per a UC2 schedule status that gives the maximum utility of 0.248. Hence, the large shift in generation from 0 to 83 MWh is observed in Figure 6.30 in period 7 for risk-seeking Gencos.

TABLE 6.14   Four Cases under Study

	Period Length (h)	24-Hour Block Contract (Nos)	12-Hour Block Contract (Nos)
Case 1	24	1	0
Case 2	24	0	2
Case 3	24	1	2
Case 4	24	0	0

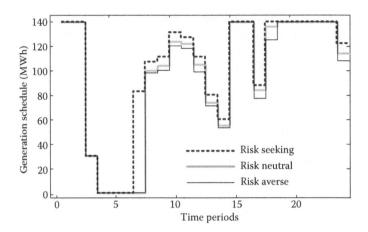

FIGURE 6.30 Optimal generation for different risk attitudes.

The effect of hedging considering contract decisions, given in Table A6.1.7, is compared in Figures 6.31 and 6.32 for risk-averse and risk-seeking Gencos, respectively. It was found that there is hardly any noticeable difference in the expected mean of different Gencos. The amount of utility that the risk-averse Genco receives from the decisions is slightly higher than that received by the risk-neutral Genco despite having a slightly lower expected profit, as can be seen from certainty equivalent (CE) plots in Figure 6.31

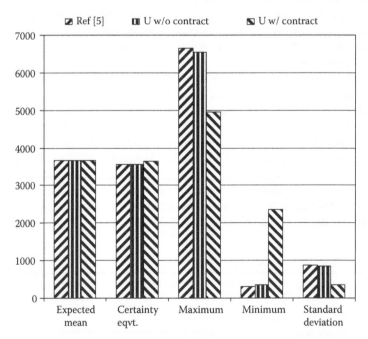

FIGURE 6.31 Comparison of risk-averse Gencos.

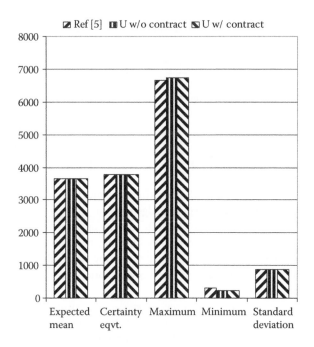

FIGURE 6.32 Comparison of risk-seeking Gencos.

and Table A6.1.7 in Appendix A6.1. This CE indicates that a risk-averse Genco will choose a decision with more certain income (higher utility) despite the lesser expected mean value. CE and hence risk premium (RP) for risk-averse producers with and without contracts indicate the effect of contracts in reducing risk. Higher values of CE (and hence lower RP) indicate that risk is considerably decreased when contracts are used. This effect can be seen from the reduction in the gap between maximum and minimum income and the reduced standard deviation leading to a more stable income. In this way, it is seen that decisions on generation as well as contracts help in hedging price uncertainty for risk-averse Gencos. It should be noted that the minimum profit of a risk-averse Genco has significantly increased due to the contracts, which comes in exchange for the reduced maximum profit.

In the case of risk-seeking producers, higher (negative) RP is indicative of increased risks and less stable income as indicated by higher standard deviation. It is observed that the performances of risk-seeking and risk-neutral Gencos are very comparable in this respect. This is because the contract price based on the expected mean is not attractive for both risk-neutral and risk-seeking Gencos and hence the only way for the Gencos to optimize their revenue is by changing the level of generation. In particular, for risk-seeking Gencos, the forward contract considered in this study plays no role in improving their objective because contracts will prevent them from reaping the benefits if the price is high.

Table 6.15 shows the comparison of optimal contract size for a risk-averse Genco in view of the level of uncertainty. The results provide an insight into the need for a proper

TABLE 6.15 Contract Size and Uncertainty

	Uncertainty	Contract Blocks	Hedging Ratio UC1	UC2	UC3
Case 1	More	h t,B(24)1	0.682	0.702	0.725
	Less	h t,B(24)1	0.739	0.757	0.777
Case 2	More	h t,B(12)1	0.437	0.461	0.512
		h t,B(12),2	0.878	0.885	0.886
	Less	h t,B(12)1	0.578	0.616	0.665
		h t,B(12),2	0.861	0.863	0.861

representation of price uncertainty while making strategies to manage it. In general, it is expected that a risk-averse Genco would hedge more in the market for higher uncertainty in price. However, cases 1 and 2 indicate that the optimal hedging ratio is higher for less uncertainty than for high uncertainty. This may be attributed to the lognormal price distribution where the probability of price reaching very high is progressively low. Hence, the utility derived by reducing the hedging ratio gives slightly better output when price uncertainty is high.

With regard to the choice of contract blocks, comparison of cases 1 and 2 in Table A6.1.7 shows that both utility and expected mean is higher for case 2. Hence, it can be said that for the hedging objective, 12-h blocks are a better choice than 24-h blocks for the forward price considered. Also, case 3, comprising both 12- and 24-h blocks, provides the same utility as case 2. This means that the choice in contract block selection does not have an added advantage with respect to utility derived in the example under consideration. However, as can be seen from Figure 6.33, the availability of multiple contract combinations resulting in the same optimal utility provides flexibility in the selection of the types of contracts.

The solution time for GAs was approximately 45 min in an Intel PIV, 2.4 GHz, and 256 MB of RAM platform. When more realistic studies are made with multiple generators with a single Genco, or multistep bid functions, the computation time is expected to increase. However, it should be possible to limit such an increase by adopting approaches suitable for these elaborate representations to achieve reasonable computation times.

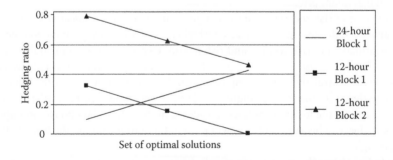

FIGURE 6.33 Set of optimal contract solutions for case 3.

6.6 Conclusions

Two major changes in the recent past, namely (i) the deregulation of power industry, and (ii) the emergence of renewable energy sources, have influenced the way power systems are operated around the world. Both these changes have introduced additional uncertainties in the system operation in different ways. Market participants now have to manage the uncertainties in the electricity price on one hand and the effects of the inherently variable nature of two most common renewable sources, wind power and solar PV power. These new features have made the task of optimal economic operation of power systems more complex where traditional UC and generation scheduling methodologies are not sufficient. This chapter presented various studies on PBUC and price-based scheduling techniques for optimization of the planning and operation of power systems under this new context.

The nature of USEP has been investigated in some detail with the intention of establishing a suitable statistical distribution to describe the behavior of electricity price in a competitive market. On the basis of graphical and statistical analyses of the historical data, the lognormal distribution was found to best describe market price behavior, though the Weibull and normal distributions also fitted reasonably well. Although price behaviors may vary from market to market, such statistical characterization of price has been used in the price-based scheduling methods described in this chapter.

General formations of the PBUC problem for competitive markets along with different solution techniques and several approaches for price-based scheduling have been presented in this chapter. Heuristic techniques in coordination with traditional UC solution methodologies are developed to obtain optimal schedules for generating units. Numerical examples have been presented to illustrate the effectiveness of various techniques.

The stochastic scheduling technique based on two-stage and multistage stochastic methods that accommodate the features of day-ahead and hour-ahead power markets, respectively, for maximizing a producer's benefit considering the stochastic nature of electricity price, has been presented in detail. Implementation of the technique was illustrated by applying the technique to the scheduling problem of a power producer with 11 generators in two different seasons. The results indicated that this stochastic self-scheduling technique is particularly effective when the uncertainty in the price is high and the price forecast is not very accurate.

Two important ways to manage the price uncertainty, namely, (i) suitable flexible bids and (ii) the use of hedging tools such as forward contracts, were also studied. The influence of these factors in short-term generation planning and corresponding profit performances of Gencos was investigated in the new scheduling scheme. Gencos risk behavior was represented by exponential utility functions. The bid functions were taken to be flexible and the simple contracts for hedging were assumed to be available. The UC task was then combined with the hedging process and the solution method based on GAs was implemented to obtain optimal scheduling. Results of numerical examples indicated that the flexible outputs and forward contracts can be used to successfully hedge against price risks in order to achieve the desired profit performance in accordance with Gencos risk behavior.

Another generation scheduling technique for price-taker Gencos was also presented, which takes into account both the market price uncertainty as well as the cost characteristics of the generating units, while ensuring a reasonable profit from the operation. The screening curve commonly used in the planning and operation of generators in the traditional operation of power systems to characterize the economic merits of the generating units were utilized in this technique to account for the costs incurred by the Gencos in their operation. This was in turn utilized to establish the desired level of bidding success in the market. The most economical way to operate a generator under this formulation is such that the generating units will run with capacity factors in the range where they are the most efficient units.

Appendix A6.1: System Data and Results

TABLE A6.1.1   Generator Unit Data

Unit No.	Output (MW)		Cost Function Coefficients			Startup Cost ($)	Minimum Up Time (h)	Minimum Down Time (h)	Minimum Cold Time (h)	Initial State (h)[a]
	Minimum	Maximum	a	b	c					
1	150	660	192.00	7.80	0.0026	440	9	9	9	9
2	160	450	264.00	17.40	0.0029	1500	6	5	5	5
3	100	400	253.09	9.00	0.0023	1000	8	5	5	8
4	140	350	212.47	13.03	0.0018	500	8	5	6	8
5	75	250	168.00	16.32	0.0014	70	4	4	5	−1
6	20	200	206.40	32.40	0.0312	250	5	5	6	−3
7	45	120	263.73	23.04	0.0084	160	4	3	3	−3
8	68.9	197	310.96	27.60	0.0031	400	5	4	4	−4
9	10	80	98.40	19.20	0.0276	120	3	1	2	−1
10	12	60	153.33	46.26	0.0384	60	1	1	1	−1
11	12	60	153.73	44.00	0.0391	60	1	1	1	−1

[a] For the initial state, " −1" means that the unit has been "off" for 1 h and "8" means that the unit has been "on" for 8 h.

TABLE A6.1.2   Generator Unit Characteristics

Unit Type	G1: Base (Coal)	G2: Intermediate (Oil)	G3: Peaking (Gas)
Rated capacity (MW)	300	200	150
Heat rate (Btu/kWh)	9000	10000	12000
Fuel cost ($/MBtu)	5.0	9.0	14.0
Plant cost ($/kW)	1000	500	250
Fixed operation and maintenance cost ($/kW/year)	20	9	1
Variable operation and maintenance cost ($/MWh)	4	6	7
Levelized fixed-charge rate/year (%)	20	20	20
Present-worth rate/year (%)	10		
Fuel price escalation/year (%)	6		
Capacity factor (%)	100		

TABLE A6.1.3 Annual Costs of the Generating Units

	Unit Type	G1: Base (Coal)	G2: Intermediate (Oil)	G3: Peaking (Gas)
Operating costs ($ × 10⁶/year)	Levelized fuel costs	201.04	268.06	357.28
	Variable levelized operation and maintenance costs	17.87	17.87	15.64
Fixed costs ($ × 10⁶/year)	Fixed levelized operation and maintenance costs	10.2	3.06	0.255
	Levelized investment costs	60	20	7.5
Total costs ($ × 10⁶/year)		289.11	308.99	398.67

TABLE A6.1.4 Characteristics of Generator Units

Unit	P_{max} (MW)	P_{min} (MW)	MDT (h)	MUT (h)	Initial Condition (h)	Initial Power (MW)	a ($)	b ($/MW)	c ($/MW²)	σ ($)	δ ($)	Time Constant (h)
1	185	54.3	4	5	5	185	143.735	11.694	0.0066	160	160	6
2	140	30.0	3	3	−4	0	50.000	13.700	0.0066	60	90	3

Source: Adapted from Shrestha, G. B., Song, K., and Goel, L. *Electric Power Systems Research*, 71, 91–98, 2004.

TABLE A6.1.5   Price Forecast and Standard Deviations for 24 h

Hour	Forecast	Standard Deviation 1	Standard Deviation 2
1	21.831	1.5	2.5
2	16.043	1.5	2.5
3	12.983	1.05	2
4	12.969	1.05	2
5	13.600	1.05	2
6	13.818	1.05	2
7	14.729	1.05	2
8	15.021	1.25	2.5
9	15.073	1.25	2.5
10	15.334	1.25	2.5
11	15.313	1.25	2.5
12	15.091	1.5	2.5
13	14.673	1.5	2.5
14	14.431	1.5	2.5
15	16.107	1.5	2.5
16	16.534	1.5	2.5
17	14.808	1.5	2.5
18	15.491	1.8	3
19	16.101	1.8	3
20	20.215	1.8	3
21	22.949	1.8	3
22	16.055	1.8	3
23	15.857	1.5	2.5
24	15.201	1.5	2.5

TABLE A6.1.6 24-Hour Generation Schedule Considering Utility Functions for Various Risk Attitudes for UC1, UC2, and UC3

		1	2	3	4	5	6	7	8	9	10	11	12	13	14	15	16	17	18	19	20	21	22	23	24
UC1	Risk neutral	140	140	30	0	0	0	0	100	104	124	122	105	74	55	140	140	84	136	140	140	140	140	140	114
	Risk averse	140	140	30	0	0	0	0	98	100	120	118	99	71	53	140	140	77	125	140	140	140	140	140	108
	Risk seeker	140	140	30	0	0	0	0	107	111	131	127	111	80	60	140	140	88	140	140	140	140	140	140	122
UC2	Risk neutral	140	140	30	0	0	0	78	100	104	124	122	105	74	55	140	140	84	136	140	140	140	140	140	114
	Risk averse	140	140	30	0	0	0	76	98	101	120	118	100	71	53	140	140	77	125	140	140	140	140	140	108
	Risk seeker	140	140	30	0	0	0	83	107	111	131	127	111	80	60	140	140	88	140	140	140	140	140	140	122
UC3	Risk neutral	140	140	30	30	30	30	78	100	104	124	122	105	74	55	140	140	84	136	140	140	140	140	140	114
	Risk averse	140	140	30	30	30	30	76	98	100	120	118	100	71	53	140	140	77	126	140	140	140	140	140	108
	Risk seeker	140	140	30	30	30	30	81	104	107	127	125	108	77	57	140	140	86	140	140	140	140	140	140	118

TABLE A6.1.7 Profits and Contract Comparisons for UC1, UC2, and UC3

	Case	Utility	CE	RP	EMV	Maximum	Minimum	SD	Contract Blocks		
									h $t,B(24)1$	h $t,B(12)1$	h $t,B(12),2$
UC1											
Risk averse	Case 1	0.6636	3628	22	3650	5005	2227	379	0.682	0.000	0.000
	Case 2	0.6645	3637	17	3654	4929	2349	339	0.000	0.437	0.878
	Case 3	0.6645	3637	18	3655	4958	2351	342	0.127	0.290	0.756
	Case 4	0.6556	3550	108	3658	6550	355	853	0.000	0.000	0.000
Risk seeking	All	0.2476	3777	−119	3658	6751	233	885	0.000	0.000	0.000
Risk neutral	All	0.6554	3548	112	3660	6653	297	868	0.000	0.000	0.000
		(0.2473)	(3774)	(−114)							
UC2											
Risk averse	Case 1	0.6639	3631	19	3650	4926	2197	354	0.702	0.000	0.000
	Case 2	0.6647	3639	16	3655	4887	2399	323	0.000	0.461	0.885
	Case 3	0.6647	3639	16	3655	4895	2383	325	0.120	0.339	0.768
	Case 4	0.6555	3549	110	3659	6625	368	860	0.000	0.000	0.000
Risk seeking	All	0.2480	3780	−121	3659	6750	248	893	0.000	0.000	0.000
Risk neutral	All	0.6553	3547	114	3661	6681	311	875	0.000	0.000	0.000
		(0.2477)	(3777)	(−116)							
UC3											
Risk averse	Case 1	0.6570	3563	14	3577	4663	2385	302	0.725	0.000	0.000
	Case 2	0.6576	3569	12	3581	4621	2505	282	0.000	0.512	0.886
	Case 3	0.6576	3569	12	3581	4623	2514	280	0.017	0.486	0.868
	Case 4	0.6478	3475	111	3586	6614	290	863	0.000	0.000	0.000
Risk seeking	All	0.2402	3708	−121	3588	6738	202	888	0.000	0.000	0.000
Risk neutral	All	0.6476	3473	115	3588	6697	233	878	0.000	0.000	0.000
		(0.2399)	(3706)	(−118)							

Note: CE, certainty equivalent; EMV, expected mean value; RP, risk premium; SD, standard deviation.

References

1. Shrestha, G.B., and Qiao, S. Statistical characterization of electricity price in competitive power markets. *International Conference on Probabilistic Methods Applied to Power Systems*, Singapore, 2010.

2. National Electricity Market of Singapore, Market Report 2006. [online]. Available at: http://www.emcsg.com/AboutUs/PublicRelations

3. National Electricity Market of Singapore, Market Report 2007. [online]. Available at: http://www.emcsg.com/AboutUs/PublicRelations

4. Stoll, H.G. *Least-Cost Electric Utility Planning*. New York: System Development & Engineering Department General Electric Company Schenectady, 1989.

5. Zhai, Q., Guan, X., and Cui, J. Unit commitment with identical units: Successive subproblem solving method based on Lagrangian relaxation. *IEEE Transactions on Power Systems*, 17(4), 1250–1257, 2002.

6. Senjyu, T., Shimabukuro, K., Uezato K., and Funabashi, T. A fast technique for unit commitment by extended priority list. *IEEE Transactions on Power Systems*, 18(2), 882–888, 2003.

7. Bjorkvoll, T., Fleten, S.-E., Nowak, M. P., Tomasgard, A., and Wallace, S. W. Power generation planning and risk management in a liberalised market. *IEEE Porto Power Tech Conference Proceedings*, Portugal, September 10–13, 2001.

8. Richter, Jr., C. W., and Sheble, G. B. Genetic algorithm evolution of utility bidding strategies for the competitive marketplace. *IEEE Transactions on Power Systems*, 13, 256–261, 1998.

9. Arroyo, J. M., and Conejo, A. J. Optimal response of a thermal unit to an electricity spot market. *IEEE Transactions on Power Systems*, 15(2), 1008–1104, 2002.

10. Richter, C. W., and Sheble, G. B. A profit based unit commitment GA for the competitive environment. *IEEE Transactions on Power Systems*, 15(2), 715–721, 2000.

11. Goldberg, D. E. *Genetic Algorithms in Search Optimization and Machine Learning*. Massachusetts: Addison-Wesley, 1989.

12. Orero S. O., and Irving M. R. Large scale unit commitment using hybrid genetic algorithm. *Electrical Power and Energy Systems*, 19(1), 45–55, 1997.

13. Swarup, K.S., and Yamashiro, S. Unit commitment solution methodology using genetic algorithms. *IEEE Transactions on Power Systems*, 17(1), 87–91, 2002.

14. Pokharel, B. K., Shrestha, G. B., Lie, T. T., and Fleten, S.-E. Profit based unit commitment in competitive markets. *Proceedings of International Conference on Power System Technology (IEEE PowerCon'2004)*, Singapore, November 21–24, 2004.

15. Pokharel, B. K., Shrestha, G. B., Lie, T. T., and Fleten, S.-E. Price based unit commitment for GenCos in deregulated markets. *IEEE Power Engineering Society General Meeting*, 1, 428–433, 2005.

16. Baillo, A., Ventosa, M., Rivier, M., and Ramos, A. Optimal offering strategies for generation companies operating in electricity spot markets. *IEEE Transactions on Power Systems*, 19(2), 745–753, 2004.

17. Plazas, M. A., Conejo, A. J., and Prieto, F. J. Multi-market optimal bidding for a power producer. *IEEE Transactions on Power Systems*, 20(4), 2041–2050, 2005.

18. Shrestha, G. B., B. K. Pokharel, Lie, T. T., and Fleten, S.-E. Price based unit commitment for bidding under price uncertainty, *IEEE Proceedings—Generation, Transmission and Distribution*, 1(4), 663–669, 2007.

19. Shrestha, G. B., Song, K., and Goel, L. An efficient stochastic self-scheduling technique for power producers in the deregulated market. *Electric Power Systems Research*, 71, 91–98, 2004.

20. Shrestha, G. B., and Qiao, S. Generation scheduling for a price taker Genco in competitive power markets. *Proceedings of the 2009 Power Systems Conference and Exposition*, Seattle, WA, USA, March, 15–18, 2009.

21. Shrestha, G. B., Qiao, S. Market oriented bidding strategy considering market price dynamics and generation cost characteristics. *IET Generation, Transmission and Distribution*, 4(2), 150–161, 2010.

22. Nogales, F. J., Contreras, J., Conejo, A. J., and Espinola, R. Forecasting next-day electricity prices by time series models. *IEEE Transactions on Power Systems*, 17(2), 342–348, 2002.

23. Neumann, J. V., and Morgenstern, O. *Theory of Games and Economic Behavior*. Princeton: Princeton University Press, 1990.

24. Conejo A. J., Nogales, F. J., Arroyo, J. M., and Bertrand, R. G. Risk-constrained self-scheduling of a thermal power producer. *IEEE Transactions on Power Systems*, 19(3), 1569–1574, 2004.

25. Conejo A. J., Nogales, F. J., and Arroyo, J. M. Price-taker bidding strategy under price uncertainty. *IEEE Transactions on Power Systems*, 17(4), 1081–1088, 2002.

26. Lu, B., and Shahidehpour, M. Short-term scheduling of combined cycle units. *IEEE Transactions on Power Systems*, 19(3), 1616–1625, 2004.

27. Lu, B., and Shahidehpour, M. Unit commitment with flexible generating units. *IEEE Transactions on Power Systems*, 20(2), 1022–1034, 2002.

28. Shrestha, G. B., Kai, S., and Goel, L. Strategic bidding for minimum power output in the competitive power market. *IEEE Transactions on Power Systems*, 16813–16818, 2001.

29. Shrestha, G. B., Pokharel, B. K., Lie, T. T., and Fleten, S.-E. Management of price uncertainty in short-term generation planning. *IEE Proceedings—Generation, Transmission and Distribution*, 2(4), 491–504, 2008.

30. Coleman, T., Branch, M. A., and Grace, A. Optimization toolbox for use with MATLAB. The Math Works Inc. [Online]. Available at: http://www-ccs.ucsd.edu/matlab/pdf_doc/optim/optim_tb.pdf

31. Huse, E. S., Wangensteen, I., and Faanes, H. H. Thermal power generation scheduling by simulated competition. *IEEE Transactions on Power Systems*, 14(2), 472–477, 1999.

7

Optimal Self-Schedule of a Hydro Producer under Uncertainty

F. Javier Díaz

Javier Contreras

7.1 Introduction

7.1.1 The Hydroelectric Unit Commitment Problem

The *classic unit commitment problem* consists of establishing, for each period of the planning horizon, the start-up and shutdown for each power unit of a power system. It is very common to find this problem integrated with the economic dispatch, so the integrated solution also computes, besides the start-up and shutdown for each power unit in each time period, the optimal generation program to the committed operating units during each period of the scheduling horizon.

Nilsson and Sjelvgren (1997b) study the start-up costs of the hydraulic units answering three basic questions: What are the causes of the start-up costs? What are the values of these costs? How does the start-up affect the short-term scheduling strategies of the generators? With regard to the causes of these costs, the authors mention the machinery wear and tear due to temperature changes during start-up, so it is necessary to perform regular maintenance to prevent life-cycle reduction. In addition to the unavailability of the units, there could be malfunctioning of the control equipment during start-up, with the corresponding technical personnel costs for reparation. Regarding the costs during start-up, the authors divide the costs into three phases depending on the units' speed: from 0% to 90% of the nominal speed, from 90% to the nominal speed, and above this speed. The authors argue that there is no common method for estimating the start-up cost, but each company uses its own way of determining it. About the third question, the start-up cost can be considered when controlling the maximum number of start-ups per day for each unit.

Li et al. (1997a) solve the thermal unit commitment problem by Lagrangian relaxation and the hydroelectric unit commitment at reservoir level by dynamic programming based on priority lists, which help to reduce the number of possible combinations. These authors combine all available units of the hydroelectric plants into an equivalent unit using an input/output aggregate curve. They reallocate the water flow from the reservoir, increasing its discharge in hours in which its marginal value is high, reducing it when it is low. Then, they update the water marginal value and repeat the allocation of hydro units until meeting the convergence criterion. Using dynamic constraints, they include issues such as the minimization of the time that units must be switched on and off, which helps to reduce the start-up and operating costs.

Li et al. (1997b) consider all available units on during the study period. Usually, this starting point produces an excess of spinning reserve. To achieve an economic operation, some units are shut down during certain periods. According to the evaluation of some economic indicators related to the relative cost savings, which unit should be shut down first is decided, then, the same is done with the other units. The procedure continues until one of the following three stopping criteria is satisfied: (i) it is not possible to obtain further reductions in cost; (ii) the unit scheduling in two consecutive iterations remains unchanged; and/or (iii) savings are zero for all remaining units.

The *hydroelectric generation unit commitment problem* is explicitly addressed in our research as it is considered an important component to optimize the operation and to evaluate the technical efficiency of a hydroelectric generation company (H-GENCO). This problem can be studied independently to determine self-programming at minimum cost of an H-GENCO in the day-ahead market (Conejo et al., 2002). However, it can also be considered as one of the subproblems of a short-term hydrothermal coordination problem solved by decomposition methods, such as Lagrangian relaxation, of a profit-maximizing company that seeks maximum technical efficiency in the operation of its generating plants.

7.1.2 Bidding Strategies in the Day-Ahead Market

With the restructuring of electric power systems, markets have evolved from vertically integrated monopolies to competitive schemes. Pool-based electricity markets have expanded into regions and countries around the world such as South America (Hammons et al., 2002), Spain (González and Basagoiti, 1999), New England (Cheung et al., 2000), and Norway (Fosso et al., 1999). As a result, market agents need to implement new strategies and methodologies to achieve higher efficiency in order to be competitive. This has led to profound changes in the procedures used by the generating companies to ensure safe and efficient operation, significantly changing the way they get their benefits.

Competition creates new markets where buyers and sellers can conduct electricity trade by auctions in the energy pool or by bilateral contracts. Therefore, cost minimization techniques are being replaced by offer algorithms made by generators to maximize their profits and, in some markets, by consumers seeking to maximize their utility functions. Prices are determined by market agents' interactions. Due to these reforms, generating companies have been forced to adapt to a new way of understanding their business. The traditional institutional framework that usually guaranteed the generating companies the recovery of their full cost has been replaced by a competitive framework in which the incomes of each company depend on its own ability to sell the energy produced by its plants. For decades, before competition, complex models were used in operation and planning, focusing on different stages of decision-making for the provision of electricity trying to minimize production costs. However, the competitive operation of a deregulated power industry is the result of the interaction of different actors with different objectives.

Our study is related to the operation of a price-taking H-GENCO owner of a system of cascaded plants along a river basin. A short-term framework is considered, focusing on the day-ahead market. Therefore, the medium- and long-term decisions may be considered as exogenous variables for the company, whose evaluation is outside the scope of this work. As the production schedule of a price-taking firm does not affect market prices, these prices can be assumed known or predicted (Szkuta et al., 1999; Angelus, 2001; Nogales et al., 2002). A company that develops bid strategies in perfectly competitive markets has incentives to bid at marginal costs, as these markets involve many small producers, and it is not possible for any of them to influence the market price. Thus, each producer must accept this price, which is fully described by the market conditions (Ladurantaye et al., 2007; Fleten and Pettersen, 2005; Conejo, Nogales, and Arroyo, 2002; Gross and Finlay, 2000).

Thermal bidding strategies that maximize the expected net incomes can show that the optimal strategy is to offer at prices reflecting the variable operating costs of the thermal plant as these variable costs are a function of the fuel costs. In hydro units the bidder may postpone its energy production if future prices are higher than at present. The hydropower plant variable costs are in fact opportunity costs depending on both future hydrological scenarios and expected load, and, most importantly, future production by other generator companies. The hydropower plant opportunity cost calculation is a complex stochastic dynamic optimization problem (Pereira and Pinto, 1991; Pereira et al., 1998).

Pereira et al. (2005) present a solution to the strategic bidding problem in short-term electricity markets using a binary expansion scheme to transform the nonlinear and nonconvex problem into a mixed-integer linear program. De la Torre et al. (2003) study the nonconvex residual demand function by a Cournot scheme using a mixed-integer programming model.

7.1.3 *Variable Head Problem* in Hydroelectric Systems

A hydroelectric generating unit has a complex operational performance. Its power output depends mainly on three variables: the net head, the water discharge to the turbine, and the turbine-generator set efficiency. The net head is a nonlinear function of several variables, which can include the water storage in the reservoir, the flow discharged to the turbine, and the water tail level (in reaction turbines). The turbine-generator set efficiency is a nonlinear function of the net head and the water discharge. Furthermore, because of technical constraints, there are upper and lower limits for the permitted discharges to the turbines. Discharges below the lower limits would cause mechanical vibrations (which would cause oscillation in the power output), cavitation, and low efficiency levels. Additionally, these units may have large prohibited operation areas that do not allow a continuous range in the generation, making the problem more complex and its solution more difficult due to the associated combinatorial nature (Finardi and da Silva, 2006).

The so-called performance curves of the hydroelectric generating units represent the power generated as a function of the water discharge. Most models related to these performance curves ignore the variable head effect in order to avoid nonlinearities and nonconcavities that could lead to local optima. However, such simplifications can lead to inaccuracies. Simple performance curves have been used, by taking approximations such as concave piecewise linear functions (Habibollahzadeh and Bubenko, 1986; Chang et al., 2001) and by modeling the so-called local best efficiency points (Nilsson and Sjelvgren, 1997b; Nilsson et al., 1998).

The variable head effects in hydropower systems have been studied, among others, by Li, Johnson and Svoboda (1997), García-González et al. (2003, 2007), Parrilla and García-González (2006), and Catalão et al. (2006, 2009). García-González et al. (2007) consider a model based on profits, with two risk aversion criteria: maximum profit Value-at Risk (VaR) and minimum Conditional Value-at-Risk (CVaR). The problem is solved by an iterative procedure with piecewise linear approximations.

Conejo et al. (2002) formulate and solve a mixed-integer linear programming model that assigns three different levels of water stored in reservoirs (low, medium, and high) to the unit performance curves; binary variables are used to select the performance curve for each of these three levels and to model each of the nonconcave curves. Borghetti et al. (2008) consider a single reservoir and a hydroelectric plant with multiple pumping units to propose an improvement of the previous approach in two steps: (i) an extension of Conejo et al. (2002) to slightly generalize their approach to a parametric number of water volumes, and (ii) an enhanced linearization with a more accurate estimation of the upper bound on the power production through a convex combination method considering both volumes and discharges. In our work, these curves are modeled by a continuous function that allows us to represent the nonconcavities where, unlike Nilsson and Sjelvgren (1997a,b), the plant operation is not restricted to the local best efficiency points, that is, the curves are appropriate to represent the global nonconcavity.

García-González et al. (2003) propose a combination of heuristics to find an initial feasible integer solution using branch and bound methods (Nilsson and Sjelvgren, 1995). Then, they do a search on the feasible solution space. García-González et al. (2007) consider the possible start-up costs and the variable head effects.

Catalão et al. (2006) analyze the short-term behavior of head-dependent reservoirs. They show that the role played by reservoirs depends not only on their relative position in the system but also on the physical data that define the hydro chain parameters. Power generation is assumed to be a function of the water discharge and the plant efficiency. For each plant, they assume that the efficiency is a linear function of the head and the latter is a function of the storage levels both in the reservoir itself and in the next one in the chain. For each reservoir, the authors assume that the water level is a linear function of the water volume stored. In this way, power generation is a nonlinear function of the water discharge and the volumes stored both in the reservoir itself and in the next one in the chain. Indefinite quadratic programming is used to solve this problem. Through a case study consisting of a system with three cascaded reservoirs, they show that the optimization process postpones power generation in the early periods to achieve high levels of storage in plants with higher technical efficiencies. Results are compared with a linear programming method, which ignores head dependency, finding increases in benefits between 2.21% and 7.86%, with a similar computational effort.

7.2 Development of an Optimization Model for Hydroelectric Operation

7.2.1 Introduction

Traditional methodological approaches applied to the analysis of hydrothermal systems usually begin with a long-term aggregate system. The system is broken down and economic signals related to hydrological and electricity price scenarios are sent to the short- and medium-term markets. In planning and programming these systems, the medium-term operation policy is a combination of hydrological and pricing stochastic forecasts, maintenance planning, and economic factors. The day-ahead hydroelectric operation programming process begins with the assessment of the mid-term operation policy to determine the amount of water to be discharged. This is based on forecasts of both the next-day market price as the water flows into the reservoirs and the company's commitments previously acquired related to its electricity sales. The result is a set of optimal water discharges for each plant in each time period, which depends on the current reservoir levels, the inflow forecasts, the electricity price forecasts in the short- and medium-term markets, and the contract portfolio that the company has undertaken.

In this work, we propose a complementary approach that considers a detailed analysis of the main features of the hydroelectric generation units. In day-ahead operation it is important to describe the relationship between the discharged water and the output power. We consider an H-GENCO that owns a cascaded reservoir system along a river basin. Beginning with the available information about the system state, in light of both hydrological and electricity price short-term forecasts, the model proposed can be useful for a company to make appropriate decisions regarding the bids to be submitted to the electricity pool in the day-ahead market. Thus, a detailed analysis of the current state and the cascaded reservoir system's physical behavior can yield useful information for the development of progressive adjustments to the available economic signals. The latter are obtained through medium- and long-term scenarios with high uncertainty, which could be gradually reduced. The problem of satisfying consumer demand is the sole responsibility of the market operator but not of the generator units whose main objective is to maximize their profits.

The technical efficiency of the hydro units is a key component to evaluate the variable head effects and to make decisions regarding the hydroelectric generating unit commitment, bidding in the day-ahead market, and self-programming, according to the economic dispatch by the market operator. In that regard, we present a model that allows us to consider the H-GENCO integrating all its plants. This methodological approach can serve as an additional input for a hydroelectric company to formulate its short-term market strategies and, possibly, to design a decision support system to improve technical and economic performances.

In traditional approaches of monopolistic centralized systems, the optimization criterion was the minimization of the total operating costs for the whole scheduling horizon. In the new competitive markets, this approach is replaced by profit maximization. In our research, we also consider technical efficiency of a hydro unit, which is modeled as a function of the net head and the water discharge. This allows us to evaluate the variable head effects in the estimation of the hydroelectric plant power output.

The remainder of this chapter is organized as follows. Section 7.2.2 refers to the hydroelectric power production function model considering variable head effects. In Section 7.2.3, a statistical regression procedure is developed to represent the hydroelectric generating unit technical efficiency as a quadratic function of the net head and the water discharge. The model constraints are formulated in Section 7.2.4. The measures of risk, VaR and CVaR, are presented in Section 7.2.5. The profit maximization objective function is defined in Section 7.2.6. The computational strategy to solve the problem is shown in Section 7.2.7. A case study of the Duero river basin in Spain is examined in Section 7.3 and relevant conclusions are presented in Section 7.4.

7.2.2 Hydroelectric Power Production Function

The power output of a hydroelectric generating unit i in time period t, expressed in megawatts (MW), can be written as in Equation 7.1, where $efi(i,t)$ represents the unit efficiency in percentage, $q(i,t)$ the water discharge to the turbine in cubic meter per second (m^3/s), and $h(i,t)$ the net head in meters (m).

$$p(i,t) = 9.81 \times 10^{-3} \times efi(i,t) \times h(i,t) \times q(i,t) \forall i \in I, \quad \forall t \in T \qquad (7.1)$$

This power function has traditionally been represented by the so-called performance curves that express the power output as a function of the water discharge. To simplify its representation and trying to avoid nonlinearities and nonconcavities that could lead to local optima, simple performance curves have been used, such as linear functions, concave piecewise linear approximations (Habibollahzadeh and Bubenko, 1986; Chang et al., 2001), or modeling the so-called local best efficiency points (Nilsson and Sjelvgren, 1997b; Nilsson et al., 1998). In all these cases, the problems associated with hydroelectric power have been addressed with linear programming techniques (Guan et al., 1999). However, these simplifications can lead to inaccuracies.

Exploiting some special features of certain constraints, the problems associated with hydroelectric power have been formulated as network flow programs, whose solution is faster than the one obtained by standard linear programming algorithms. Brännlund et al. (1986) use a network flow technique combined with dynamic programming. Different techniques and methodologies have been proposed to get greater benefits in hydroelectric generation (Alley, 1977; Guan et al., 1995). Approaches such as capacity curves, dynamic programming, Lagrangian relaxation, and heuristic methods have been used. Some more complex and sophisticated models address the problem of optimality more realistically. Deng et al. (2004) present a stochastic programming framework; Legalov and Palamarchuk (2005) propose a dynamic programming model; Valenzuela and Mazumdar (2001) apply stochastic optimization to maximize generator profits; Dentcheva et al. (1996) use a recursive procedure with primal and dual methodologies in a mixed-integer linear programming model; and Correa et al. (2007) incorporate interior point methods and transmission network constraints.

Hydroelectric power is a nonlinear and nonconvex function of the water discharge (Finardi and da Silva, 2006; Siu et al., 2001; Wang et al., 2004). Siu et al. (2001) develop a model of the British Columbia Hydro Power Authority system in Canada and apply it to a 2700 MW plant with 10 units of four different types. Through an example, this model shows that allocating all available units in a plant could produce a 15% efficiency loss or an energy equivalent to 80 MWh. Wang et al. (2004) use a function for each hydraulic head and quadratic approximations for the relationship between power and discharge. Ferrer (2004) considers the reservoir gross head as a cubic polynomial of the water volume and a quadratic function of the water discharge; the unit efficiency is assumed to be a second-degree concave polynomial of the water discharge, and the power output a fourth-degree polynomial.

Although the literature related to the problem is abundant, models that explicitly consider technical efficiency of hydroelectric generating units have not been found. In this chapter, the performance curves of these units are represented through continuous nonlinear functions to deal with nonconcavities where, unlike Nilsson and Sjelvgren (1997b), the operation of the units is not limited to the best local efficiency points, that is, the curves are suitable to represent global nonconcavity. Therefore, our study develops a mixed-integer nonlinear programming (MINLP) model of the operation of an H-GENCO that owns a cascaded reservoir chain along a river basin in the context of a short-term market. In this way, it is possible to get a better estimate of the generated power and to achieve significant water savings when considering the technical efficiency in the operation of hydroelectric plants.

7.2.3 Technical Efficiency of the Generating Units as a Function of the Net Head and the Water Discharge

A hydroelectric generating unit has a complex operational performance. Its power generation depends on three variables: the net head, the water discharge to the turbines, and the technical efficiency of the turbine-generator set. The net head is a nonlinear function of variables such as the gross head (associated with the reservoir water volume), the water discharge, and the water level in the reservoir tail. Moreover, the turbine efficiency is a nonlinear function of the net head and the water discharge.

Many related works consider the net head, the water discharge, and the technical efficiency as parameters with average values for the scheduling horizon, leading to the formulation of the power output as a linear function of the water discharge (Velásquez et al., 1999; Campo and Restrepo, 2004; Catalão et al., 2006, 2009). Considering these two magnitudes as variables, the net head and the technical efficiency of the units, could complicate the problem of calculating the generated power of a hydroelectric plant. However, this can have significant effects, as it allows us to get more accurate and useful results for making appropriate decisions in short-term electricity markets.

Our research begins with the information available from the manufacturers usually denoted as efficiency graphs at different levels, depending on the net head and the water discharge. The graphs are known as "Hill diagrams," such as the one illustrated in Figure 7.1 adapted from Díaz (2009) and Finardi (2003). In this figure, for any pair (net head, water discharge) associated with any point of the feasible operating region, it is possible to read the turbine efficiency (and the power output). A reaction turbine is considered; in this case, the efficiency of a Francis turbine is represented as a quadratic function of the net head and the water discharge; in this way, the adjusted model using multiple nonlinear regression is presented in Equation 7.2, while ignoring the subscripts and using the following notation:

$$\eta = \beta_0 + \beta_1 h + \beta_2 q + \beta_3 h^2 + \beta_4 q^2 + \beta_5 hq \tag{7.2}$$

where η represents the efficiency of a hydroelectric generating unit in percentage (%), h the net head (m), q the discharge of water to the turbine (m^3/s), and β_i ($i = 0, 1, 2, \ldots, 5$) the parameters of the regression model (to estimate).

The procedure used to represent the technical efficiency as a quadratic function of the net head and the water discharge is based on the development of a multivariable nonlinear statistical regression analysis described as follows. It starts by reading 378 points from the curves in Figure 7.1 using the text and image viewer GSview, version 4.7, for Postscript files. Then, an algorithm programmed in the statistical package R, version 2.4.1, is used to transform the data to the original scale. With these 378 records, the regression model is adjusted in expression 7.2 to estimate the β_i ($i = 0, 1, 2, \ldots, 5$) parameters and the fitted model of expression 7.3 is obtained. The statistical tests t and F are fulfilled, giving an adjusted R^2 equal to 0.82, meaning that the model explains approximately 82% of the data variability in the corresponding sampling

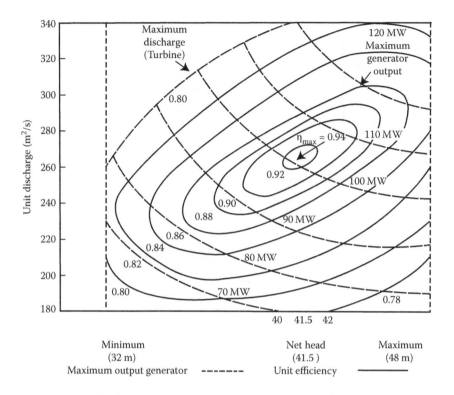

FIGURE 7.1 "Hill diagram" of a Francis turbine: Technical efficiency as a net head and water discharge function. (Adapted from Finardi, E. C., Alocação de unidad hidrelétricas em sistemas hidrotérmicos utilizando relaxação Lagrangeana e programação quadrática sequencial, PhD Thesis, Universidade Federal de Santa Catarina, Florianópolis, Brazil, 2003.)

intervals. According to the values found in the p, t, and F tests, all the proposed parameters are significantly different from zero. Figure 7.2 shows the graphs of the sampled points and the fitted model.

$$E(\eta/h,q) = -1.37 + 0.0773h + 0.00502q - 0.00131h^2 - 0.0000191q^2 + 0.000121hq \quad (7.3)$$

From the above results, it can be concluded that expression 7.3 is a good adjusted model to estimate the efficiency of the hydroelectric generating units as a quadratic function of the net head, in meters (m), and the water discharge, in cubic meters per second (m³/s). Thus, for any couple (net head, water discharge) associated with any point of the feasible operating region, the above expression can be used to estimate the turbine efficiency, from which it is possible to calculate the power output, in megawatts (MW), using Equation 7.1. The water discharge to the unit i at time t is considered as the decision or control variable, and the net head is a variable that depends on the volume of water stored in the reservoir.

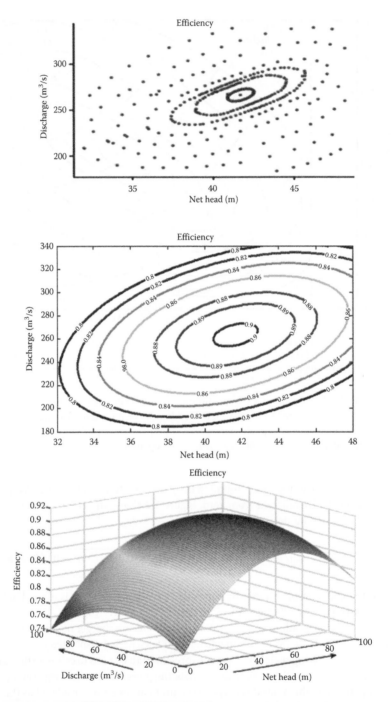

FIGURE 7.2 Efficiency of a Francis turbine as a quadratic function of the net head and the water discharge. Sampling points (first section). Fitting model (second and third sections).

After statistical validation of the results obtained by multiple nonlinear regressions, the adjusted model is subject to mathematical validation considering the efficiency as the above quadratic function of the net head and water discharge. Using mathematical analysis applied to this function, its optimal point is found, which coincides with the design point of the turbine, as expected, and as detailed below.

Every turbine is designed to operate with nominal or design parameters: net head (hd) and water discharge (qd). The pair (hd,qd) is known as the design point. In the Hill diagram of Figure 7.1 these parameters take approximate values of 41.67 m and 262.43 m³/s, respectively, where the turbine gets its maximum efficiency or its optimal performance, close to 94%. Moreover, for any point of the feasible operation region associated with the pair (h,q), where h represents the net head, and q the water discharge, it is possible to read the turbine efficiency. Operating far from these parameters will make the electrical power production process inefficient.

Díaz (2009) proposes as a working hypothesis, the existence of a trajectory of maximum relative efficiency, which depends on the level reached by the net head, for which the optimal discharge of maximum relative efficiency is possible to obtain from the Hill diagram with the aforementioned net head. The optimal discharge, for a given net head, can be found, by optimization, by equating to zero the partial derivative of the efficiency function with respect to the discharge, $\delta\eta/\delta q = 0$ (for which h is considered a constant), and solving the resulting equation for the discharge, q, as a function of the net head, h. In the case of expression 7.3, expression 7.4 is obtained, from which q can be expressed as a linear function of h, as in expression 7.5.

$$\delta\eta/\delta q = \beta_2 + 2\beta_4 q + \beta_5 h \qquad (7.4)$$

$$q = -\left(\beta_2/2\beta_4\right) - \left(\beta_5/2\beta_4\right) \times h \qquad (7.5)$$

The technical efficiency of a Francis turbine is approximated by a bivariable quadratic model, as specified in Equation 7.3, from which a linear function that allows us to obtain the maximum relative efficiency discharge associated with a given net head is obtained. For this couple (net head, water discharge), it is possible to calculate both the turbine efficiency, and the power output, that is, the power level at maximum relative efficiency can be obtained. This discharge of maximum relative efficiency provides useful information for a price-taker hydropower company to make decisions regarding the commitment of its generating units, bidding, and self-programming after the market operator performs economic dispatch.

Using a procedure similar to that used to find the discharge at maximum relative efficiency for a given net head, we equate to zero the partial derivative of the efficiency, this time with respect to the net head, $\delta\eta/\delta h = 0$, considering the discharge q as constant:

$$\delta\eta/\delta h = \beta_1 + 2\beta_3 h + \beta_5 q \qquad (7.6)$$

From the above analysis, a system of two simultaneous linear equations with two variables is obtained, Equations 7.4 and 7.6, whose solution must reproduce the design point of the Hill diagram. For the numerical example of the case study, a system of two simultaneous linear equations with two variables, h and q, is formulated as shown below.

$$q = 131.4136 + 2.1675 \times h$$

$$h = 29.5038 + 0.0462 \times q$$

The solution of this system, except for rounding errors, replicates the design point, observed in the Hill diagram; see Figure 7.1 (hd, qd) = (41.67, 262.43). The advantage of having an analytic function of this type is the possibility of developing a mathematical procedure for decision-making. For example, to obtain the optimal discharge in the short-term (hourly, daily, etc.) is possible, given a good approximation of the net head. Thus, we are also able to obtain the daily or hourly operation at maximum efficiency of the generation units.

7.2.4 Constraints on the Hydroelectric Generation System

The constraints of a hydropower system are represented by expressions 7.7 through 7.28. Then, they will be represented in blocks.

7.2.4.1 Water Balance Equations

The equality constraints Equation 7.7 presented below represent the continuity equations of the reservoirs. In each reservoir, the water volume in the current period is calculated as the volume in the previous period, plus the total inflow to the reservoir (both the natural inflow and the contributions from upstream reservoirs), less the reservoir total outflow, formed by both the water discharge to the turbines and the spillage. The M constant, equal to 0.0036, is a conversion factor from m³/s to Hm³/h. The water travel time from a reservoir to the next one downstream is one hour. The transition matrix MATtra(i,j) is composed of binary values which are equal to 1 if the reservoir i receives flows from reservoir j upstream, and 0 otherwise.

$$\mathrm{vol}(i,t) = \mathrm{vol}(i,t-1) + W(i,t) + M * \left\{ \sum_{j \in J} \mathrm{MATtra}(i,j) \right.$$
$$\left. *[q(j,t-1) + \mathrm{ver}(j,t-1)] - q(i,t) - \mathrm{ver}(i,t) \right\} \tag{7.7}$$

7.2.4.2 Constraints on the Reservoirs

In the following, the limits of the stored water volume, the net head, and the spillage are shown. In Equation 7.8, the volume of reservoir i during time period t must remain between its lower and upper limits. In Equation 7.9, the net head is bounded by its minimum and maximum values. In Equation 7.10, the reservoir spillage associated to plant i is bounded by its maximum capacity, VERmax(i).

$$\mathrm{VOL\,min}(i) \leq \mathrm{vol}(i,t) \leq \mathrm{VOL\,max}(i) \quad \forall i \in I \tag{7.8}$$

$$\mathrm{Hmin}(i) \leq h(i,t) \leq \mathrm{Hmax}(i) \quad \forall i \in I \tag{7.9}$$

$$\mathrm{ver}(i,t) \leq \mathrm{VER\,max}(i) \quad \forall i \in I, \quad \forall t \in T \tag{7.10}$$

7.2.4.3 Constraints on the Hydroelectric Generating Units

The limits of the water discharge to the turbines, the power generation, and the volume discharged during the scheduling horizon are shown in constraints 7.11 through 7.13. In Equation 7.11, as established by the $v(i,t)$ binary variables, the water discharge to the turbine associated with plant i in period t must remain within its upper and lower bounds, if the turbine is on, then $v(i,t) = 1$; but if the turbine is off, then $v(i,t) = 0$, and the discharge is also equal to 0. Expression 7.12 is similar to expression 7.11, where the power is bounded by its minimum and maximum limits. In Equation 7.13, the total discharged volume in cubic hectometers (Hm³) is calculated in each reservoir for the entire scheduling horizon.

$$Qmin(i) * v(i,t) \leq q(i,t) \leq Qmax(i) * v(i,t) \quad \forall i \in I, \ \forall t \in T \quad (7.11)$$

$$Pmin(i) * v(i,t) \leq p(i,t) \leq Pmax(i) * v(i,t) \quad \forall i \in I, \ \forall t \in T \quad (7.12)$$

$$VOLdes(i) = M * \sum_{t \in T} q(i,t) \quad \forall i \in I \quad (7.13)$$

7.2.4.4 Computation of the Net Head

In this section, the net head for each reservoir in each time period is calculated. In Equation 7.14 the water level in the upstream reservoir of plant i for each time period t, $Nsup(i,t)$, is computed as a quadratic function of the stored water volume, whose parameters, $a0(i)$, $a1(i)$, and $a2(i)$, are assumed to be known. In Equation 7.15 the lower level, $Ninf(i,t)$ is estimated as a function of the elevation tail downstream of plant i, depending on the water discharge q during time period t. In Equation 7.16 the hydraulic losses, $per(i,t)$, are calculated, and in Equation 7.17 the net head is computed.

$$Nsup(i,t) = a0(i) + a1(i) * vol(i,t) + a2(i) * vol^2(i,t) \quad \forall i \in I, \ \forall t \in T \quad (7.14)$$

$$Ninf(i,t) = b_0 + b_1 * q(i,t) + b_2 * q(i,t)^2 \quad (7.15)$$

$$per(i,t) = C * q^2(i,t) \quad (7.16)$$

$$h(i,t) = Nsup(i,t) - Ninf(i,t) - per(i,t) \quad (7.17)$$

7.2.4.5 Scaling the Hill Diagram

It is assumed that each plant operates with a Francis turbine, or its equivalent, represented by a Hill diagram similar to the one in Figure 7.1, which, from this point, will be considered as associated with a standard turbine, whose values of net head and water discharge will be referred to as "hest" and "qest," respectively. However, it is considered that each turbine has its own design parameters, net head and water discharge, h and q, respectively, associated with its own Hill diagram. For this reason, a scaling of the Hill diagram associated with each of the individual turbines is required to match the measurement scale of the latter with the corresponding one of a standard turbine. This can be accomplished by a linear transformation of the data in each of the axes, net head and water discharge.

A linear transformation from the values $h(i,t)$ to their corresponding ones in the scale of the standard turbine, hest(i,t), is performed to scale the axis of the net head in each reservoir i, according to the linear function 7.18, where the slope, mhest(i), is calculated as in Equation 7.19. Using the first equation, and making $h(i,t) = \text{Hmin}(i)$, hest$(i,t) = \text{Hmin}(0)$ is obtained, verifying that the minimum net head of turbine i, Hmin(i) is associated with the corresponding minimal head of a standard turbine, Hmin(0). This correspondence is also valid for the maximum heads, where Hmax(i) is associated with Hmax(0), which can be verified by making $h(i,t) = \text{Hmax}(i)$ to find hest$(i,t) = \text{Hmax}(0)$, using both Equations 7.18 and 7.19.

$$\text{hest}(i,t) = \text{Hmin}(0) + \text{mhest}(i) * [h(i,t) - \text{Hmin}(i)] \quad \forall i \in I, \quad \forall t \in T \qquad (7.18)$$

$$\text{mhest}(i) = [\text{Hmax}(0) - \text{Hmin}(0)] / [\text{Hmax}(i) - \text{Hmin}(i)] \quad \forall i \in I \qquad (7.19)$$

Similar to scale the axis of the water discharge in each reservoir i, a transformation of the values $q(i,t)$ to their corresponding ones in the scale of the standard turbine qest(i,t) is performed by the function 7.20. The latter is a nonlinear function because of the presence of the $v(i,t)$ binary variables whose appearance will be explained below. The slope mqest(i) is calculated as in Equation 7.21. Likewise the net head case, we replace $q(i,t)$ by both its minimum and maximum values, Qmin(i) and Qmax(i), in turbine i. It is possible to verify that these values are associated with the corresponding minimum and maximum discharges of the standard turbine, Qmin(0) and Qmax(0), respectively. However, there is a problem when the turbine is off and, therefore, its discharge is zero. This change of scale for the discharge values works provided that the discharge is within the limits specified by the technical parameters of the turbine, Qmin(i) and Qmax(i), when the turbine is on; however, when the turbine is off, the discharge $q(i,t)$ is 0, making the factor $[q(i,t) - \text{Qmin}(i)]$ negative, as well as the product of this factor by the positive constant mqest(i). This product, subtracted from Qmin(0), would yield a value of qest(i,t) lower than the minimum discharge Qmin(0), which would cause an error for positive values of this latter parameter. To solve this difficulty, analyzing the constraints 7.11, the equivalence $q(i,t) = 0 \leftrightarrow v(i,t) = 0$ can be derived, showing the double implication as it is done below. If $q(i,t) = 0$, then Qmin$(i)*v(i,t) = 0$, considering the left side of the constraints 7.11, which implies that $v(i,t) = 0$, as Qmin(i) is a positive constant. On the other hand, if $v(i,t) = 0$, then, using the constraints 7.11, we obtain $0 \leq q(i,t) \leq 0$, which implies that $q(i,t) = 0$. Therefore, the water discharge to the standard turbine can be represented in general as shown in Equation 7.20, and it is valid in both cases, when the turbine is on, $v(i,t) = 1$, and when it is off, $v(i,t) = 0$.

$$\text{qest}(i,t) = \{\text{Qmin}(0) + \text{mqest}(i) * [q(i,t) - \text{Qmin}(i)]\} * v(i,t) \quad \forall i \in I, \quad \forall t \in T \quad (7.20)$$

$$\text{mqest}(i) = [\text{Qmax}(0) - \text{Qmin}(0)] / [\text{Qmax}(i) - \text{Qmin}(i)] \quad \forall i \in I \qquad (7.21)$$

7.2.4.6 Calculation of the Technical Efficiency of the Generation Units and the Power Output

The changes of scale in the net head and water discharge axes allow us to calculate the technical efficiency of the generating units using Equation 7.3, as shown in Equation 7.22, and the power output using Equation 7.1, as shown in Equation 7.23.

$$
\begin{aligned}
\mathrm{efi}(i,t) = \big\{ &\beta_0 + \beta_1 \mathrm{hest}(i,t) + \beta_2 \mathrm{qest}(i,t) + \beta_3 \mathrm{hest}^2(i,t) + \beta_4 \mathrm{qest}^2(i,t) \\
&+ \beta_5 \mathrm{hest}(i,t)^* \mathrm{qest}(i,t) \big\} * v(i,t) \quad \forall i \in I, \quad \forall t \in T
\end{aligned}
\tag{7.22}
$$

$$
p(i,t) = 0.00981 * h(i,t) * q(i,t) * \mathrm{efi}(i,t) \quad \forall i \in I, \quad \forall t \in T
\tag{7.23}
$$

7.2.4.7 Modeling of the Start-up and Shutdown of Units

The nonnegative variables used in the model are shown in Equations 7.24 and 7.25. The constraints 7.26 and 7.27 use binary variables as specified in Equation 7.28 in order to model the dynamics of the start-up and shutdown of the plants during the scheduling horizon (Brännlund et al., 1986). The $z(i,t)$ variables may seem superfluous, as they appear only in these two constraints. However, numerical simulations have shown their ability to significantly reduce the computation time (Conejo et al., 2002). The discharge, $q(i,t)$, standard discharge, $\mathrm{qest}(i,t)$, efficiency, $\mathrm{efi}(i,t)$, and power, $p(i,t)$ are equal to 0 if plant i is off (offline) in period t, that is, if $v(i,t) = 0$.

$$
\mathrm{efi}(i,t), h(i,t), \mathrm{hest}(i,t), p(i,t), q(i,t), \mathrm{qest}(i,t), \mathrm{ver}(i,t), \mathrm{vol}(i,t) \geq 0 \quad \forall i \in I, \forall t \in T
\tag{7.24}
$$

$$
\mathrm{COSenc}(i), \mathrm{INGope}(i), \mathrm{INGvfa}(i), \mathrm{VOLdes}(i) \geq 0 \, \forall i \in I
\tag{7.25}
$$

$$
y(i,t) - z(i,t) = v(i,t) - v(i,t-1) \quad \forall i \in I, \quad \forall t \in T
\tag{7.26}
$$

$$
y(i,t) + z(i,t) \leq 1 \quad \forall i \in I, \quad \forall t \in T
\tag{7.27}
$$

$$
v(i,t), y(i,t), z(i,t) \in \{0,1\} \quad \forall i \in I, \quad \forall t \in T
\tag{7.28}
$$

7.2.5 Measures of Risk: VaR and CVaR

VaR is a measure computed as the maximum profit value such that the probability of the profit being lower than or equal to this value is lower than or equal to $1 - \delta$:

$$
\mathrm{VaR} = \max\{x/p(B \leq x) \leq 1 - \delta\}
\tag{7.29}
$$

Usually, δ ranges between 0.9 and 0.99 (Conejo et al., 2008). In this study, δ is considered equal to 0.95.

CVaR is the expected profit not exceeding a measure ζ called VaR:

$$\text{CVaR} = E\left(B/B \leq \zeta\right) \tag{7.30}$$

Mathematically, CVaR can be defined as

$$\max \zeta - \frac{1}{1-\delta} \sum_{n=1}^{N} \rho_n \eta_n \tag{7.31}$$

subject to

$$-B_n + \zeta - \eta_n \leq 0 \tag{7.32}$$

$$\eta_n \geq 0 \tag{7.33}$$

In Equation 7.32, η_n is equal to zero if scenario n has a profit greater than ζ. For the remaining scenarios, η_n is equal to the difference between ζ and the corresponding profit.

For stochastic problems, VaR has the difficulty of requiring the use of binary variables for its modeling. Instead, CVaR computation does not require the use of binary variables and it can be modeled by the simple use of linear constraints. CVaR represents an appropriate approach to address risk management for a hydropower producer. However, previous MINLP approaches (Catalão et al., 2010; Díaz et al., 2011) did not consider risk management.

The concepts of VaR and CVaR are illustrated in Figure 7.3.

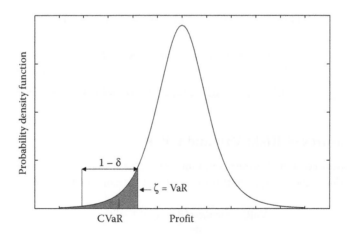

FIGURE 7.3 Value-at-Risk and conditional Value-at-Risk.

7.2.6 Objective Function: Profit Maximization and CVaR-Based Profit Maximization

The purpose of any H-GENCO in an electricity market is to maximize its own profit, calculated as the difference between its incomes and its total operating costs. The production costs are negligible for an H-GENCO. The most significant ones are the start-up costs of the units that are mainly caused by the additional maintenance required for the mechanical equipment and the malfunction of the control equipment. Thus, the profit maximization objective function can be expressed as in Equation 7.34, where the total profit of the system is the sum of the profits of each of its plants, and is composed of three terms: (i) the operating income, (ii) the income from the future value of the water stored in the reservoir at the end of the programming period as defined by a medium- or long-term model, and (iii) the start-up cost, according to the plant type (Nilsson and Sjelvgren, 1997a,b).

In Equation 7.35, the operating income of each plant belonging to the H-GENCO is calculated as the sum of the electricity prices multiplied by the power output in all time periods. In Equation 7.36, the income from the future water value is obtained as the volume stored in the last period multiplied by a constant that represents the future water value of reservoir i, VFA(i). In Equation 7.37, the start-up costs for the scheduling horizon are represented as the sum of the respective start-up costs for the time periods when the unit is on, that is, when $y(i,t) = 1$.

$$\text{Max}\,Z = \sum_{i \in I} [\text{INGope}(i) + \text{INGvfa}(i) - \text{COSenc}(i)] \qquad (7.34)$$

where

$$\text{INGope}(i) = \sum_{t \in T} [L(t) * p(i,t)] \quad \forall i \in I \qquad (7.35)$$

$$\text{INGvfa}(i) = \text{vol}(i,\text{NT}) * \text{VFA} \quad \forall i \in I \qquad (7.36)$$

$$\text{COSenc}(i) = \sum_{t \in T} [\text{SU}(i) * y(i,t)] \quad \forall i \in I \qquad (7.37)$$

In the short-term hydro scheduling problem under consideration, the objective function takes into account all the price scenarios at once weighted by their occurrence probability. The problem can be formulated to maximize

$$J = \sum_{n=1}^{N} \rho_n B_{n^+} \alpha \left[\zeta - \frac{1}{1-\delta} \sum_{n=1}^{N} \rho_n \eta_n \right] \qquad (7.38)$$

In Equation 7.38, ρ_n is the probability associated with scenario n, α is the positive weighting factor to achieve an appropriate trade-off between profit and risk, and ζ is the VaR at a confidence level of δ. B_n is the benefit for each price scenario, given by

$$B_n = \lambda_{tn} \sum_{i=1}^{I} P_{it} \qquad (7.39)$$

where λ_{tn} is the energy price for scenario n at period t, and P_{it} is the power generation of plant i during period t.

7.2.7 Strategy for Solving the Hydroelectric Operation Scheduling Problem

Some nonlinear phenomena are often approximated using simplifying assumptions to find their solution by the application of linear programming techniques or some of their extensions, such as separable linear programming, successive linear programming, or quadratic programming. However, these assumptions may be very far from the true behavior of the real system and, therefore, their results may be unreliable. In other cases, dynamic programming techniques, or some of their variants, are applied. Nonlinear programming techniques have great potential, although, in practice, they are rarely used.

The operation scheduling problem of an H-GENCO and its corresponding optimization model, such as the one formulated in this chapter, is a large-scale MINLP, as it incorporates a number of nonlinear constraints, including binary decision variables and a long time horizon. Due to the complexity of the model formulated, the existence of an efficient method to find a global solution may not be guaranteed. Thus, the goal is to find a local optimal solution of good quality, using MINLP, adapted to the structure of the problem at hand.

The feasible region of a mathematical programming problem can be expressed in terms of equality and inequality constraints, as presented in the models developed throughout this chapter. The equality constraints are considered associated with the water balance in each reservoir in each time period and to the definition of the volumes discharged during the scheduling horizon, VOLdes(i). Furthermore, the model includes equality constraints associated with some definitions for each reservoir in each time period, such as the upper and lower levels of water storage, Nsup(i,t) and Ninf(i,t), respectively, the head loss in the pipelines, per(i,t), the net head, $h(i,t)$, the discharge, $q(i,t)$, the technical efficiency, efi(i,t), the power produced, $p(i,t)$, and the equations defining the scaling variables of the turbines, hest(i,t), mhest(i,t), qest(i,t), and mqest(i,t). The inequality constraints are represented for each reservoir or hydroelectric plant associated and each time period by the operating limits for the water volume, vol(i,t), the net head, $h(i,t)$, the spillage, ver(i,t), the water discharge, $q(i,t)$, and the power generated, $p(i,t)$.

The model can be viewed in the context of MINLP where there are nonlinear functions, such as those used for calculating the water level in the reservoir upstream of the plant i, Nsup(i,t), in Equation 7.14, the level of the tail elevation at the discharge point, Ninf(i,t), in Equation 7.15, the hydraulic losses, per(i,t), in Equation 7.16, the standard discharge, qest(i,t), in Equation 7.20, the technical efficiency of the units, efi(i,t), in Equation 7.22, and the power output, $p(i,t)$, in Equation 7.23. In addition, there are binary variables in Equations 7.11 and 7.12 to set lower and upper limits for the water discharge, in Equation 7.20 for the power output in order to calculate the standard discharge, qest(i,t), in Equation 7.22 to estimate the technical efficiency, efi(i,t), and in Equations 7.26 and 7.27 to model the dynamics of start-up and shutdown of units. Binary variables appear in a linear function in Equation 7.20 and in a nonlinear one in Equation 7.22.

The model is implemented in GAMS (Brooke et al., 2000) and it is solved using the SBB solver for branch and bound methods and the CONOPT optimizer for nonlinear programming.

7.3 Case Study: Hydroelectric Operation Scheduling for a Cascaded Reservoir System Along the Duero River in Spain

The aim of this study, as stated in the introduction, is to obtain the optimal hydroelectric generation resulting from the energy offers to the day-ahead market. In this section, the results of a realistic case study are presented and discussed. The system comprises eight cascaded plants along the Duero river basin in Spain, whose data are shown in Table 7.1, as taken from Conejo et al. (2002) for a period of one week of the year 2008. The information and relevant assumptions are completed and updated, as presented below, to formulate and solve the model as an MINLP. The natural water inflow to each reservoir is assumed constant during the time horizon—in our case, 1 week divided into 168-h periods. The initial and final reservoir water volumes are known at the beginning of the study period. The latter ones, as well as the future water value, VFA, are usually obtained by medium- or long-term planning models (see Conejo et al., 2002).

The MINLP approach, which not only considers head-dependency and technical efficiency of the units, but also price uncertainty and risk management, has been applied to a case study based on one important Spanish cascaded hydro system. This approach has been developed and implemented in GAMS (Brooke et al., 2000) and solved using the

TABLE 7.1 Hydro System Data

Plant	Qmin(i) (m³/s)	Qmax(i) (m³/s)	VOLini(i) (Hm³)	VOLmin(i) (Hm³)	VOLmax(i) (Hm³)	W(i,t) (Hm³/h)	SU(i) (€)	Pmax(i) (MW)	VFA(i) (€/Hm³)
1	2	62	100	6	225	0.051	110	28.62	84.0
2	5	163	80	6	162	0.058	150	69.52	221.2
3	14	464	790	6	1200	0.603	200	139.05	630.0
4	19	662	33	6	66	0.051	250	116.38	900.2
5	18	628	13	6	26	0.051	350	186.66	854.0
6	14	479	1200	6	2586	0.199	1500	833.28	651.0
7	29	985	50	6	115	0.500	2000	1159.63	1338.4
8	30	1028	90	6	181	0.048	1000	550.90	1397.2

optimization solver packages SBB and CONOPT. Numerical simulation has been performed on a 600-MHz-based processor with 256 MB of RAM memory.

For the sake of simplicity, we develop a single scenario for a fixed water final volume, where, without loss of generality, we assume that the final water volume is equal to the initial one for each reservoir. Therefore, the production schedule is indifferent to the stored water future value, VFA(i), at the end of the period previously defined. A proportional change in all VFA(i) would not affect the production schedule (volume discharged, VOLdes, and operative income, INGope), although the revenue from future water value, INGvfa, and the profit could change. Figure 7.4 shows the spatial coupling between reservoirs.

In the studies by Catalão et al. (2010) and Díaz et al. (2011), the energy prices are considered as deterministic input data. Instead, several price scenarios that could be obtained from price forecasting are considered in our work, taking the actual prices of the last 20 weeks of 2010 of the Iberian electricity market from www.omel.es. The prices related to these 20-week scenarios are shown in Figure 7.5 over the 168-h horizon. The probability of occurrence of each generated scenario is, in this case, 1/20, this is 0.05.

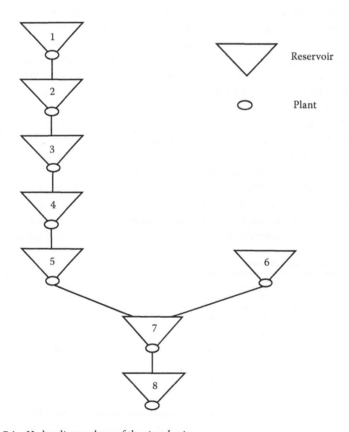

FIGURE 7.4 Hydraulic topology of the river basin.

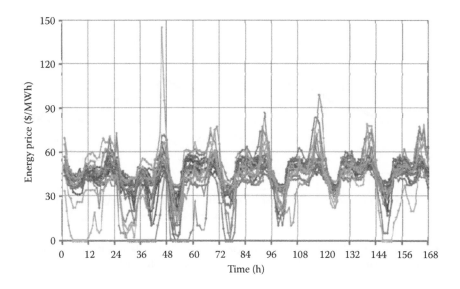

FIGURE 7.5 Energy price scenarios.

The storage targets for the short-term time horizon, which are established by medium-term planning studies, may be represented either by a penalty on water storage or by a previously determined future cost function (Uturbey and Simões Costa, 2007).

The expected profit versus profit standard deviation is presented in Figure 7.6, considering six values for α. This figure provides the maximum achievable expected profit for

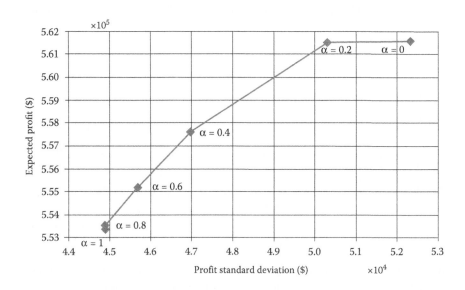

FIGURE 7.6 Expected profit versus profit standard deviation.

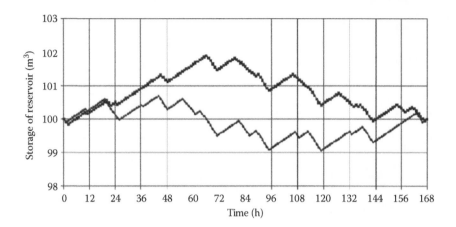

FIGURE 7.7 Optimal reservoir storage trajectories. The dark line denotes the results obtained using a risk level $\alpha = 0$, whereas the clear one denotes the results obtained using a risk level $\alpha = 1$.

each risk level or, alternatively, the minimum achievable risk level for each expected profit. An analysis of this figure, known as efficient frontier or Markowitz frontier, reveals that, for a risk-neutral producer ($\alpha = 0$), the expected profit is approximately $561,700 with a standard deviation of $52,300. On the other hand, a risk-averse producer ($\alpha = 1$) expects to achieve a lower profit of $553,200 with a lower standard deviation of $44,800.

For a detailed analysis we have selected plant 1, for which the water storage and water discharge are presented. The evolution of the reservoir volume associated with plant 1 and its discharge are shown in Figures 7.7 and 7.8, respectively. In particular,

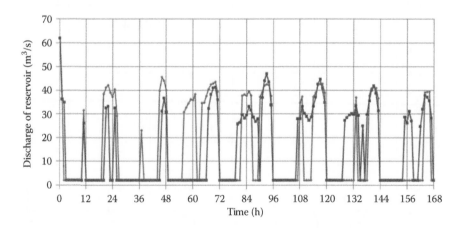

FIGURE 7.8 Optimal plant discharge trajectories. The dark line denotes the results obtained using a risk level $\alpha = 0$, whereas the clear one denotes the results obtained using a risk level $\alpha = 1$.

the optimal reservoir storage trajectories are shown in Figure 7.7 and the optimal plant discharge trajectories are shown in Figure 7.8. The dark lines denote the results obtained using a risk level $\alpha = 0$, while the clear ones denote the results obtained using a risk level $\alpha = 1$. Risk makes a different behavior possible, especially for the first reservoir, implying that, for a risk-neutral producer, the influence of the head change effect is more relevant.

The results in these two figures are consistent. The risk-neutral producer aims at discharging mostly during peak hours, obtaining maximum profit. Instead, by assuming higher values for the risk penalty factors, the number of online hours tends to decrease. Analyzing these figures, it can be verified that the risk level corresponding to $\alpha = 0$ implies a higher expected profit than for $\alpha = 1$. However, $\alpha = 0$ is riskier than $\alpha = 1$, because a financial loss can occur under some scenarios. Thus, a risk-averse investor would prefer $\alpha = 1$ because it gives almost the same expected profit level and exhibits lower financial risk. Hence, our model allows the decision-maker to obtain solutions according to the desired risk exposure level.

7.4 Conclusions

The main motivation of this paper is to provide a price-taker H-GENCO with a short-term scheduling tool in a pool-based electricity market. In this study, the technical efficiency of hydro generating units is a key component for the analysis and the evaluation of the variable head effects, the hydroelectric generating unit commitment, and the bidding design in the day-ahead market. The technical efficiency estimation is a quadratic function of the net head and water discharge resulting from a Hill diagram supplied by the turbine manufacturer. A general formulation of the unit performance curve as a continuous nonlinear and nonconcave function, incorporating the technical efficiency, is provided to overcome the inaccuracies of the discrete approaches.

The hydroelectric generating unit characteristics are modelled in detail in order to obtain a good approximation to the relation between the net head, the water discharge, and the technical efficiency. The mathematical functions formulated allow the adequate treatment of nonlinear and nonconcave unit performance curves. We have tested our models with case studies of the Spanish electric system related to a chain of reservoirs along the Duero river basin.

The main achievements associated to this work are listed below:

1. A systematic and complete characterization of the operation of hydropower systems in short-term competitive markets.
2. The estimation of the production function of a hydroelectric generating unit in a plant with a regulating reservoir based on the information available and considering that the efficiency of the turbine-generator set has a complex operational behavior due to the interdependency of several variables involved in the electricity generation process.
3. A realistic case study to test our methodology.

References

Alley, W. T. 1977. Hydroelectric plant capability curves, *IEEE Trans. Power App. Syst.*, 3, 999–1003.

Angelus, A. 2001. Electricity price forecasting in deregulated markets, *Electricity J.*, 14, 32–41.

Borghetti, A., D'Ambrosio, C., Lodi, A., and Martello, S. 2008. An MILP approach for short-term hydro scheduling and unit commitment with head-dependent reservoir, *IEEE Trans. Power Syst.*, 23(3), 1115–1124.

Brännlund, H., Bubenko, J. A., Sjelvgren, D., and Andersson, N. 1986. Optimal short term operation planning of a large hydrothermal power system based on a nonlinear network flow concept, *IEEE Trans. Power Syst.*, 1(4), 1154–1161.

Brooke, A., Kendrick, D., Meeraus, A., and Raman, R. 2000. *GAMS/Cplex 7.0 User Notes*. Washington, DC: GAMS Development Corp.

Campo, R. and Restrepo, P. 2004. Estudio de optimalidad del programa MPODE, *Mundo Eléctrico*, 18(54), 32–35.

Catalão, J. P. S., Mariano, S. J. P. S., Mendes, V. M. F., and Ferreira, L. A. F. M. 2006. Parameterisation effect on the behavior of a head-dependent hydro chain using a non-linear model. *Electr. Power Syst. Res.*, 76, 404–412.

Catalão, J. P. S., Mariano, S. J. P. S., Mendes, V. M. F., and Ferreira, L. A. F. M. 2009. Scheduling of head-sensitive cascaded hydro systems: A nonlinear approach, *IEEE Trans. Power Syst.*, 24(2), 337–346.

Catalão, J. P. S., Pousinho, H. M. I., and Mendes, V. M. F. 2010. Mixed-integer nonlinear approach for the optimal scheduling of a head-dependent hydro chain, *Electr. Power Syst. Res.*, 80(8), 935–942.

Chang, G. W., Aganagic, M. J., Waight, G., Medina, J., Burton, S., Reeves, T., and Christoforidis, M. 2001. Experiences with mixed integer linear programming based approach on short-term hydro scheduling, *IEEE Trans. Power Syst.*, 16(4), 743–749.

Cheung, K., Shamsollahi, P., Sun, P., Milligan, J., and Potishnak, M. 2000. Energy and ancillary service dispatch for the interim ISO New England electricity market, *IEEE Trans. Power Syst.*, 15(3), 968–974.

Conejo, A. J., Arroyo, J. M., Contreras, J., and Villamor, F. A. 2002. Scheduling of a hydro producer in a pool based electricity market, *IEEE Trans. Power Syst.*, 17(4), 1265–1272.

Conejo, A. J., Nogales, F. J., and Arroyo, J. M. 2002. Price-taker bidding strategy under price uncertainty, *IEEE Trans. Power Syst.*, 17(4), 1081–1088.

Conejo, A. J., García-Bertrand, R., Carrión, M., Caballero, A., and Andrés, A. 2008. Optimal involvement in futures markets of a power producer, *IEEE Trans. Power Syst.*, 23(2), 701–711.

Correa, C. A., Bolaños, R. A., and Ruiz, A. G. 2007. Methods of nonlinear interior point applied to hydrothermal dispatch problem, *Scientia Technica*, 13, 91–96.

De la Torre, S., Conejo, A. J., and Contreras, J. 2003. Simulating oligopolistic pool-based electricity markets: A multiperiod approach, *IEEE Trans. Power Syst.*, 18(4), 1547–1555.

Deng, S., Shen, Y., and Sun, H. 2004. Stochastic co-optimization for hydro-electric power generation, Power Systems Engineering Research Center (PSERC). Available at: http://www.pserc.wisc.edu/documents/publications/papers/2007_general_publications/

Dentcheva, D., Gollmer, R., Moller, A., Romisch, W., and Schultz, R. 1996. Solving the unit commitment problem in power generation by primal and dual methods, Report of the German Federal Ministry of Education, Science, Research and Technology (BMBF). Berlin: Institut fuer Mathematik, Humboldt-Universitaet.

Díaz, F. J. 2009. La eficiencia técnica como un nuevo criterio de optimización para la generación hidroeléctrica a corto plazo, *Dyna*, 76(157), 91–100.

Díaz, F. J., Contreras, J., Muñoz, J. I., and Pozo, D. 2011. Optimal scheduling of a price-taker cascaded reservoir system in a pool-based electricity market, *IEEE Trans. Power Syst.*, 26(2), 604–615.

Ferrer, A. B. 2004. Applicability of deterministic global optimization to the short-term hydro-thermal coordination problem, PhD Thesis, Universitat Politècnica de Catalunya, Barcelona, Spain.

Finardi, E. C. 2003. Alocação de unidad hidrelétricas em sistemas hidrotérmicos utili-zando relaxação Lagrangeana e programação quadrática sequencial, PhD Thesis, Universidade Federal de Santa Catarina, Florianópolis, Brazil.

Finardi, E. C. and da Silva, E. L. 2006. Solving the hydro unit commitment problem via dual decomposition and sequential quadratic programming, *IEEE Trans. Power Syst.*, 21(2), 835–843.

Fleten, S. and Pettersen, E. 2005. Constructing bidding curves for a price-taking retailer in the Norwegian electricity market, *IEEE Trans. Power Syst.*, 20(2), 701–778.

Fosso, O., Gjelsvik, A., Haugstad, A., Mo, B., and Wagensteen, I. 1999. Generation scheduling in a deregulated system. The Norwegian case, *IEEE Trans. Power Syst.*, 14(2), 75–81.

García-González, J., Parrilla, E., Barquín, J., Alonso, J., Saiz-Chicharro, A., and González, A. 2003.Under-relaxed iterative procedure for feasible short-term scheduling of a hydro chain, in *Proceedings of the IEEE Power Tech*, Bologna, Italy, June 23–26.

García-González, J., Parrilla, E., and Mateo, A. 2007. Risk-averse profit-based optimal scheduling of a hydro-chain in the day-ahead electricity market, *Eur. J. Operat. Res.*, 181(3), 1354–1369.

González, J. and Basagoiti, P. 1999. Spanish power exchange market and information sys-tem. Design, concepts, and operating experience, in *Proceedings of the 21st IEEE International Conference Power Industry Computer Applications*, Santa Clara, CA, USA, May 16–21.

Gross, G. and Finlay, D. 2000. Generation supply bidding in perfectly competitive electric-ity markets, *Comput. Math. Org. Theor.*, 6, 83–98.

Guan, X., Luh, P. B., and Zhang, L. 1995. Non-linear approximation method in Lagrangian relaxation-based algorithms for hydrothermal scheduling, *IEEE Trans. Power Syst.*, 10(2), 772–778.

Guan, X., Svoboda, A., and Li, C. 1999. Scheduling hydro power systems with operating restricted zones and discharge ramping constraints, *IEEE Trans. Power Syst.*, 14(1), 126–131.

Habibollahzadeh, H. and Bubenko, J. A. 1986. Applications of decomposition techniques to short-term operation planning of hydro thermal power systems, *IEEE Trans. Power Syst.*, 1(1), 41–47.

Hammons, T. J., Rudnick, H., and Barroso, L. A. 2002. Latin America: Deregulation in a hydro-dominated market, *Hydro Review Worldwide*, 10(4), 20–27.

Ladurantaye, D., Gendreau, M., and Potvin, J. 2007. Strategic bidding for price- taker hydroelectricity producers *IEEE Trans. Power Syst.*, 22(4), 2187–2203.

Legalov, D. T. and Palamarchuk, S. I. 2005. Hydro generation scheduling with electricity price calculation, in *Proceedings of the IEEE Power Tech*, St. Petersburg, Russia, June 27–30, Vol. 1, pp. 1–6.

Li, C., Johnson, R. B., and Svoboda, A. J. 1997b. A new unit commitment method, *IEEE Trans. Power Syst.*, 12(1), 113–119.

Li, C., Svoboda, A. J., Li, T. C., and Johnson, R. B. 1997a. Hydro unit commitment in hydro-thermal optimization, *IEEE Trans. Power Syst.*, 12(2), 764–769.

Nilsson, O. and Sjelvgren, D. 1995. Mixed-integer programming applied to short term planning of a hydro-thermal system, in *Proceedings of the IEEE Power Industry Computer Applications*, Salt Lake City, USA, pp. 158–163.

Nilsson, O. and Sjelvgren, D. 1997a. Variable splitting applied to modeling of start-up costs in short term hydro generation scheduling, *IEEE Trans. Power Syst.*, 12(2), 770–775.

Nilsson, O. and Sjelvgren, D. 1997b. Hydro unit start-up costs and their impact on the short term scheduling strategies of Swedish power producers, *IEEE Trans. Power Syst.*, 12(1), 38–44.

Nilsson, O., Söder, L. and Sjelvgren, D. 1998. Integer requirements modeling of spinning reserve in short term scheduling of hydro systems, *IEEE Trans. Power Syst.*, 13(3), 959–964.

Nogales, F. J., Contreras, J., Conejo, A. J., and Espínola, R. 2002. Forecasting next-day electricity prices by time series models, *IEEE Trans. Power Syst.*, 17(2), 342–348.

Parrilla, E. and García-González, J. 2006. Improving the B & B search for large-scale hydro-thermal scheduling weekly problems, *Int. J. Elect. Power Energy Syst.*, 28, 339–348.

Pereira, M. V. and Pinto, L. 1991. Multistage stochastic optimization applied to energy planning, *Math. Program.*, 52, 359–375.

Pereira, M. V., Campodónico, M., and Kelman, R. 1998. Long-term hydro scheduling based on stochastic models, in *Proceedings of EPSOM '98*, Zurich, Switzerland, September 23–25.

Pereira, M. V., Granville, S., Fampa, M. C., Dix, R., and Barroso, L. A. 2005. Strategic bidding under uncertainty: A binary expansion approach, *IEEE Trans. Power Syst.*, 20(1), 180–188.

Siu, T. K., Nash, G., and Shawwash, Z. K. 2001. A practical hydro dynamic unit commitment and loading model, *IEEE Trans. Power Syst.*, 16(2), 301–306.

Szkuta, B. R., Sanabria, L. A., and Dillon, T. S. 1999. Electricity price short-term forecasting using artificial neural networks, *IEEE Trans. Power Syst.*, 14(3), 851–857.

Uturbey, W. and Simões Costa, A. 2007. Dynamic optimal power flow approach to account for consumer response in short term hydrothermal coordination studies, *IET Gen. Trans. Dist.*, 1(3), 414–421.

Valenzuela, J. and Mazumdar, M. 2001. Probabilistic unit commitment under a deregulated market, Part II: New features in unit commitment models, in B. F. Hobbs, M. H. Rothkopf, R. P. O'Neill, and H.-p Chao (eds), *The Next Generation of Electric*

Power Unit Commitment Models, International Series in Operations Research & Management Science, Vol. 36. Dordrecht: Kluwer Academic Publishers, pp. 139–152.

Velásquez, J. M., Restrepo, P. J., and Campo, R. 1999. Dual dynamic programming. A note on implementation, *Water Resources*, 35, 34–44.

Wang, J., Yuan, X., and Zhang, Y. 2004. Short-term scheduling of large-scale hydropower systems for energy maximization, *J. Water Resour. Plng. Mgmt.*, 130(3), 198–205.

8

Hydrothermal Producer Self-Scheduling

Christos K. Simoglou

Pandelis N. Biskas

Anastasios
G. Bakirtzis

8.1 Introduction

Over the last 20 years the world's largest developing countries have experienced a restructuring process in their power sectors, with the target of improving fair competition and decreasing end-consumer prices. In this new environment, large volumes of electric energy are being traded in the day-ahead market. Energy producers offer energy and reserves on this market on the basis of their ability to produce energy/contribute to reserves for a specific period on the following day. The day-ahead markets, usually also called spot markets owing to the very short-term forward delivery, are organized as sealed-bid auctions, with voluntary or mandatory participation.

In a day-ahead energy auction, generating units submit supply offers and load representatives submit demand bids for all market products (energy and reserves) for each trading interval (hour or half-hour) of the next day to the system (or market) operator. Supply offers and demand bids can be either "priced," in the form of a set of price quantity (€/MWh–MWh) pairs, or "nonpriced," in the form of quantity (MWh), only offers or bids. The system operator processes supply offers and demand bids and computes the market clearing price (MCP) for all products as well as the trading volumes. The objective of the optimization is the maximization of the social welfare (or equivalently the minimization of the total production cost minus the consumer utility) along with the satisfaction of the system- and/or zonal-wide energy demand and reserves requirements, unit operating constraints, and system interzonal constraints.

On the other hand, an electricity producer faces everyday the fundamental problem of self-scheduling his own thermal and hydropower generating units so as to maximize his total profits from his participation in the day-ahead energy and reserves markets. The solution of the producer self-scheduling problem provides the desired commitment program as well as the energy and reserve contribution of each producer's generating unit, for all periods of the next day. Once the optimal self-schedule is computed, the electricity producer can design a suitable bidding strategy, such as those presented in References [1,2], that will indirectly lead the ISO to issue dispatch schedules close to the producer optimal self-schedules.

Many different optimization methods have been proposed in the literature for the solution of the above problems, such as dynamic programming [3], Lagrangian relaxation [4,5], network flow programming [6], mixed integer programming [7], and metaheuristic techniques [8,9]. An extensive review of the literature for solving hydrothermal coordination and self-scheduling problems can be found in Reference [10]. Considering thermal generating units only, indicative practical mixed-integer linear programming (MILP)-based approaches that are suitable for both traditional (conventional unit commitment) and competitive (self-scheduling) environments are given in References [11–16]. The short-term scheduling of hydro units is particularly addressed in References [17–22], whereas the combination of thermal and hydro subsystems in a single portfolio for a generation company is presented in References [23,24].

In an uncertain market environment, several models have been presented for the solution of the producer self-scheduling problem and the development of optimal offering strategies in day-ahead electricity markets under a stochastic programming framework [25–29]. In these models the producer is assumed to be risk-neutral, as the objective is the maximization of the expected profits, without including any specific measure accounting for risk. The risk effect under different measures (variance, downside risk, Conditional Value-at-Risk—CVaR, etc.) is modeled in References [30–35], whereas a detailed overview of risk assessment methods on energy trading is given in Reference [36].

In this chapter, a hydrothermal electricity producer owning thermal and hydropower generating units as well as pumped storage plants (henceforth, "Producer") and participating in the day-ahead energy and reserves markets is considered. The Producer self-scheduling problem is a profit maximization problem, which is formulated and solved as an MILP, considering the co-optimization of the Producer participation in the energy as well as primary (up/down), secondary (up/down), and tertiary (spinning/nonspinning) reserves markets in a pool-based framework. The mathematical formulation provided is suitable for the self-scheduling of a hydrothermal Producer acting either as a price-taker or as a price-maker in the day-ahead energy and reserves markets. Forward bilateral contracts with end-consumers are considered and their effect on the Producer self-schedule and profits is examined. Uncertainty of market conditions is also modeled within a two-stage stochastic programming framework, while the CVaR metric accounting for risk management is also incorporated. Postprocessing techniques are applied for the construction of the generating units optimal offer curves for the day-ahead electricity market.

The objective and the mathematical formulation of the hydrothermal Producer self-scheduling problem in both cases (price-taker/price-maker) are further described in the following sections. Initially, the deterministic and, subsequently, the stochastic hydrothermal Producer self-scheduling problems are described. For the reader's convenience, the explanation of all symbols that appear in the following sections for the detailed mathematical formulation of the optimization problem and are not defined in the text is given in Section 8.7.

The hydrothermal Producer self-scheduling problem addressed in this chapter is formulated as a constrained optimization problem and its solution can be found through high-level commercially available software such as GAMS [37], which allows for a compact and precise representation of large-scale and complex optimization problems. Such a computational environment also allows for the use of state-of-the-art solvers such as CPLEX, which is a high-performance solver suitable for linear and mixed-integer linear programming. GAMS/CPLEX is adopted for the modeling and solution of the present optimization problem.

8.2 Objective of the Deterministic Hydrothermal Producer Self-Scheduling Problem

8.2.1 Price-Taker Producer Objective

First, we consider a price-taker hydrothermal Producer, that is, a producer that has no capability of altering the MCPs of market products (energy and reserves). Although hourly clearing prices depend on the bidding behavior of all market participants, they are assumed to be known and used as input parameters in the Producer's optimization problem. Appropriate forecasting methods such as artificial neural networks, time-series analysis, and so on, can be used to predict the hourly clearing prices of all market products for the next day [38–43].

In the case of a price-taker Producer, the self-scheduling problem is modeled as a mathematical optimization problem, as follows:

$$Max\left\{\sum_{i\in I}\sum_{t\in T}\left[\lambda_t^E\cdot p_{it}+\sum_{m\in M}\left(\lambda_t^m\cdot r_t^m\right)-\sum_{i\in I}c_{it}\left(p_{it}\right)\right]\right\} \tag{8.1}$$

subject to

$$\mathbf{x}_{it}=\left[p_{it},u_{it},y_{it},z_{it}....\right]\in\Pi_i \quad \forall i\in I, \quad t\in T \tag{8.2}$$

where, λ_t^{pr} is the forecast clearing price of market product pr (pr = E: energy, pr = m: reserves) in hour t (in €/MWh), p_{it} is the power output of generating unit i accepted by the ISO in hour t in the day-ahead energy auction (in MW), r_{it}^m is the contribution of unit i in reserve type m during hour t (in MW), and $c_{it}\,(p_{it})$ is the total production cost of thermal unit i in hour t at level p_{it} (in €/h).

The Producer profit during the planning period, represented in Equation 8.1, is the Producer revenue from the energy market and the various reserves markets minus the Producer units' total operating cost. For the sake of simplicity, the cost of hydro units and the cost of the generating units for providing reserves are neglected.

As the MCPs are fixed, the market revenue—the product of the clearing price times the Producer's energy production/reserves contribution—is a linear function of the Producer's energy production/reserves contribution, respectively, which are the main decision variables in this approach.

Constraints 8.2 represent the operating constraints of the Producer units, that is, start-up and shutdown procedures, minimum up/down times, minimum/maximum power output restrictions, ramp-rate limits, fuel limitations, and so on. These constraints can be expressed as linear constraints on the continuous and the binary decision variables associated with the operation of the unit, denoted by the vector decision variable $\mathbf{x}_{it}$.

All constraints related to the operation of the Producer's thermal and hydro generating units are described in detail in Sections 8.3.2 and 8.3.4, respectively.

8.2.2 Price-Maker Producer Objective

In the case that the Producer acts as a price-maker (Stackelberg monopolist) in all day-ahead markets (energy and reserves), his daily market revenues are no longer calculated on exogenous input parameters (e.g., MCPs) but the influence of his decisions on the resulting MCPs is appropriately taken into account. The Producer's strategic behavior is explicitly modeled through the use of residual demand curves for all market products, also known as price-quota curves. The residual demand curve is a stepwise monotonically nonincreasing function that expresses how the MCP of each market product (energy or reserves) changes as the Producer's quota (total accepted quantity by the ISO) for this product changes. The residual demand curve is further described in Section 8.2.2.1.

In the case of a price-maker Producer, the aim of the Producer is to compute not only his quota and the corresponding energy and reserves clearing prices (from the respective hourly residual demand curves) but also the commitment scheduling, that is, the desired contribution of each unit in energy and reserves, so that the total daily profit from his participation in the day-ahead energy and reserves markets is maximized, as in the case of the price-taker Producer.

The objective function of the price-maker Producer self-scheduling problem is formulated as follows:

$$Max \left\{ \sum_{t \in T} \left[\pi_t^E \left(\xi_t^E \right) \cdot \xi_t^E + \sum_{m \in M} \pi_t^m \left(\xi_t^m \right) \cdot \xi_t^m - \sum_{i \in I} c_{it} \left(p_{it} \right) \right] \right\} \tag{8.3}$$

where $\pi_t^{pr} \left(\xi_t^{pr} \right)$ is the residual demand curve of market product pr (energy and reserves) in hour t, expressed as a stepwise monotonically nonincreasing function that represents the MCP as a function of the Producer quota, ξ_t^{pr}.

Owing to the products of variables in Equation 8.3, the optimization problem is nonlinear. An alternative linear formulation requires additional continuous and binary decision variables as well as the incorporation of additional constraints 8.5 through 8.9 presented in Section 8.2.2.1 in the model, so that the nonlinear objective function 8.3 is transformed to the equivalent linear expression given by

$$Max \left\{ \sum_{t \in T} \left[\sum_{b \in B^E} \left[\pi_{bt}^E \cdot \left(d_{bt}^E + w_{bt}^E \cdot D_{bt}^{E,\min} \right) \right] \right. \right.$$
$$\left. \left. + \sum_{m \in M} \sum_{b \in B^m} \left[\pi_{bt}^m \cdot \left(d_{bt}^m + w_{bt}^m \cdot D_{bt}^{m,\min} \right) \right] - \sum_{i \in I} c_{it} \left(p_{ito} \right) \right] \right\} \tag{8.4}$$

Similarly with the previous case, the hydrothermal Producer's profit during the scheduling period represented in Equation 8.3, Equation 8.4 is computed as the difference between its revenue from the day-ahead energy and reserves markets minus the Producer units' total operating cost. Constraints 8.2, denoting the Producer units' feasible operating region, must also be included in the optimization problem, as already mentioned in Section 8.2.1.

8.2.2.1 Residual Demand Curve

In a given hour, a Producer acting as a price-maker in all market products (energy and reserves) contributes a specific amount of energy and reserves in order to serve the load demand and the corresponding system reserve requirements, though he is capable of altering the respective MCPs to his own benefit. The amount of energy or reserves that the Producer desires to be accepted by the market operator in the day-ahead auction is called quota. The curve that expresses how the MCP of a market product changes as the Producer quota of this product changes is called price-quota curve or residual demand curve [16]. Each residual demand curve is formed as the difference of the aggregated demand curve for each market product and the aggregated competitors' offer curve for the same product. A typical residual demand curve for energy is shown in Figure 8.1.

The residual demand curve is formed as a stepwise monotonically nonincreasing function of the Producer's quota. The hourly residual demand curves that the Producer faces can be estimated based on forecasts of the system load demand/reserve requirements and the competitors' energy/reserve offers for every hour of the scheduling horizon [2]. The mathematical formulation of the price-taker Producer self-scheduling problem is a special case of the price-maker Producer problem. In this special case, the residual demand curve of any hour of the scheduling horizon consists of one step only, with offer quantity equal to the total load demand/reserve requirement and offer price equal to the forecast product MCP.

Figure 8.2 shows the variables and parameters needed for the linearization of the price-maker Producer revenue as a function of its quota of a given market product (energy/reserves). The stepwise nature of the hourly residual demand curve allows for the linear formulation of the Producer revenue using the continuous variables d_{bt} and the binary variables w_{bt}. The gray shaded area denotes the Producer revenue for a given market product. Constraints 8.5 through 8.9 that follow must be incorporated to the self-scheduling problem formulation, as they are necessary for the linearization of the problem objective 8.4.

$$\xi_t^E = \sum_{i \in I} P_{it} - \sum_{h \in H} p_{ht}^A \quad \forall t \in T \tag{8.5}$$

$$\xi_t^m = \sum_{i \in I} r_{it}^m \quad \forall t \in T, \ m \in M \tag{8.6}$$

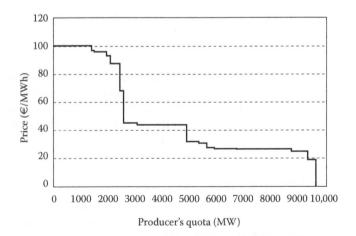

FIGURE 8.1   Residual demand curve for energy.

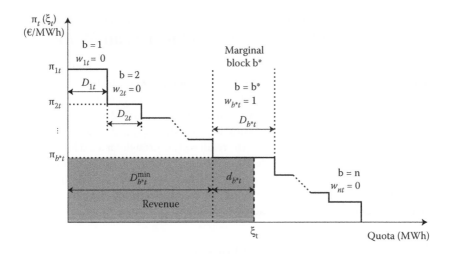

FIGURE 8.2 Linear formulation of the residual demand curve.

$$\xi_t^{pr} = \sum_{b \in B^{pr}} \left(d_{bt}^{pr} + w_{bt}^{pr} \cdot D_{bt}^{pr,min} \right) \quad \forall t \in T, \; pr \in PR \tag{8.7}$$

$$0 \leq d_{bt}^{pr} \leq w_{bt}^{pr} \cdot D_{bt}^{pr} \quad \forall b \in B^{pr}, \; t \in T, \; pr \in PR \tag{8.8}$$

$$\sum_{b \in B^{pr}} w_{bt}^{pr} = 1 \quad \forall t \in T, \; pr \in PR \tag{8.9}$$

Equation 8.5 defines the Producer energy quota, ξ_t^E, in hour t, which is equal to the summation of the power output of thermal units and hydroplants minus the pumping load of pumped storage plants, and it is required in order to compute the energy MCP from the residual demand curve [16], $\pi_t^E(\xi_t^E)$, used in Equation 8.3 to compute the Producer revenue.

Similarly, Equation 8.6 defines the Producer quota of reserve type m, ξ_t^m, in hour t, which is equal to the summation of the contribution of thermal units and hydroplants in reserve type m, and it is required so as to compute the reserve type m MCP from the residual demand curve, $\pi_t^m(\xi_t^m)$, used in Equation 8.3 to compute the Producer revenue. Given that tertiary reserve can be provided by units either as spinning (3S) or nonspinning (3NS) reserve, the following equation must be added to the model:

$$r_{it}^3 = r_{it}^{3S} + r_{it}^{3NS} \quad \forall i \in I, \; t \in T \tag{8.10}$$

Constraints 8.7 express the linear formulation of the Producer quota of each market product pr (energy and reserves) in every hour t as a function of variables d_{bt}^{pr} and w_{bt}^{pr}. Constraints 8.8 denote that the blocks of the price quota curve of product pr accepted by the ISO in every hour are upper-bounded positive values. Constraints 8.8 and 8.9 together state that only one variable d_{bt}^{pr} is different from zero in every hour.

8.3 Formulation of the Deterministic Hydrothermal Producer Self-Scheduling Problem as an MILP

In this section, the deterministic hydrothermal Producer self-scheduling problem is formulated as an MILP. A detailed modeling of the operating constraints of all generating units is presented separately for the thermal and hydro subsystems in Sections 8.3.1 through 8.3.4, respectively. In Section 8.3.6, an extension of the hydrothermal Producer self-scheduling problem formulation is provided, in which the hydrothermal Producer is considered to play a dominant role in the retail sector through forward bilateral contracts with end-consumers. The modeling of the uncertainties of the day-ahead market conditions is deferred to Section 8.4, where a two-stage stochastic programming model is formulated for the efficient self-scheduling of the Producer and the development of optimal offer curves for the day-ahead energy market.

8.3.1 Thermal Unit Operating Phases

Figure 8.3 shows the different operating phases of a thermal unit. After being reserved ($u_{it} = 0$) for T_i^{off} hours ($T_i^{off} \geq DT_i$), the unit starts up at hour t_1 ($y_{it} = 1$) and remains committed ($u_{it} = 1$) until it is shut down at hour t_5 ($z_{it} = 1$). Once committed, the unit follows four consecutive operation phases: (i) synchronization, (ii) soak, (iii) dispatchable, and (iv) desynchronization, denoted by binary variables u_{it}^{syn}, u_{it}^{soak}, u_{it}^{disp}, and u_{it}^{des}, respectively (Figure 8.3). The first two phases comprise the unit start-up phase. During the dispatchable phase, the unit can receive dispatch instructions to vary its power output between its technical minimum and its nominal power output according to its ramp-rate limits, and contributes to the system reserves. During the desynchronization phase, the unit power output follows a predefined shutdown sequence.

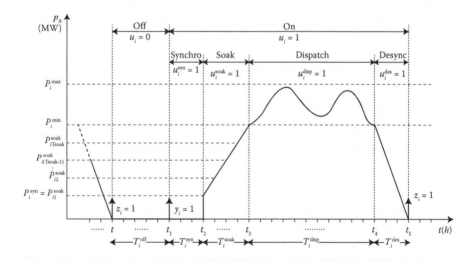

FIGURE 8.3 Operating phases of a thermal unit.

TABLE 8.1 Thermal Unit Start-up Modeling

Start-up Type	Prior Reservation Time (h) $\underline{T}_i^\ell \leq T_i^{off} < \bar{T}_i^\ell$	Synchronization Time (h)	Soak Time (h)	Start-up Cost (€)
Hot	$0 \leq T_i^{off} < T_i^w$	$T_i^{syn,h}$	$T_i^{soak,h}$	SUC_i^h
Warm	$T_i^w \leq T_i^{off} < T_i^c$	$T_i^{syn,w}$	$T_i^{soak,w}$	SUC_i^w
Cold	$T_i^c \leq T_i^{off} < T^-$	$T_i^{syn,c}$	$T_i^{soak,c}$	SUC_i^c

The accurate modeling of the unit start-up sequence requires special attention. Three start-up types are modeled, $\ell \in L = \{h, w, c\}$ (h: hot, w: warm, and c: cold), each with distinct synchronization time, soak time, and start-up cost, as shown in Table 8.1. Both $T_i^{syn,\ell}$ and $T_i^{soak,\ell}$ depend on the thermal unit's prior reservation time T_i^{off}. In this Table, T^- represents a large number of hours in the past (larger than the maximum reservation time to cold start all units) and is further explained in Appendix A8.1.

The thermal unit start-up sequence consists of two phases: (i) synchronization phase and (ii) soak phase.

Once a type ℓ start-up decision is taken, $y_{it}^\ell = 1$ (see Figure 8.3), the thermal unit enters the synchronization phase that lasts for $T_i^{syn,\ell}$ hours and during which the power output of the unit is 0 MW. Subsequently, the unit enters the soak phase that lasts for $T_i^{soak,\ell}$ hours and during which the unit operates between the synchronization load, P_i^{syn}, and the technical minimum power output, P_i^{min}. A detailed description of the unit's soak phase ramp-up sequence is given in Section 8.3.2.3.

Once a thermal unit enters a hot, warm, or cold start-up phase, it should complete the start-up sequence, and enter the dispatchable phase for at least 1 h, or as long as needed to satisfy the minimum up time requirement before shutting down.

8.3.2 Thermal Unit Constraints

8.3.2.1 Start-Up-Type Constraints

$$y_{it}^\ell \leq \sum_{\tau=t-\bar{T}_i^\ell+1}^{t-\underline{T}_i^\ell} z_{i\tau} \quad \forall i \in I, \ t \in T, \ \ell \in L, \quad \text{where } \tau \in T^- \tag{8.11}$$

$$y_{it} = \sum_{\ell \in L} y_{it}^\ell \quad \forall i \in I, \ t \in T \tag{8.12}$$

Constraints 8.11 select the correct start-up type of the ith unit, depending on the unit's prior reservation time as described by the first two columns of Table 8.1. This is achieved in Equation 8.11 by constraining the type ℓ start-up variable of unit i during hour t, y_{it}^ℓ, to be zero, unless there was a prior shutdown of the unit in the time interval

$(t - \overline{T}_i^\ell, t - \underline{T}_i^\ell]$. Constraints 8.12 ensure that only one start-up type per start-up is selected. Planning horizon extended to the past, T^-, is further explained in Appendix A8.1.

8.3.2.2 Synchronization Phase Constraints

$$u_{it}^{\text{syn},\ell} = \sum_{\tau=t-T_i^{\text{syn},\ell}+1}^{t} y_{i\tau}^\ell \quad \forall i \in I, \ t \in T, \ \ell \in L, \ \text{where } \tau \in T^- \tag{8.13}$$

$$u_{it}^{\text{syn}} = \sum_{\ell \in L} u_{it}^{\text{syn},\ell} \quad \forall i \in I, \ t \in T \tag{8.14}$$

Constraints 8.13 ensure that unit i enters the synchronization phase immediately following start-up (see Figure 8.3). The duration of the synchronization phase, $T_i^{\text{syn},\ell}$, depends on the start-up type, ℓ. This is achieved in Equation 8.13 by turning on the type ℓ synchronization phase binary variable, $u_{it}^{\text{syn},\ell}$, whenever there is a type ℓ start-up of the unit in the prior $T_i^{\text{syn},\ell}$ hours. Constraints 8.14 ensure that only one synchronization type per start-up is selected.

8.3.2.3 Soak Phase Constraints

$$u_{it}^{\text{soak},\ell} = \sum_{\tau=t-T_i^{\text{syn},\ell}-T_i^{\text{soak},\ell}+1}^{t-T_i^{\text{syn},\ell}} y_{i\tau}^\ell \quad \forall i \in I, \ t \in T, \ \ell \in L, \ \text{where } \tau \in T^- \tag{8.15}$$

$$u_{it}^{\text{soak}} = \sum_{\ell \in L} u_{it}^{\text{soak},\ell} \quad \forall i \in I, \ t \in T \tag{8.16}$$

Constraints 8.15 ensure that thermal unit i should enter a soak phase following its synchronization (see Figure 8.3). The duration of the soak phase depends on the unit i start-up type (hot, warm, cold). Constraints 8.16 ensure that only one soak phase type per start-up is selected.

In general, during the soak phase, the power output of a thermal unit follows a predefined sequence of megawatt values, depending on the start-up type, ℓ, as shown in Figure 8.4:

$$\left\{ P_{is}^{\text{soak},\ell}, s = 1, ..., T_i^{\text{soak},\ell} \right\} \tag{8.17}$$

Thus, during the soak phase, the power output of the unit is constrained by

$$p_{it}^{\text{soak}} = \sum_{\ell \in L} \sum_{\tau=t-T_i^{\text{syn},\ell}-T_i^{\text{soak},\ell}+1}^{t-T_i^{\text{syn},\ell}} y_{i\tau}^\ell \cdot P_{i,t-T_i^{\text{syn}}-\tau+1}^{\text{soak},\ell} \quad \forall i \in I, \ t \in T, \ \ell \in L, \ \text{where } \tau \in T^- \tag{8.18}$$

Two special cases are considered.

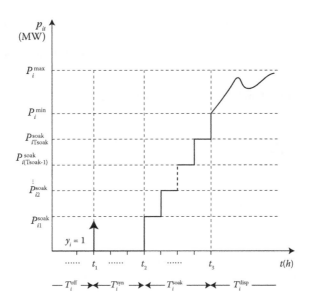

FIGURE 8.4 Soak phase of a thermal unit with predefined sequence of megawatt values.

In the first case, during the soak phase, the thermal unit power output increases linearly from its synchronization load, P_i^{syn}, to the unit technical minimum, $P_i^{\min}$ (Figure 8.3). In this case, the soak phase megawatt sequence of Equation 8.17 is defined by

$$P_{is}^{\text{soak},\ell} = P_i^{\text{syn}} + (s-1) \cdot \frac{P_i^{\min} - P_i^{\text{syn}}}{T_i^{\text{soak},\ell}}, \quad s = 1, ..., T_i^{\text{soak},\ell} \tag{8.19}$$

In the second case, the thermal unit power output during the soak phase is fixed to the constant value P_i^{soak} (in MW). In this case, P_i^{soak} can be factored out of the second sum of Equation 8.18, and using Equation 8.15, Equation 8.18 is simplified to

$$p_{it}^{\text{soak}} = \sum_{\ell \in L} \bar{P}_i^{\text{soak}} \cdot u_{it}^{\text{soak},\ell} \quad \forall i \in I, \quad t \in T \tag{8.20}$$

8.3.2.4 Desynchronization Phase Constraints

$$u_{it}^{\text{des}} = \sum_{\tau=t+1}^{t+T_i^{\text{des}}-1} z_{i\tau} \quad \forall i \in I, \quad t \in T \tag{8.21}$$

$$p_{it}^{\text{des}} = \sum_{\tau=t}^{t+T_i^{\text{des}}-1} z_{i\tau} \cdot (\tau - t) \cdot \frac{P_i^{\min}}{T_i^{\text{des}}} \quad \forall i \in I, \quad t \in T \tag{8.22}$$

where $\tau \in T^+$. Planning horizon extended to the future, T^+, is further explained in Appendix A8.1.

Constraints 8.21 ensure that thermal unit i should operate in a desynchronization phase lasting T_i^{des} hours before its shutdown (see Figure 8.3). The thermal unit power output during the desynchronization process decreases linearly from its technical minimum power output to 0 MW (see Equation 8.22).

8.3.2.5 Minimum Up/Down Time Constraints

$$\sum_{\tau=t-UT_i+1}^{t} y_{i\tau} \leq u_{it} \quad \forall i \in I,\ t \in T,\ \ell \in L, \quad \text{where } \tau \in T^- \tag{8.23}$$

$$\sum_{\tau=t-DT_i+1}^{t} z_{i\tau} \leq 1 - u_{it} \quad \forall i \in I,\ t \in T,\ \ell \in L, \quad \text{where } \tau \in T^- \tag{8.24}$$

Constraints 8.23 and 8.24 enforce the minimum up/down time constraints, respectively; that is, unit i must remain committed (decommitted) at hour t if its start-up (shutdown) occurred during the previous $UT_i - 1$ ($DT_i - 1$) hours [44].

8.3.2.6 Logical Status of Commitment

$$u_{it} = u_{it}^{\text{syn}} + u_{it}^{\text{soak}} + u_{it}^{\text{disp}} + u_{it}^{\text{des}} \quad \forall i \in I,\ t \in T \tag{8.25}$$

$$y_{it} - z_{it} = u_{it} - u_{i(t-1)} \quad \forall i \in I,\ t \in T \tag{8.26}$$

$$y_{it} + z_{it} \leq 1 \quad \forall i \in I,\ t \in T \tag{8.27}$$

Constraints 8.25 ensure that only one at most of the binary variables corresponding to the different commitment states of unit i can equal 1 in every hour. Constraints 8.26 model the logic of the start-up and shutdown statuses change. Constraints 8.27 require that unit i may not be started up and shut down simultaneously in a given hour.

8.3.2.7 Power Output Constraints

$$u_{it}^{\text{AGC}} \leq u_{it}^{\text{disp}} \quad \forall i \in I,\ t \in T \tag{8.28}$$

$$0 \leq r_{it}^{1+} \leq R_i^1 \cdot u_{it}^{\text{disp}} \quad \forall i \in I,\ t \in T \tag{8.29}$$

$$0 \leq r_{it}^{1-} \leq R_i^1 \cdot u_{it}^{\text{disp}} \quad \forall i \in I,\ t \in T \tag{8.30}$$

$$0 \leq r_{it}^{2+} \leq 15 \cdot RU_i^{\text{AGC}} \cdot u_{it}^{\text{AGC}} \quad \forall i \in I,\ t \in T \tag{8.31}$$

$$0 \leq r_{it}^{2-} \leq 15 \cdot RD_i^{AGC} \cdot u_{it}^{AGC} \quad \forall i \in I, \quad t \in T \tag{8.32}$$

$$0 \leq r_{it}^{3S} \leq R_i^{3S} \cdot u_{it}^{\text{disp}} \quad \forall i \in I, \quad t \in T \tag{8.33}$$

$$r_{it}^{3NS} \leq R_i^{3NS} \cdot u_{it}^{3NS} \quad \forall i \in I, \quad t \in T \tag{8.34}$$

$$r_{it}^{3NS} \geq P_i^{\min} \cdot u_{it}^{3NS} \quad \forall i \in I, \quad t \in T \tag{8.35}$$

$$u_{it}^{3NS} \leq 1 - u_{it} \quad \forall i \in I, \quad t \in T \tag{8.36}$$

$$p_{it} - r_{it}^{2-} \geq 0 \cdot u_{it}^{\text{syn}} + p_{it}^{\text{soak}} + p_{it}^{\text{des}} + P_i^{\min} \cdot \left(u_{it}^{\text{disp}} - u_{it}^{AGC} \right)$$
$$+ P_i^{\min,AGC} \cdot u_{it}^{AGC} \quad \forall i \in I, \, t \in T \tag{8.37}$$

$$p_{it} + r_{it}^{2+} \leq 0 \cdot u_{it}^{\text{syn}} + p_{it}^{\text{soak}} + p_{it}^{\text{des}} + P_i^{\max} \cdot \left(u_{it}^{\text{disp}} - u_{it}^{AGC} \right)$$
$$+ P_i^{\max,AGC} \cdot u_{it}^{AGC} \quad \forall i \in I, \, t \in T \tag{8.38}$$

$$p_{it} - r_{it}^{1-} - r_{it}^{2-} \geq 0 \cdot u_{it}^{\text{syn}} + p_{it}^{\text{soak}} + p_{it}^{\text{des}} + P_i^{\min} \cdot u_{it}^{\text{disp}} \quad \forall i \in I, \, t \in T \tag{8.39}$$

$$p_{it} + r_{it}^{1} + r_{it}^{2+} + r_{it}^{3S} \leq 0 \cdot u_{it}^{\text{syn}} + p_{it}^{\text{soak}} + p_{it}^{\text{des}} + P_i^{\max} \cdot u_{it}^{\text{disp}}$$
$$+ \left(P_i^{\min} - P_i^{\max} \right) \cdot z_{i\left(t + T_i^{\text{des}}\right)} \quad \forall i \in I, \, t \in T \tag{8.40}$$

where $t + T_i^{\text{des}} \in T^+$.

Constraints 8.28 state that unit i may provide automatic generation control (AGC) if and only if it is on dispatch. Constraints 8.29 through 8.34 set the upper limits of primary up/down, secondary up/down, and tertiary spinning/nonspinning reserves, respectively. As shown by constraints 8.28 through 8.33, unit i may contribute in synchronized reserves if and only if it is dispatchable, whereas during the synchronization, soak, and desynchronization phases the contribution of the synchronized reserves is equal to zero. Constraints 8.35 enforce the tertiary nonspinning reserve contribution to be greater than the minimum power output of unit i. Constraints 8.36 state that unit i may provide tertiary nonspinning reserve if and only if it is offline.

Constraints 8.37 through 8.40 define the limits of the power output of thermal unit i in every commitment state. The first three terms on the right-hand side of Equations 8.37 through 8.40 constrain the output of the unit during synchronization, soak, and desynchronization. If unit i is on synchronization in hour t (i.e., $u_{it}^{\text{syn}} = 1$), the power output will be equal to 0, whereas the terms p_{it}^{soak} and p_{it}^{des} are defined in Equations 8.18 or 8.20 and 8.22, respectively. The last term of the right-hand side of Equation 8.40 ensures that the unit will operate at its technical minimum power output the hour before entering the desynchronization phase; this term must be omitted for fast-start units.

8.3.2.8 Ramp-Up/Down Constraints

$$p_{it} - p_{i(t-1)} \le RU_i \cdot 60 + N \cdot \left(u_{it}^{\text{syn}} + u_{it}^{\text{soak}}\right) \quad \forall i \in I,\ t \in T \tag{8.41}$$

$$p_{i(t-1)} - p_{it} \le RD_i \cdot 60 + N \cdot \left(z_{it} + u_{it}^{\text{des}}\right) \quad \forall i \in I,\ t \in T \tag{8.42}$$

Constraints 8.41 and 8.42 introduce the effect of ramp-rate limits on the power output. Note that N is a large constant, so that constraints 8.41 and 8.42 are relaxed when unit i is in the synchronization, soak, or desynchronization phase.

8.3.2.9 Fuel Limitations

$$\sum_{t \in T} p_{it} \le \bar{F}_i \quad \forall i \in I \tag{8.43}$$

Constraints 8.43 establish an upper bound on the sum of the energy production of a thermal generating unit i during the scheduling horizon due to fuel limitations, for example, shortage of lignite/natural gas/oil reserves.

8.3.3 Total Thermal Unit Production Cost

The total production cost of a thermal unit includes the unit's start-up-type-dependent start-up cost, SUC_i^ℓ (see Figure 8.5), shutdown cost, SDC_i, and the hourly operating cost

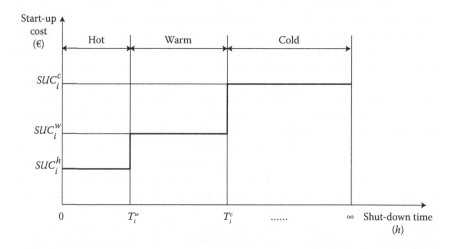

FIGURE 8.5   Start-up cost from hot, warm, or cold standby until load with synchronization.

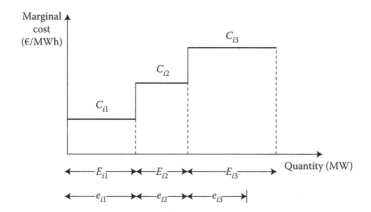

FIGURE 8.6 Thermal generating unit stepwise marginal cost function (convex cost function).

defined by the unit's no-load cost, NLC_i, and the stepwise marginal cost function, $(E_{if}, C_{if}), f \in F^i$, shown in Figure 8.6.

The general case of a stepwise marginal cost function of the units is modeled with the addition of new variables mc_{ift} and e_{ift}, defined as follows [1,11]:

F^i : Set of steps of the stepwise marginal cost function of unit j, $F^i = \{1,2,...,F^i\}$.

e_{ift} : Portion of step f of the ith unit's marginal cost function loaded in hour t (in MWh).

mc_{ift} : Binary variable indicating unit i output higher than or equal to step f during hour t (needed for nonconvex cost function only).

The total production cost of a thermal unit is modeled as follows:

$$c_{it}(p_{it}) = \sum_{\ell \in L} SUC_i^\ell \cdot y_{it}^\ell + SDC_i \cdot z_{it} + NLC_i \cdot \left(u_{it} - u_{it}^{syn}\right)$$
$$+ \sum C_{if} \cdot e_{ift} \quad \forall i \in I, \, t \in T \tag{8.44}$$

In the general case of nonconvex cost functions, the following constraints must be added to the model:

$$\sum_{f \in F^i} e_{ift} = p_{it} \quad \forall i \in I, \, t \in T \tag{8.45}$$

$$mc_{i(f+1)t} \cdot E_{if} \leq e_{ift} \leq mc_{ift} \cdot E_{if} \quad \forall i \in I, \, f \in F^i, \, t \in T \tag{8.46}$$

$$mc_{i(f+1)t} \leq mc_{ift} \quad \forall i \in I, \, f \in F^i, \, t \in T \tag{8.47}$$

If the unit's marginal cost function is nondecreasing (convex cost function; see Figure 8.6), the use of the additional binary variables mc_{ift} is avoided and constraints 8.46 and 8.47 should be replaced by constraint 8.48 in the model.

$$0 \leq e_{ift} \leq E_{if} \quad \forall i \in I, \ f \in F^i, \ t \in T \tag{8.48}$$

The start-up cost used in Equation 8.44 is discretized in three levels, one for each start-up type (see Figure 8.5). However, a more detailed start-up cost model, which allows hourly resolution of the prior reservation time, such as the one presented in References [45,46], may also be used.

It should be noted that the start-up cost, SUC_i, includes expenses up to the end of the synchronization phase; the unit operating cost during the soak phase is computed separately, as the sum of the unit's no-load cost and the integral of the unit's marginal cost function, as shown in Equation 8.44. Similarly, the shutdown cost, SDC_i, involves all expenses needed for the unit's shutdown besides the unit operating cost during the desynchronization phase.

Example 8.1 describes the optimal self-scheduling (commitment and dispatch schedule) of a thermal Producer owning a single combined-cycle gas turbine (CCGT) and acting as a price-taker in all day-ahead markets (energy and reserves).

Example 8.1: Optimal Self-Scheduling of a Thermal Producer

We consider a thermal Producer with one typical CCGT who participates as a price-taker in all day-ahead markets (energy and reserves) responding to the respective forecast clearing price curves for all market products.

Table 8.2 presents the main technoeconomic data of the CCGT, whereas Table 8.3 shows the forecast MCPs for energy. Prices for primary up/down, secondary up/down, and tertiary spinning reserves are considered to be constant during the whole scheduling period and equal to 20, 16, and 8€/MWh, respectively. The price for tertiary nonspinning reserve is considered to be equal to 5€/MWh during peak-load hours (i.e., hours 10–16 and 21–23), and zero during off-peak hours. Note that in the time interval before the market horizon ($t = 0$), the CCGT was operating at its nominal power output and had been synchronized for 20 h.

The size of the CCGT is small compared to the system requirements and, thus, coupling constraints 8.84 through 8.89 are relaxed in this example. In addition, fuel limitations as in Equation 8.43 are inactive and a convex

---

TABLE 8.2 Combined Cycle Unit Data

						Hot		Warm		Cold			Marginal Cost Range (€/MWh)	
P^{max} (MW)	P^{min} (MW)	UT (h)	DT (h)	T^w (h)	T^c (h)	$T^{syn,h}$ (h)	$T^{soak,h}$ (h)	$T^{syn,w}$ (h)	$T^{soak,w}$ (h)	$T^{syn,c}$ (h)	$T^{soak,c}$ (h)	T^{des} (h)	Min.	Max.
476	250	4	3	5	12	0	1	1	1	3	3	2	55.0	59.0

TABLE 8.3 Forecast Market Clearing Prices (Energy)

Hour	Price (€/MWh)
1	52.34
2	32.21
3	31.52
4	27.85
5	27.85
6	30.64
7	48.19
8	55.78
9	64.11
10	76.45
11	85.32
12	88.12
13	95.16
14	106.75
15	108.89
16	65.18
17	59.17
18	52.23
19	52.23
20	87.54
21	112.45
22	117.89
23	89.32
24	66.00

cost function is considered. Therefore, the optimization problem in this example consists of the problem objective 8.1 subject to constraints 8.11 through 8.42, 8.44 and 8.45, and 8.48.

Figure 8.7 illustrates the optimal self-schedule of the CCGT. Owing to the low-energy MCP of the early morning hours, the CCGT is desynchronized from 3:00 to 6:00 p.m., following a 2-h desynchronization phase (hours 1–2). Although the MCP is lower than the combined cycle unit's variable cost during hours 1–2, the CCGT cannot be shut down immediately, as it must first follow a 2-h desynchronization phase. In addition, although the MCP in hours 7–8 is lower than the combined cycle unit's variable cost, it is more profitable for the CCGT to perform a hot start-up in hour 7 than to perform a warm start-up in hour 9 (see Tables 8.1 and 8.3), due to the higher start-up cost in the latter case (see Figure 8.5). In hours 8–24, the combined cycle unit is in the dispatchable phase and is capable of contributing to different types of reserves.

In hours 8–10, 16–19, and 24, the CCGT's opportunity cost for providing reserves (defined as the difference between the forecast energy MCP and the unit's marginal cost) is lower than the respective reserves prices and so the Producer decides to operate its CCGT strictly within its power output limits in order to contribute to different types of reserves. On the contrary, during hours 11–15

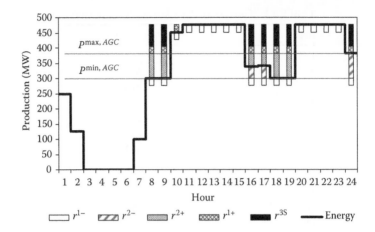

FIGURE 8.7 Energy and reserves contribution of a typical CCGT unit (476 MW) in the day-ahead energy and reserves markets.

and 20–23, the combined cycle units, opportunity cost for providing reserves is much higher than all reserves prices, making it more profitable for the Producer to operate its CCGT at its nominal power output, without contributing to any type of upward reserves. The total daily profit of the CCGT is equal to €135,008.

8.3.4 Hydro Unit Constraints

This section presents the constraints imposed on the operation of hydroplants and pumped storage plants over the scheduling horizon.

Given that hydro units are fast-start units, they can be committed/decommitted in a few minutes and, thus, they do not follow the start-up and shutdown procedures, including synchronization, soak, and desynchronization phases, presented in Section 8.3.2. As a result, all commitment (binary) variables corresponding to these phases (i.e., $u^{syn}_{jt} = u^{soak}_{jt} = u^{des}_{jt} = 0$) as well as all related constraints 8.11 through 8.22 standing for thermal units are omitted for hydroplants. Additionally, the total operating cost (including start-up and shutdown costs) of a hydroplant is considered to be equal to zero.

8.3.4.1 Hydro Intertemporal Constraints

Constraints 8.23 through 8.27 and 8.41 and 8.42 enforcing the expressions of minimum up/down time, logical status of commitment, and ramp-rate limits for thermal units are also valid for hydroplants, by substituting index "*i*" with index "*j*" ("*j*" denotes hydroplants) and taking into account that $u^{syn}_{jt} = u^{soak}_{jt} = u^{des}_{jt} = 0$.

8.3.4.2 Hydro Power Output Constraints

$$u^{AGC}_{jt} \leq u^{disp}_{jt} \quad \forall j \in J,\ t \in T \tag{8.49}$$

$$0 \leq r^{2+}_{jt} \leq 15 \cdot RU^{AGC}_{j} \cdot u^{AGC}_{jt} \quad \forall j \in J,\ t \in T \tag{8.50}$$

$$0 \leq r_{jt}^{2-} \leq 15 \cdot RD_j^{AGC} \cdot u_{jt}^{AGC} \quad \forall j \in J, \, t \in T \tag{8.51}$$

$$0 \leq r_{jt}^{3S} \leq R_j^{3S} \cdot u_{jt}^{disp} + p_{jt}^{A} \quad \forall j \in J, \, t \in T \tag{8.52}$$

$$r_{jt}^{3NS} \leq P_j^{max} \cdot u_{jt}^{3NS} \quad \forall j \in J, \, t \in T \tag{8.53}$$

$$r_{jt}^{3NS} \geq P_j^{min} \cdot u_{jt}^{3NS} \quad \forall j \in J, \, t \in T \tag{8.54}$$

$$u_{jt}^{3NS} \leq 1 - u_{jt} \quad \forall j \in J, \, t \in T \tag{8.55}$$

$$p_{jt} - r_{jt}^{2-} \geq P_j^{min} \cdot \left(u_{jt}^{disp} - u_{jt}^{AGC} \right) + P_j^{min,AGC} \cdot u_{jt}^{AGC} \quad \forall j \in J, \, t \in T \tag{8.56}$$

$$p_{jt} + r_{jt}^{2+} \leq P_j^{max} \cdot \left(u_{jt}^{disp} - u_{jt}^{AGC} \right) + P_j^{max,AGC} \cdot u_{jt}^{AGC} \quad \forall j \in J, \, t \in T \tag{8.57}$$

$$p_{jt} + r_{jt}^{2+} + r_{jt}^{3S} \leq P_j^{max} \cdot u_{jt}^{disp} \quad \forall j \in J, \, t \in T \tag{8.58}$$

$$\sum_{t \in T} y_{jt} \leq M \quad \forall j \in J \tag{8.59}$$

Constraints 8.49 through 8.55 enforce the reserves contribution limits of hydroplants and are similar to constraints 8.28 through 8.36 standing for thermal units, with the exception of constraints 8.29 and 8.30 that are omitted in the case of hydro units, as these units do not contribute to primary up/down reserves due to a transient phenomenon that takes place when the water flow through a penstock changes suddenly, known as the "water hammer effect" [47]. As shown in Equation 8.52, the pumping load of a pumped storage plant can also contribute to tertiary spinning reserve.

Constraints 8.56 through 8.58 enforce the power output plus/minus the corresponding reserves contribution of a hydroplant j to be within the respective nominal limits. Given that hydro units are not involved in synchronization, soak, and desynchronization phases, constraints 8.56 through 8.58 are much simpler than the respective power output constraints 8.37 through 8.40 standing for thermal units.

Finally, constraints 8.59 enforce the maximum number of hydroplant start-ups during the daily scheduling horizon. In this chapter, M is equal to 2 to coincide with the number of peaks within the scheduling period.

8.3.4.3 Hydro Prohibited Operating Zones

$$\sum_{z \in Z} u_{jt}^{z} \cdot P_j^{min,z} \leq p_{jt} \leq \sum_{z \in Z} u_{jt}^{z} \cdot P_j^{max,z} \quad \forall j \in J, \, t \in T \tag{8.60}$$

$$\sum_{z \in Z} u_{jt}^{z} = u_{jt}^{disp} \quad \forall j \in J, \, t \in T \tag{8.61}$$

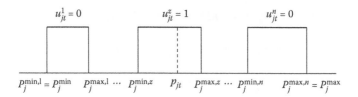

FIGURE 8.8 Hydroplant permissible operating zones.

A hydroplant may have prohibited operating zone(s) due to oscillations and physical limitations of unit components [48]. For an online hydroplant with $n-1$ prohibited operating zone(s), the total operating region is divided into n discrete subregions, named permissible operating zones, and the hydroplant must operate in one of these subregions (see Figure 8.8).

Constraints 8.60 and 8.61 ensure that the power output of a hydroplant j will be within the bounds of only one permissible operating zone in every hour t that hydroplant j is in dispatchable phase.

8.3.4.4 Discrete Pumping

$$p_{ht}^A = \sum_{n \in N} g_{ht}^n \cdot P_{A,h}^n \quad \forall h \in H,\, t \in T \tag{8.62}$$

$$g_{ht}^n + u_{ht}^{\text{disp}} \le 1 \quad \forall n \in N^h,\, h \in H,\, t \in T \tag{8.63}$$

Constraints 8.62 ensure that the pumping load of a pumped storage plant h is equal to the sum of the online units' pumping load. In this modeling, the rotation frequency (r/min) of a pumping unit turbine n is constant, resulting in a discrete pumping load, $P_{A,h}^n$. Recently, variable frequency motors are also used for pumping, allowing for continuous pumping load variation.

Constraints 8.63 prohibit simultaneous pumping and generation mode operation [17].

8.3.4.5 Hydro Energy Limitations

$$E_j^{\min} \le \sum_{t \in T} p_{jt} \le E_j^{\max} \quad \forall j \in J \tag{8.64}$$

Constraints 8.64 establish lower and upper bounds on the sum of the energy production (in MWh) of a hydroplant j during the scheduling horizon. In fact, the upper production bound is selected to be consistent with a medium-term planning policy, whereas the lower bound is usually imposed in order to avoid spillage as well as due to requirements of irrigation, navigation, recreation, and so on. These constraints comprise a simple "energy-only" modeling of the hydro resources management.

8.3.4.6 Reservoir Dynamics

Apart from the simple hydro energy limitations given by Equation 8.64, the hydraulic coupling among cascaded hydroplants is explicitly formulated. In addition, a possible time delay between the discharge of a hydroplant and the resulting inflows of the downstream reservoir is also taken into account. Finally, forecast net water inflows are assumed to be known.

$$
v_{jt} = v_{j(t-1)} + M \cdot \left[I_{jt} - q_{jt} - sp_{jt} + q_{jt}^A \right]
$$
$$
+ M \cdot \sum_{j^u \in J^u} \left[q_{j^u(t-d_{ju})} + sp_{j^u(t-d_{ju})} - q_{j^u(t-d_{ju})}^A \right] \quad \forall j \in J,\ t \in T \qquad (8.65)
$$

$$
0 \le v_{jt} \le V_j^{\max} \quad \forall j \in J,\ t \in T \qquad (8.66)
$$

$$
v_{j0} = V_j^{\text{ini}} \quad \forall j \in J \qquad (8.67)
$$

$$
v_{jT} = V_j^{\text{fin}} \quad \forall j \in J \qquad (8.68)
$$

where $q_{jt}^A = e_j^A \cdot sc_j^A \cdot p_{jt}^A$ denotes the water flow (in m³/s) from the lower to the upper reservoir in the case of a pumped storage plant while operating in pump mode.

Constraints 8.65 represent the hourly reservoir water balance, taking into account the operation (including pumping) of the upstream hydroplants of a given hydroplant and the related time delay, d_{ju}, whereas constraints 8.66 state the reservoir stored volume limits.

Constraints 8.67 and 8.68 define the initial and final (target) reservoir volumes, respectively. Initial and final reservoir contents are usually obtained by medium-term planning studies and, in general, the final water content of each reservoir is different from its initial value.

The incorporation of constraints 8.65 through 8.68 in the model formulation usually requires that the simple energy-only modeling given by Equation 8.64 be disregarded.

8.3.4.7 Variable Head Modeling

As already mentioned in the introduction, to formulate precisely the self-scheduling problem of a hydro subsystem, it is essential to model the generation characteristics of hydroplants describing the relationship between the head of the associated reservoir, the water discharged, and the hydroplant power output. This relationship is often represented by a set of nonlinear unit performance curves, which comprise the well-known Hill diagram (see Figure 8.9).

Due to the nonlinear nature of the head effect, most optimization models dealing with daily operation of the hydro system neglect the effect of the reservoir head variation on the hydroplants operating schedule (commitment and dispatch) or deal with

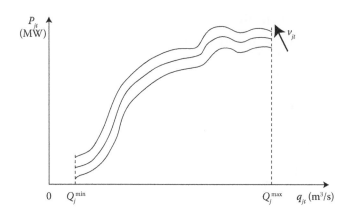

FIGURE 8.9   Hill diagram.

simplifications, such as the concave piecewise linear approximation or the modeling of the best local efficiency points.

An early attempt to provide an explicit 0/1 MILP approach for the solution of the head effect problem under the hydrothermal Producer self-scheduling framework is given in Reference [18], where the set of curves of the Hill diagram is reduced to three curves according to two levels of the stored water in the reservoir, and each curve is approximated by a piecewise linear function. This approach is extended in Reference [19], where a more sophisticated modeling is introduced through an enhanced linearization technique based on two-dimensional considerations for the nonlinear relationship among water volume, discharge, and power output. Although this approach leads to a more precise representation of the head effect, the computational burden caused and the computing time required are considerably increased.

This paragraph presents a simplified version of the 0/1 MILP approach presented in Reference [18] for the modeling of the relationship between the performance curves and the variation of the water volume of the reservoir. In this model, the power output of a hydroplant *j* is considered to be a linear function of the water discharge for a given curve (see Figure 8.10) instead of a piecewise linear function [18]. However, this simplified model is extended to take into account the capability of the hydroplants to operate under AGC and/or contribute in the various types of reserves, as already described in the previous paragraphs. Figure 8.10 illustrates the linearization of the performance curves of a hydroplant *j*, whereas Figure 8.11 shows how the power output limits of a hydroplant *j* are modified according to the water content of the reservoir (reservoir volume zone). Note that the specific consumption of a hydroplant *j* [in (m³/s)/MW] decreases and, subsequently, the slope of the hydroplant performance curves increases with the increase in the reservoir water content.

Although in this model no additional binary variables are used for the linearization of the nonconvex performance curves, the present formulation leads to increased computational burden and subsequent slow execution times due to the complex modeling of the operating constraints of hydroplants (AGC and reserves modeling).

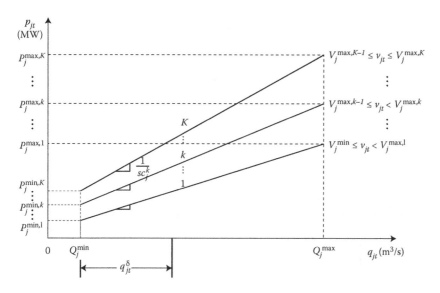

FIGURE 8.10 Performance curves of hydroplant j.

The general mixed-integer linear formulation for the discretization of the reservoir in K water volume zones is given by the following constraints:

$$v_{jt} \geq \sum_{k=1}^{K-1} V_j^{\max,k} \cdot \left(x_{jt}^k - x_{jt}^{k+1} \right) \quad \forall j \in J, \; t \in T \qquad (8.69)$$

$$v_{jt} \leq \sum_{k=1}^{K} V_j^{\max,k} \cdot \left(x_{jt}^{k-1} - x_{jt}^k \right) \quad \forall j \in J, \; t \in T \qquad (8.70)$$

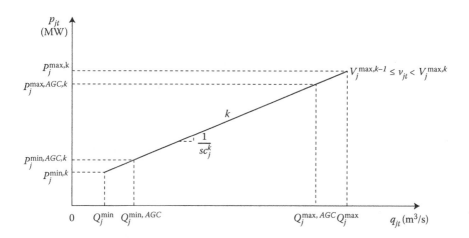

FIGURE 8.11 Power output limits of hydroplant j.

TABLE 8.4   Reservoir Volume Zones Discretization

Reservoir Volume Zone	x_{jt}^1	x_{jt}^2	x_{jt}^{k-1}	x_{jt}^k	x_{jt}^{k+1}
1	0	0	0	0	0
2	1	0	0	0	0
3	1	1	0	0	0
k	1	1	1	0	0
k + 1	1	1	1	1	0

$$x_{jt}^k \geq x_{jt}^{k+1} \quad \forall k = 1, \ldots, K-1, \; j \in J, \; t \in T \tag{8.71}$$

In the above formulation, binary variables x_{jt}^k are used for the selection of the correct curve according to the reservoir volume. Table 8.4 presents the permissible combinations of values 0–1 for all binary variables involved, as required by constraints 8.71 and considering that always $x_{jt}^0 = 1$ and $x_{jt}^K = 0$.

The mixed-integer linear formulation of the hydroplant performance curves is defined by the following constraints:

$$q_{jt} = Q_j^{\min} \cdot \left(u_{jt}^{\text{disp}} - u_{jt}^{AGC} \right) + Q_j^{\min,AGC} \cdot u_{jt}^{AGC} + q_{jt}^\delta \quad \forall j \in J, \; t \in T \tag{8.72}$$

$$q_{jt} \leq Q_j^{\max} \cdot \left(u_{jt}^{\text{disp}} - u_{jt}^{AGC} \right) + Q_j^{\max,AGC} \cdot u_{jt}^{AGC} \quad \forall j \in J, \; t \in T \tag{8.73}$$

$$p_{jt} \leq P_j^{\min,k} \cdot \left(u_{jt}^{\text{disp}} - u_{jt}^{AGC} \right) + P_j^{\min,AGC,k} \cdot u_{jt}^{AGC} + \frac{1}{sc_j^k} \cdot q_{jt}^\delta$$
$$+ N \cdot \left((k-1) - \sum_{m=1}^{k-1} x_{jt}^m + \sum_{m=k}^{K-1} x_{jt}^m \right) \quad \forall k \in K, \; j \in J, \; t \in T \tag{8.74}$$

$$p_{jt} \geq P_j^{\min,k} \cdot \left(u_{jt}^{\text{disp}} - u_{jt}^{AGC} \right) + P_j^{\min,AGC,k} \cdot u_{jt}^{AGC} + \frac{1}{sc_j^k} \cdot q_{jt}^\delta$$
$$- N \cdot \left((k-1) - \sum_{m=1}^{k-1} x_{jt}^m + \sum_{m=k}^{K-1} x_{jt}^m \right) \quad \forall k \in K, \; j \in J, \; t \in T \tag{8.75}$$

$$p_{jt} - r_{jt}^{2-} \geq P_j^{\min,k} \cdot \left(u_{jt}^{\text{disp}} - u_{jt}^{AGC} \right) + P_j^{\min,AGC,k} \cdot u_{jt}^{AGC}$$
$$- N \cdot \left((k-1) - \sum_{m=1}^{k-1} x_{jt}^m + \sum_{m=k}^{K-1} x_{jt}^m \right) \quad \forall k \in K, \; j \in J, \; t \in T \tag{8.76}$$

$$p_{jt} + r_{jt}^{2+} \leq P_j^{\max,k} \cdot \left(u_{jt}^{\text{disp}} - u_{jt}^{AGC} \right) + P_j^{\max,AGC,k} \cdot u_{jt}^{AGC}$$
$$+ N \cdot \left((k-1) - \sum_{m=1}^{k-1} x_{jt}^m + \sum_{m=k}^{K-1} x_{jt}^m \right) \quad \forall k \in K, \; j \in J, \; t \in T \tag{8.77}$$

$$p_{jt} + r_{jt}^{2+} + r_{jt}^{3S} \le P_j^{\max,k} \cdot u_{jt}^{disp} + N \cdot \left((k-1) - \sum_{m=1}^{k-1} x_{jt}^m + \sum_{m=k}^{K-1} x_{jt}^m \right)$$

$$\forall k \in K, \ j \in J, \ t \in T \tag{8.78}$$

$$r_{jt}^{3NS} \ge P_j^{\min,k} \cdot u_{jt}^{3NS} - N \cdot \left((k-1) - \sum_{m=1}^{k-1} x_{jt}^m + \sum_{m=k}^{K-1} x_{jt}^m \right)$$

$$\forall k \in K, \ j \in J, \ t \in T \tag{8.79}$$

$$r_{jt}^{3NS} \le P_j^{\max,k} \cdot u_{jt}^{3NS} + N \cdot \left((k-1) - \sum_{m=1}^{k-1} x_{jt}^m + \sum_{m=k}^{K-1} x_{jt}^m \right)$$

$$\forall k \in K, \ j \in J, \ t \in T \tag{8.80}$$

Constraints 8.72 and 8.73 denote that the water discharge of hydroplant *j* in each hour is equal to the minimum water discharge (operating under *AGC* or not) plus the portion of the water discharged needed to obtain the desirable power output and cannot exceed the maximum water discharge of the hydroplant. Constraints 8.74 and 8.75 define the hydroplant power output in relation to the reservoir volume zone and taking into account the possible operation of the hydroplant under *AGC*. It should be noted that N is a large constant, so that constraints 8.74 and 8.75 are relaxed for all possible combinations of binary variables x_{jt}^k other than those corresponding to the appropriate reservoir volume zone (see Table 8.4).

Similarly, constraints 8.76 through 8.78 enforce the power output plus/minus the corresponding reserves contribution of a hydroplant *j* to be within the nominal limits corresponding to the correct reservoir volume zone. Finally, constraints 8.79 and 8.80 define the tertiary nonspinning reserve contribution limits for hydroplant *j*, as these limits strongly depend on the current reservoir volume zone (see also Figure 8.10).

Note that, in the case of the variable head modeling, constraints 8.76 through 8.78, 8.79, and 8.80 should substitute the initial power output and reserves contribution constraints 8.56 through 8.58, 8.53, and 8.54, respectively, in the model formulation.

The incorporation of the variable head modeling in the hydrothermal Producer self-scheduling problem increases significantly the computational burden due to the inclusion of the additional binary variables x_{jt}^k as well as the complex constraints 8.72 through 8.80. Thus, a simplified hydro model that does not take into account the variable head modeling as well as the minimum water discharge of a hydroplant is given by constraints 8.53, 8.54, 8.56 through 8.58, and constraints 8.81 through 8.83 below:

$$p_{jt} = \frac{1}{sc_j^{av}} q_{jt} \quad \forall j \in J, \ t \in T \tag{8.81}$$

$$0 \le q_{jt} \le Q_j^{\max} \quad \forall j \in J, \ t \in T \tag{8.82}$$

In this case, the power output of a hydroplant j is considered to vary proportionally to the water discharge, where the average specific consumption is considered to be equal to the mean value of the specific consumptions that correspond to all reservoir volume zones, as follows:

$$sc_j^{av} = \frac{\sum_{k \in K} sc_j^k}{K} \quad \forall j \in J \tag{8.83}$$

This simplified model is suitable for the daily self-scheduling problem formulation of hydroplants with large reservoirs, as the daily variation of the head of the upstream reservoir has little effect on the energy production of hydroplants with large dams. On the contrary, when a longer scheduling period is considered and/or the hydro subsystem involves hydroplants associated with small reservoirs, the variable head modeling already discussed leads to more accurate results.

In each case, for the sake of simplicity, the specific consumption of the pumped storage plants while operating in pump mode, sc_h^A, is considered to be equal to the respective average specific consumption while operating in generation mode, sc_h^{av}, that is, no variable head modeling is considered for the pumping operation.

In the numerical application section, an indicative test case illustrating the usefulness of the variable head modeling for a hydro system comprising both large and small hydroplants/reservoirs is analytically discussed.

8.3.5 System Constraints

$$\sum_{i \in I} p_{it} \leq D_t \quad \forall t \in T \tag{8.84}$$

$$\sum_{i \in I} r_{it}^{1+} \leq RR_t^{1+} \quad \forall t \in T \tag{8.85}$$

$$\sum_{i \in I} r_{it}^{1-} \leq RR_t^{1-} \quad \forall t \in T \tag{8.86}$$

$$\sum_{i \in I} r_{it}^{2+} \leq RR_t^{2+} \quad \forall t \in T \tag{8.87}$$

$$\sum_{i \in I} r_{it}^{2-} \leq RR_t^{2-} \quad \forall t \in T \tag{8.88}$$

$$\sum_{i \in I} r_{it}^{3S} + \sum_{i \in I} r_{it}^{3NS} \leq RR_t^{3} \quad \forall t \in T \tag{8.89}$$

The system (coupling) constraints 8.84 through 8.89 are necessary only in the case of a hydrothermal Producer acting as a price-taker and reflect the fact that, no matter how high the energy and the reserves prices are, the energy and the reserves that the system operator will purchase from the Producer will not exceed the corresponding system requirements. These coupling constraints may be relaxed if the size of the Producer is small compared to the system requirements. In this case, the price-taker Producer problem may be decomposed by unit.

In the case of the price-maker hydrothermal Producer these constraints are redundant, as the maximum quantity of energy and reserves of the Producer that will be purchased by the system operator is defined indirectly by the hourly residual demand curves, as already discussed in Section 8.2.2.1.

8.3.6 Hydrothermal Producer also Acting as a Retailer

In this section, the hydrothermal Producer is considered to act as a dominant price-maker Producer as well as a dominant retail service provider (henceforth, "Retailer") through forward bilateral contracts with end-consumers. Therefore, the Producer is considered to be a power company with a dominant role in both the production and retail sectors of an electricity market. As a Retailer, the Producer buys energy from the day-ahead market at the hourly MCPs and sells it to end-consumers at predefined quantities and prices in order to fulfill daily contractual obligations. Following a specific rule in several electricity markets (such as the Greek electricity market), the Retailer is also charged for the procurement of reserves in proportion to the forward contracted quantity of energy.

The incorporation of forward bilateral contracts along with the procurement cost of the associated reserves in the optimization target of the self-scheduling problem seriously affect the self-scheduling program of the power company that acts as a price-maker in the production sector. Numerical results show that the power company's incentive to withhold capacity and increase the short-term electricity prices is significantly reduced by the presence of forward bilateral contracts and the procurement cost of the associated reserves.

In the case that the price-maker Producer acts also as a Retailer, the objective function of the self-scheduling problem 8.3 is transformed as follows:

$$Max\left\{\sum_{t\in T}\left[\pi_t^E\left(\xi_t^E\right)\cdot\xi_t^E + \sum_{m\in M}\pi_t^m\left(\xi_t^m\right)\cdot\xi_t^m\right.\right.$$

$$\left.\left. + CQ_t^E\cdot\left(CP_t^E - \pi_t^E\left(\xi_t^E\right)\right) - \sum_{m\in M}CQ_t^m\cdot\pi_t^m\left(\xi_t^m\right) - \sum_{i\in I}c_{it}\left(p_{it}\right)\right]\right\} \qquad (8.90)$$

or, equivalently:

$$Max\left\{\sum_{t\in T}\left[\sum_{b\in B^E}\left[\pi_{bt}^E\cdot\left(d_{bt}^E + w_{bt}^E\cdot D_{bt}^{E,\min}\right)\right] + \sum_{m\in M}\sum_{b\in B^m}\left[\pi_{bt}^m\cdot\left(d_{bt}^m + w_{bt}^m\cdot D_{bt}^{m,\min}\right)\right]\right.\right.$$

$$\left.\left. + CQ_t^E\cdot\left(CP_t^E - \pi_t^E\left(\xi_t^E\right)\right) - \sum_{m\in M}CQ^m\cdot\pi_t^m\left(\xi_t^m\right) - \sum_{i\in I}c_{it}\left(p_{it}\right)\right]\right\} \qquad (8.91)$$

The power company's profit during the scheduling period in Equation 8.90 or, equivalently, Equation 8.91 is computed as the difference between its revenue from the day-ahead energy and reserves markets as well as the forward bilateral contracts for energy delivery minus the procurement cost for energy and reserves and the units' total operating cost.

As a Retailer, the power company is charged for the procurement of reserves in proportion to its forward contracted quantity of energy, as follows:

$$CQ_t^m = \frac{CQ_t^E}{D_t} \cdot RR_t^m \qquad (8.92)$$

The optimization problem of the power company is also subject to the system and unit operating constraints 8.5 through 8.83, already described in the previous sections.

8.3.7 Model Summary

To sum up, the mathematical formulation of the deterministic hydrothermal Producer self-scheduling problem presented in Sections 8.2 through 8.3.6 can be classified in different categories regarding the problem objective as well as the generating units operating constraints and system constraints, as follows:

1. *Price-taker Producer*: In the case of a price-taker Producer, the objective function to be maximized is given by Equation 8.1. System constraints 8.84 through 8.89 must also be added to the model formulation.
2. *Price-maker Producer*: In the case of a price-maker Producer, objective function 8.4 is used along with constraints 8.5 through 8.10 needed for the linearization of the respective residual demand curves.
3. *Price-maker Producer and Retailer*: In the case of a price-maker Producer also acting as a Retailer, objective function 8.4 must be replaced by function 8.91 and Equation 8.92 should also be added to the model.

Regarding the operating constraints of generating units, different constraints must be included in the model formulation for the thermal and hydro subsystems, respectively, as follows:

i. *Thermal System*: Constraints 8.11 through 8.48 are necessary for the thermal system modeling in all cases.
ii. *Hydro System*: Three models with regard to the water resources management, the reservoir dynamics, and the modeling of the head effect can be considered in ascending order of complexity, as follows:
   a. *Energy-only Model*: The simplest hydro model comprises the energy-only modeling, whereas no head variation dependence is explicitly formulated. In this case, constraints 8.23 through 8.27, 8.41, 8.42, and 8.49 through 8.64 must be included in the model formulation.
   b. *Hydraulic Coupling Model*: This model differentiates from the energy-only model in that constraints 8.64 must be replaced by constraints 8.65 through

8.68 denoting the reservoir dynamics and Equations 8.81 through 8.83 must be taken into account. Therefore, this model comprises the following constraints: 8.23 through 8.27, 8.41, 8.42, 8.49 through 8.63, 8.65 through 8.68, and 8.81 through 8.83.

c. *Full Hydro Model:* In this model, the reservoir dynamics as well as the relationship between the head of the associated reservoir, the water discharged, and the hydroplant power output is taken into account. In this case, the following constraints constitute the hydro system model: 8.23 through 8.27, 8.41, 8.42, 8.49 through 8.52, 8.55, 8.59 through 8.63, and 8.65 through 8.80.

8.4 Uncertainty Modeling

As already mentioned, in all previous sections of this chapter, the solution of the hydrothermal Producer self-scheduling problem is based on the Producer's deterministic knowledge of the day-ahead market conditions, regarding either the MCPs (in the case that the Producer acts as a price-taker) or the system requirements and competitors' offers for all market products (in the case that the Producer acts as a price-maker). In each case, the Producer's estimates exhibit some degree of uncertainty, which, if not appropriately addressed, may lead to a subsequent offering strategy that will yield suboptimal solutions regarding both the generating units' commitment and Producer profits from its participation in day-ahead markets.

To take into account market uncertainty, as the uncertain input parameters may usually be described through probability functions, a common practice is to approximate the probability distribution of the input parameters by a number of scenarios with associated probabilities of occurrence. For instance, random parameter θ (e.g., MCP in the price-taker Producer problem) can be represented by $\theta(\omega)$, $\omega \in \Omega = \{1, 2, \ldots, \Omega\}$, where ω is the scenario index and Ω is the number of scenarios. Each realization $\theta(\omega)$ is associated with a probability of occurrence $prob_\omega$, defined as

$$prob_\omega = P(\omega \mid \theta = \theta(\omega)), \quad \text{where} \sum_{\omega \in \Omega} prob_\omega = 1 \tag{8.93}$$

One option is to solve the deterministic optimization problem already described in the previous sections by substituting the forecast input parameters (MCPs or system requirements and competitors' offers) by their respective expected values. Although this option may seem efficient, it usually does not result in the best possible outcome.

A more sophisticated alternative option is to formulate the hydrothermal Producer self-scheduling problem in a stochastic programming framework. The basic idea of the stochastic programming is that (optimal) decisions should be based on data available at the time the decisions are made and should not depend on future observations. In other words, the optimization of the decision-making under uncertainty is closely related to the concept of hedging against all possible future realizations of the random parameters.

In this section, the deterministic hydrothermal Producer self-scheduling problem is appropriately converted to a stochastic optimization problem and the associated formulation is presented. For the reader's convenience, a brief introduction on the basics of the stochastic programming is given in the immediately following section. Further information on stochastic optimization can be found in References [49,50].

8.4.1 Stochastic Programming Basics

Each stochastic programming problem is defined in a number of stages. Each stage denotes a point in time where related decisions are made or uncertainty is revealed. In general, stochastic programming problems are grouped in two categories depending on the number of stages considered, namely *two-stage* and *multistage stochastic programming problems*. As the hydrothermal Producer self-scheduling problem is later formulated as a two-stage stochastic problem, in this section the two-stage stochastic programming problems are only described. The multistage stochastic programming problems are formulated similarly [28,49].

In a two-stage stochastic programming framework, variables that represent decisions that are made before the uncertainty is revealed are known as *first-stage* or *here-and-now* variables. Decision variables representing decisions that are made once the uncertainty is revealed are known as *second-stage* or *wait-and-see* variables. If uncertainty is modeled through a number of scenarios, second-stage variables are dependent on each scenario, whereas first-stage variables are identical for all scenarios (*nonanticipativity*) [49].

In general, a stochastic programming problem can be represented by a scenario tree. The scenario tree consists of a set of nodes and branches. The nodes represent the points where decisions are made or uncertainty is revealed (stages). Each node, except the root node, has a single predecessor and may have several successors. The first node, where first-stage decisions are made, is called the root. Nodes without any successors are called leaves. The nodes connected to the root node represent second-stage decisions, and in a two-stage stochastic programming problem are also the leaves of the tree. The branches represent different realizations of random parameters. Figure 8.12 illustrates a simple scenario tree for a two-stage stochastic problem.

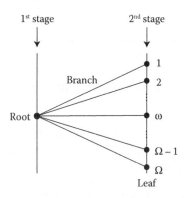

FIGURE 8.12   Two-stage scenario tree.

In a two-stage stochastic programming problem, an optimal policy is devised against all possible future realizations of a random parameter, $\theta(\omega)$. This optimal strategy consists of taking an initial decision, x (first-stage decision), before the uncertainty is revealed, observing the realization of the random parameter, $\theta(\omega)$, and then taking a corrective decision, $y(\omega)$ (second-stage decision), tailored to the realization of the random parameter (recourse action). Note that the corrective decision, $y(\omega)$, depends also on the initial decision, x.

The general formulation of a two-stage stochastic linear programming problem with recourse is as follows:

$$\min\left\{c^T x + E\left[\min q(\omega)^{\mathrm{T}} y(\omega)\right]\right\} \tag{8.94}$$

subject to

$$Ax = b \tag{8.95}$$

$$W(\omega)y(\omega) + T(\omega)x = h(\omega) \tag{8.96}$$

$$x \in X, y(\omega) \in Y \tag{8.97}$$

In the case that some of the variables, x, $y(\omega)$, are integers, Equations 8.94 through 8.97 represent a stochastic MILP problem. In Equations 8.94 through 8.97, decision variables x are first-stage variables, whereas decision variables $y(\omega)$ are second-stage variables.

Note that the use of the stochastic programming approach even under a limited number of scenarios usually leads to a dramatic increase in the size of the problem to be solved and may even lead to intractability, due to the enormous number of equations and variables simultaneously considered. Therefore, the respective deterministic problem to be converted into a stochastic one should be as simple as possible, in order to yield feasible solutions.

In the following section, the hydrothermal Producer self-scheduling problem is explicitly formulated as a two-stage stochastic programming problem with recourse.

8.4.2 Formulation of the Hydrothermal Producer Self-Scheduling Problem as a Two-Stage Stochastic Programming Problem with Recourse

In this section, the hydrothermal Producer self-scheduling problem is formulated as a two-stage stochastic programming problem with recourse. For consistency, we consider a price-maker Producer that acts also as a Retailer through forward bilateral contracts with the end-consumers, thus forming a dominant power company, as already described in Section 8.3.6.

To reduce the problem size, the Producer is considered to participate in the energy market only (no reserves markets are considered). In addition, regarding the hydro model, head dependencies are ignored.

As the Producer acts as a price-maker in the day-ahead energy market only, the uncertainty of the market conditions is limited in the uncertainty of the respective residual demand curves. Under the two-stage stochastic programming formulation of the Producer self-scheduling problem, the aim of the Producer is to devise its optimal offering strategy for its own generating units. In other words, instead of computing a single-step offer curve for each hour and each generating unit (deterministic problem), the Producer is now capable of deriving a multistep offer curve for each hour and each generating unit, taking into account the uncertain behavior of competitors through scenario-dependent residual demand curves.

The formulation of this problem as a two-stage stochastic programming problem lies in the fact that the commitment decisions for each generating unit are considered to be scenario-independent, and, therefore, are unique for all possible scenarios (first-stage decisions), whereas the power output of each generating unit depends on the realization of each scenario (second-stage decisions).

In the proposed two-stage stochastic model, the risk associated with the profit variability is explicitly taken into account through the CVaR, which has already been analytically described in Chapter 2.

In Section 8.4.2.1 the construction of the scenario-dependent residual demand curves is described, whereas in Section 8.4.2.2 the explicit mathematical formulation of the aforementioned two-stage stochastic programming problem is presented. In Section 8.4.2.3 the risk-constrained formulation of the Producer self-scheduling problem is presented.

8.4.2.1 Representing Uncertainty of the Residual Demand Curves

The formulation of the Producer's residual demand curves entails a high degree of uncertainty, regarding the estimates of the system load demand and the competitors' energy offers.

In some electricity markets, (such as the Greek one), the system load forecast and the forecast for the energy injection from renewable energy sources (which is considered as nonpriced energy injection and, therefore, is subtracted from the total system load demand to be served by the conventional generating units) that the market operator uses for the clearing of the day-ahead market become publicly available few hours before the participants submit their offers. Therefore, we consider that the Producer has perfect knowledge of the net system load demand (defined as the difference between the total system load and the renewable energy sources injection) before computing its own self-schedule and, subsequently, submitting its own offer curves.

The competitors' energy offers can be estimated on the basis of the Producer's gathered information on competitors (through press releases, etc.) and on ISO-released information on past market outcomes; in some electricity markets the energy offers of all participants become public information some time after the market clearing [51,52]. In addition, the Producer may wish to incorporate uncertainties due to unpredictable events, such as the competitor unit outages.

Hence, we assume that the Producer faces Ω possible series of 24 residual demand curves. Considering only one realization (scenario) ω, one can compute the total hourly Producer quota, ξ_t^ω, as well as the power output of each Producer generating unit, p_{it}^ω, for each hour of the scheduling horizon, resulting in the vectors of quantities $(\xi_1^\omega, \xi_2^\omega, ..., \xi_{24}^\omega)$ and $(p_{i1}^\omega, p_{i2}^\omega, ..., p_{i24}^\omega)$ for each realization ω, respectively. Obviously, for each realization ω, the Producer quota is equal to the sum of the generating unit power output, as required by Equation 8.5, if all variables are augmented with superscript ω.

As already mentioned, each hourly quota corresponds to a resulting MCP from the respective residual demand curve. Thus, the vector of quantities for each realization ω, $(\xi_1^\omega, \xi_2^\omega, ..., \xi_{24}^\omega)$ results in a vector of respective clearing prices $\left(\pi_1^\omega(\xi_1^\omega), \pi_1^\omega(\xi_2^\omega), ..., \pi_{24}^\omega(\xi_{24}^\omega)\right)$ (see Figure 8.13).

In the case that the Producer participates in a financial market, where the Producers are allowed to submit a single offer curve for their entire quota (portfolio offer), each pair $\left(\xi_t^\omega, \pi_t^\omega(\xi_t^\omega)\right)$ can be considered as the offer decided by the Producer for hour t and realization ω. Conversely, in the case that a physical electricity market is considered, each generating unit is obliged to submit its own energy offer curve. In this chapter, we consider a physical electricity market and, therefore, each Producer unit's energy offer for hour t and realization ω is represented by the quantity (MWh)–price (€/MWh) pair $\left(p_{it}^\omega, \pi_t^\omega(\xi_t^\omega)\right)$.

Following the logic presented in Reference [2], if we focus on a particular hour t, it is concluded that the Ω quantities and the Ω prices decided by the Producer for each generating unit i, although corresponding to different market outcomes, constitute the offer curve that the generating unit should submit to the market for that hour.

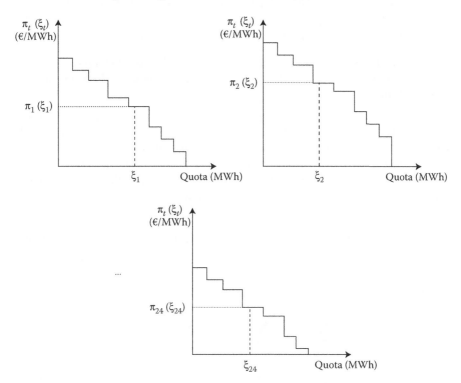

FIGURE 8.13 Twenty-four residual demand curves for each realization ω.

To guarantee that the offer curve decided by the model is nondecreasing (as required by most day-ahead energy markets), the following condition must hold for each pair of offers $\left\{\left(p_{it}^{\omega},\pi_t^{\omega}(\xi_t^{\omega})\right),\ \left(p_{it}^{\omega'},\pi_t^{\omega'}(\xi_t^{\omega'})\right)\right\}$ submitted for hour t for each generating unit i:

$$\left(p_{it}^{\omega}-p_{it}^{\omega'}\right)\cdot\left(\pi_t^{\omega}\left(\xi_t^{\omega}\right)-\pi_t^{\omega'}\left(\xi_t^{\omega'}\right)\right)\geq 0 \quad \forall i\in I,\ t\in T,\ \omega,\omega'\in\Omega,\ \omega'>\omega \quad (8.98)$$

The above nonlinear expression is equivalent to the following group of linear constraints [2]:

$$p_{it}^{\omega}-p_{it}^{\omega'}\geq -x_t^{\omega\omega'}\cdot M^p \quad \forall i\in I,\ t\in T,\ \omega,\omega'\in\Omega,\ \omega'>\omega \qquad (8.99)$$

$$p_{it}^{\omega'}-p_{it}^{\omega}\geq -(1-x_t^{\omega\omega'})\cdot M^p \quad \forall i\in I,\ t\in T,\ \omega,\omega'\in\Omega,\ \omega'>\omega \qquad (8.100)$$

$$\pi_t^{\omega}\left(\xi_t^{\omega}\right)-\pi_t^{\omega'}\left(\xi_t^{\omega'}\right)\geq -x_t^{\omega\omega'}\cdot M^{\pi} \quad \forall i\in I,\ t\in T,\ \omega,\omega'\in\Omega,\ \omega'>\omega \qquad (8.101)$$

$$\pi_t^{\omega'}\left(\xi_t^{\omega'}\right)-\pi_t^{\omega}\left(\xi_t^{\omega}\right)\geq -(1-x_t^{\omega\omega'})\cdot M^{\pi} \quad \forall i\in I,\ t\in T,\ \omega,\omega'\in\Omega,\ \omega'>\omega \qquad (8.102)$$

where M^p is a big quantity, M^{π} is a big price, and $x_t^{\omega\omega'}$ is a binary variable. If $x_t^{\omega\omega'}=0$, then $p_{it}^{\omega}\geq p_{it}^{\omega'}$ and $\pi_t^{\omega}\left(\xi_t^{\omega}\right)\geq\pi_t^{\omega'}\left(\xi_t^{\omega'}\right)$. If $x_t^{\omega\omega'}=1$, then $p_{it}^{\omega}\leq p_{it}^{\omega'}$ and $\pi_t^{\omega}\left(\xi_t^{\omega}\right)\leq\pi_t^{\omega'}\left(\xi_t^{\omega'}\right)$. In each case, the nondecreasing nature of the offer curves is assured.

8.4.2.2 Mathematical Formulation of the Hydrothermal Producer Self-Scheduling Problem as a Two-Stage Stochastic Programming Problem

As already mentioned in the previous section, in the presence of uncertainty in the residual demand curves, the Producer is considered to face Ω possible series of 24 residual demand curves (scenarios), each with associated probability $prob_{\omega}$.

In the case that the price-maker Producer acts also as a Retailer, the objective function of the proposed stochastic model results from the respective objective functions of the deterministic self-scheduling problem 8.90 as follows:

$$Max\left\{\sum_{\omega\in\Omega}prob_{\omega}\cdot\left\{\sum_{t\in T}\left[\pi_t^{\omega}\left(\xi_t^{\omega}\right)\cdot\xi_t^{\omega}+CQ_t^E\cdot\left(CP_t^E-\pi_t^{\omega}\left(\xi_t^{\omega}\right)\right)-\sum_{i\in I}c_{it}^{\omega}\left(p_{it}^{\omega}\right)\right]\right\}\right\} \qquad (8.103)$$

or, equivalently,

$$Max\left\{\sum_{\omega\in\Omega}prob_{\omega}\cdot\left\{\sum_{t\in T}\left[\sum_{b\in B}\left[\pi_{bt}^{\omega}\cdot\left(d_{bt}^{\omega}+w_{bt}^{\omega}\cdot D_{bt}^{min,\omega}\right)\right]\right.\right.\right.$$
$$\left.\left.\left.+CQ_t^E\cdot\left(CP_t^E-\pi_t^{\omega}\left(\xi_t^{\omega}\right)\right)-\sum_{i\in I}c_{it}^{\omega}\left(p_{it}^{\omega}\right)\right]\right\}\right\} \qquad (8.104)$$

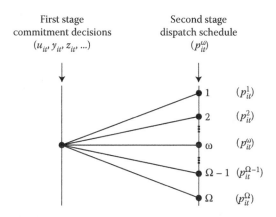

FIGURE 8.14   Scenario tree.

As in the stochastic optimization problem only the energy product is considered, superscript "*E*" denoting the energy market product in the deterministic formulation is omitted from the above expressions to simplify notation.

The stochastic optimization problem of the power company is also subject to the unit operating constraints considered for the respective deterministic problems. Therefore, constraints 8.11 through 8.48 (thermal system) as well as 8.23 through 8.27, 8.41, 8.42, 8.49 through 8.63, 8.65 through 8.68, and 8.81 through 8.83 (hydro system) must be also added to the model formulation. In these constraints, all variables except for the commitment binary ones are augmented with superscript "ω," representing their realization in scenario ω (second-stage variables). The commitment binary variables (y_{it}, z_{it}, u_{it}, ...) remain scenario-independent, as they are unique for all scenario realizations (first-stage variables) (nonanticipativity) [49]. For consistency, in the stochastic model the above problem constraints are denoted as 8.11′ through 8.48′ (thermal system) and 8.23′ through 8.27′, 8.41′, 8.42′, 8.49′ through 8.63′, 8.65′ through 8.68′, and 8.81′ through 8.83′ (hydro system).

Figure 8.14 represents the scenario tree for the aforementioned two-stage stochastic programming problem.

8.4.2.3 Risk Modeling

In this paragraph, the risk associated with the profit variability is explicitly taken into account in the problem formulation through the CVaR at the α-confidence level (α-CVaR).

As already described in Chapter 2, in a profit maximization problem with discrete probability distribution, α-CVaR is defined as the expected profit of the $(1 - a) \cdot 100\%$ worst scenarios (the ones with lowest profit). This risk measure can be easily incorporated within an optimization problem using linear terms only (no further binary variables are needed).

The risk-constrained formulation of the aforementioned stochastic optimization problem is as follows [35]:

$$Max\left\{\sum_{\omega\in\Omega}prob_\omega\cdot\left\{\sum_{t\in T}\left[\pi_t^\omega\left(\xi_t^\omega\right)\cdot\xi_t^\omega+CQ_t\cdot\left(CP_t-\pi_t^\omega\left(\xi_t^\omega\right)\right)-\sum_{i\in I}c_{it}^\omega\left(p_{it}^\omega\right)\right]\right.\right.$$
$$\left.\left.+\beta\cdot\left(\eta-\frac{1}{1-\alpha}\sum_{\omega\in\Omega}prob_\omega\cdot s_\omega\right)\right\}\right\} \qquad (8.105)$$

or, equivalently,

$$Max\left\{\sum_{\omega\in\Omega}prob_\omega\cdot\left\{\sum_{t\in T}\left[\sum_{b\in B}\left[\pi_{bt}^\omega\cdot\left(d_{bt}^\omega+w_{bt}^\omega\cdot D_{bt}^{min,\omega}\right)\right]\right]+CQ_t\cdot\left(CP_t-\pi_t^\omega\left(\xi_t^\omega\right)\right)\right.\right.$$
$$\left.\left.-\sum_{i\in I}c_{it}^\omega\left(p_{it}^\omega\right)\right\}+\beta\cdot\left(\eta-\frac{1}{1-\alpha}\sum_{\omega\in\Omega}prob_\omega\cdot s_\omega\right)\right\} \qquad (8.106)$$

subject to problem constraints 8.11′ through 8.48′ (thermal system), 8.23′ through 8.27′, 8.41′, 8.42′, 8.49′ through 8.63′, 8.65′ through 8.68′, and 8.81′ through 8.83′ (hydro system), linear constraints 8.98 through 8.102 plus the following ones needed for the linear formulation of the CVaR:

$$\eta-\sum_{t\in T}\left\{\left[\sum_{b\in B}\left[\pi_{bt}^\omega\cdot\left(d_{bt}^\omega+w_{bt}^\omega\cdot D_{bt}^{min,\omega}\right)\right]\right]+CQ_t\cdot\left(CP_t-\pi_t^\omega\left(\xi_t^\omega\right)\right)-\sum_{i\in I}c_{it}^\omega\left(p_{it}^\omega\right)\right\}$$
$$\leq s_\omega\quad\forall\omega\in\Omega \qquad (8.107)$$

$$s_\omega\geq 0\quad\forall\omega\in\Omega \qquad (8.108)$$

where $\alpha\in(0,1)$ is an input parameter denoting the confidence level, $\beta\in[0,\infty)$ is a weighting factor for the incorporation of risk in the expected profit objective function, η represents the Value-at-Risk (VaR), and s_ω is an auxiliary continuous nonnegative variable defined as the maximum between zero and the difference between the VaR and the profit of scenario ω.

The expected profit of the Producer is computed as follows:

$$\sum_{\omega\in\Omega}prob_\omega\cdot\left\{\sum_{t\in T}\left[\pi_t^\omega\left(\xi_t^\omega\right)\cdot\xi_t^\omega+CQ_t\cdot\left(CP_t-\pi_t^\omega\left(\xi_t^\omega\right)\right)-\sum_{i\in I}c_{it}^\omega\left(p_{it}^\omega\right)\right]\right\} \qquad (8.109)$$

The objective function to be maximized, Equation 8.105 or, equivalently, Equation 8.106, consists of the expected profit of the Producer and the CVaR of the profit multiplied by the weighting parameter $\beta\in[0,\infty)$. In fact, this weighting parameter enforces the trade-off between profit and risk; the higher the value of β, the more risk-averse the Producer. In the case that $\beta = 0$, risk is not considered and the Producer is risk-neutral.

In Section 8.5.2.2, it is shown how the variability of the Producer profit can be significantly reduced in the presence of forward bilateral contracts with the end-consumers.

8.5 Numerical Application

In this section, the numerical application of the aforementioned optimization models is presented. In Section 8.5.1, the numerical application of the deterministic self-scheduling problem already described in Sections 8.2 and 8.3 is presented, whereas the respective numerical applications for the stochastic self-scheduling problem, already discussed in Section 8.4, is presented in Section 8.5.2.

8.5.1 Deterministic Problem Application

In this section, the proposed deterministic optimization model for the solution of the hydrothermal Producer self-scheduling problem is applied to the daily self-scheduling of the dominant power company of the Greek interconnected power system in both the production and retail sectors. Currently, this power company owns a majority of the conventional thermal generating units and the total hydroelectric production. The power company is considered to act as a price-maker Producer in the day-ahead energy and reserves markets and as a Retailer through forward bilateral contracts with the end-consumers.

For the sake of simplicity, in all test cases presented in this numerical application, the fuel limitations of thermal generating units are relaxed (see Constraint 8.43), whereas convex cost functions are considered for all thermal units (constraints 8.46 and 8.47 are disregarded). In addition, no prohibited operating zones of hydroplants are considered (see Constraints 8.60 and 8.61) and no time delay between the discharge of a hydroplant and the resulting inflows of the downstream reservoir is taken into account (see Equation 8.65).

In all test cases, the hydraulic coupling model is considered. In the first two test cases presented in Sections 8.5.1.2.1 and 8.5.1.2.2, the effect of the variable head modeling on the hydroplants self-scheduling is neglected. The variable head effect is examined through an indicative test case presented in Section 8.5.1.2.3, where the full hydro model is adopted.

8.5.1.1 System and Units Data

In this case study, the dominant power company of the Greek power system owns 30 thermal units and 13 hydroplants, including 2 pumped storage plants. The total installed thermal and hydro capacity is 7404 and 2934 MW, respectively.

A summary of the thermal system can be found in Table 8.5. Table 8.6 presents an overview of the hydro system with the main technoeconomic data of hydroplants [8]. There are six river basins (A–F) and four blocks of cascaded hydroplants/reservoirs (A–D). The maximum pumping load of the pumped storage plants is given in parentheses. The average specific consumption for all hydroplants is computed as the mean value

TABLE 8.5 Thermal System Overview (Deterministic Case)

Unit Type	Number of Units	Installed Capacity (MW)	Marginal Cost Range (€/MWh)	Start-up Cost Range (€/MWh)
Base load (lignite-fired)	20	4770	29.0–33.8	16,200–87,000
Intermediate load (CCGTs)	3	1404	54.0–61.5	11,300–33,000
Peak load (OCGTs, oil)	7	1230	82.5–108.0	1200–3700

Note: CCGT, combine-cycle gas turbine; OCGT, open-cycle gas turbine.

of the respective values given in Reference [8] and is used in the first two test cases where the simplified hydro model is adopted. The forecast net water inflows for all reservoirs are considered to be constant over the whole scheduling period, whereas the minimum volume of water stored in all reservoirs is equal to zero.

The load demand and the system requirements in all types of reserves are typical curves of the Greek power system [53]. The load demand curve in combination with the anticipated competitors' energy offers are used for the construction of the Producer's hourly residual demand curves for energy. As already mentioned, these curves are nonincreasing and, in this case study, consist of 20 steps. The sum of the competitors' energy offer quantities at each hour is considered to be constant and equal to 2400 MW, whereas the respective competitors' offer prices vary from 62.0 to 100.0 €/MWh. The residual demand curve for energy that the price-maker Producer faces in hour 9 is shown in Figure 8.15. As the Producer acts also as a price-maker in all reserves markets, the hourly residual demand curves for all types of reserves are formulated similarly, on the basis of the respective competitors' reserves offers, and consist of 5–8 steps. The competitors' offer prices for all reserves types vary from 0.02 to 60 €/MWh.

TABLE 8.6 Hydro System Overview (Deterministic Case)

Hydroplant/ Reservoir	River	P^{max} (MW)	P^{min} (MW)	sc^{av} [(m³/s)/MW]	I (m³/s)	V^{max} (Hm³)	V^{ini} (Hm³)	V^{fin} (Hm³)
#1	A	375	20	0.854	11.44	1158.6	447.0	441.9
#2	A	315 (220)	20	1.851	0.18	16.0	7.0	16.0
#3	A	108	20	2.914	0.13	10.0	5.0	7.9
#4	B	437	20	1.231	40.58	3320.9	1584.9	1588.4
#5	B	320	20	1.542	0.40	52.0	15.5	15.5
#6	B	150	20	3.168	0.04	11.9	1.5	0.0
#7	C	210	20	0.174	2.11	144.3	82.6	82.7
#8	C	300	20	1.749	4.65	344.8	181.7	182.1
#9	C	34	10	14.875	0.04	4.1	1.8	1.6
#10	D	375 (250)	20	1.083	12.80	563.0	500.0	492.0
#11	D	116	20	1.940	1.31	69.6	51.2	60.5
#12	E	64	10	0.517	0.28	46.2	10.8	9.2
#13	F	130	20	0.214	3.80	299.0	148.3	147.4

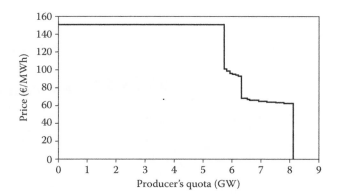

FIGURE 8.15 Producer's residual demand curve for energy (hour 9).

In the case that the hourly system load demand/reserves requirements exceeds the total amount of energy/reserves offered by the competitors, the quantity of the first step of the corresponding hourly residual demand curve is equal to the excess load demand/reserve requirement and the price is equal to the price cap of energy or reserves, respectively. The price cap for energy and reserves have been set equal to 150 and 100 €/MWh, respectively.

It should also be noted that base-load and peak-load thermal units do not operate in AGC mode (and do not contribute to secondary up/down reserve), whereas hydro units do not contribute to primary up/down reserve.

8.5.1.2 Test Results

In this section, three different test cases are examined. In the first case (case D.1), the Producer participates in the day-ahead energy and all reserves markets, determining the degree of involvement in each market described in the introduction. In this case, the effect of the variable head modeling on the hydroplants self-scheduling is neglected.

In the second case (case D.2), the Producer participates in all energy and reserves markets acting simultaneously as a Retailer that has agreed to deliver CQ_t^E (MWh) at a contract price of CP_t^E (€/MWh) through forward bilateral contracts. The procurement cost for reserves is also taken into account, according to Equation 8.92.

In the third case (case D.3), the Producer participates in the day-ahead energy and all reserves markets, whereas the variable head modeling is explicitly taken into account for several hydroplants and its effect on the Producer self-schedule and profits is thoroughly analyzed.

8.5.1.2.1 Energy and Reserves Markets

In the first case, the Producer participates as a Stackelberg monopolist in the day-ahead energy and all reserves markets facing hourly residual demand curves for all market products.

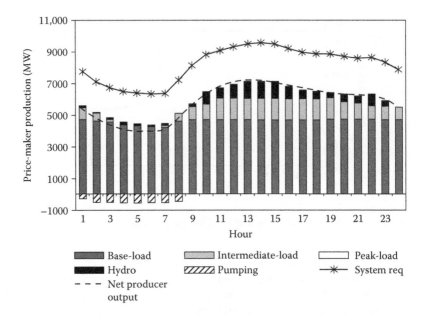

FIGURE 8.16   Price-maker optimal generating schedule per unit type (case D.1).

Figure 8.16 illustrates (a) the hourly system load demand, (b) the hourly power generation from various fuel types, and (c) the net hourly Producer generation (total generation minus total pumping).

During the entire 24-h scheduling period, the net Producer's generation profile follows the shape of the hourly system load demand curve, as it is equal to the system load minus the constant hourly total amount of energy offered by the competitors' units (2400 MW) (see Figure 8.18). In other words, the Producer withholds capacity either by holding certain intermediate-load units offline (hours 3–9) or by enforcing them to operate below their nominal power output, so that the energy MCP is raised to the price cap (150 €/MWh) (see Figure 8.18).

It should be noted that the pumped storage plants operate in pump mode during the low-demand period (i.e., hours 1–8) only. In these hours, base-load and intermediate-load units (low-cost and medium-cost units, respectively) increase their output so as to maintain the constant difference (=2400 MW) between the net Producer's generation and the hourly system load, as described in the previous paragraph (see Figure 8.16). The water stored in this period is mainly released during the high-demand period (i.e., hours 9–23), resulting in an increased production of hydroplants and substituting the operation of the high-cost peak-load units in these hours. In fact, despite the major operation of hydroplants in peak-load hours, these units also operate in a low-power output during low-demand hours (i.e., hours 1–7), so as to provide secondary up/down reserve in these hours and increase profits. Note that all peak-load units are offline during the entire scheduling period in this case.

8.5.1.2.2 *Energy Market, Reserves Markets, and Forward Bilateral Contracts*

In the second case, the dominant Producer not only participates in the energy and reserves markets but also acts as a Retailer through forward bilateral contracts with customers. In this case, the power company's hourly contracted quantity of energy is considered to be equal to the hourly system load demand; that is, the power company is the only energy Retailer in the market. For the sake of simplicity, the forward contract price is considered to be constant during the scheduling horizon and equal to 100 €/MWh, whereas the adoption of any daily contract price profile is straightforward.

The aim of this case study is to show how the presence of forward bilateral contracts as well as the procurement cost of the associated reserves force the power company to increase its total hourly energy production as well as its hourly contribution in all types of reserves as much as possible, in order to induce lower MCPs (or system marginal prices—SMPs) for all market products (given that $\pi_t^{pr}(\xi_t^{pr})$ is a stepwise monotonically nonincreasing function of the Producer quota of each market product, ξ_t^{pr}), so as to reduce the cost of closing its short position in the day-ahead market and maximize its total profits. Table 8.7 presents the Producer's optimal generating schedule per unit type in this case.

TABLE 8.7   Optimal Generating Schedule per Unit Type (Case D.2)

Hour	System Load (MW)	Base-Load Unit Production (MW)	Intermediate-Load Unit Production (MW)	Peak-Load Unit Production (MW)	Hydroplant Production (MW)	Total Pumping (MW)
1	7737	4770	1350	0	21	0
2	7124	4770	1355	0	49	−250
3	6736	4770	1300	0	0	−470
4	6494	4770	1300	0	0	−470
5	6379	4770	1350	0	11	−470
6	6350	4770	1350	0	11	−470
7	6371	4770	1350	0	21	−470
8	7182	4770	1350	0	21	−250
9	8124	4770	1364	164	26	0
10	8867	4770	1404	473	420	0
11	9109	4770	1404	482	653	0
12	9329	4770	1404	482	874	0
13	9485	4770	1404	482	1029	0
14	9565	4770	1404	482	1110	0
15	9489	4770	1404	482	1033	0
16	9207	4770	1404	482	751	0
17	8947	4770	1404	482	491	0
18	8868	4770	1404	482	412	0
19	8852	4770	1404	482	396	0
20	8704	4770	1395	303	436	0
21	8618	4770	1395	303	350	0
22	8699	4770	1395	303	431	0
23	8326	4770	1395	240	121	0
24	7889	4770	1350	0	21	0

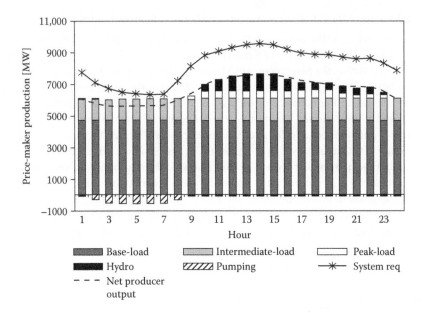

FIGURE 8.17   Price-maker optimal generating schedule per unit type (case D.2).

Figure 8.17 illustrates (a) the hourly system load demand, (b) the optimal generating schedule per unit type, and (c) the hourly net Producer generation (total generation minus total pumping). The difference between the net Producer generation and the system load is covered by the competitors' energy offers (see Figure 8.18).

As shown in Table 8.7 and Figure 8.17, the base-load lignite units operate at their nominal power output during the whole scheduling period, whereas the intermediate-load units are fully loaded only during the high-demand hours (i.e., hours 10–19). In addition, the peak-load units are committed only during the peak-load hours (i.e., 9–23), so that the Producer quota increases and, subsequently, a significant reduction in the energy SMP is induced. Similarly with case D.1, the pumped storage plants operate in pump mode during the low-demand period (i.e., hours 2–8) only, whereas hydro-plants also operate at a low-power output during low-demand hours (i.e., hours 1–2, 5–9, and 24), so as to have the capability to provide secondary up reserve in these hours and increase profits (see Figure 8.20).

Figures 8.19 and 8.20 illustrate the primary-up and secondary-up reserve contribution of the Producer, respectively. Similar to the energy residual demand curve, the gap between the system requirement and the total contribution is covered by the competitors' reserve offers.

Figures 8.17 and 8.18 illustrate that, with the exception of the low-demand hours 1–8 and 24, during the entire scheduling period, the net Producer's generation profile follows the shape of the hourly system load demand curve, as it is reduced by a constant hourly amount of energy offered by the competitors (1800 MW). The resulting MCP of energy is remarkably reduced as compared to that in case D.1 and remains in the range

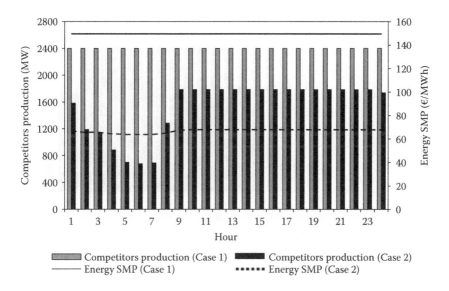

FIGURE 8.18 Competitors' production and energy system marginal prices (cases D.1 and D.2).

of 64.0–68.0 €/MWh during the entire scheduling period (see Figure 8.18), thus leading to a remarkable decrease in the cost of energy supply for the power company.

A similar scheduling strategy is followed by the Producer for all types of reserves in order to lead all reserves clearing prices to the lowest possible value (0.02 €/MWh). Figure 8.19 shows that, during low-demand hours, as intermediate-load units do not operate at their maximum power output (see Tables 8.5 and 8.7), they can contribute to

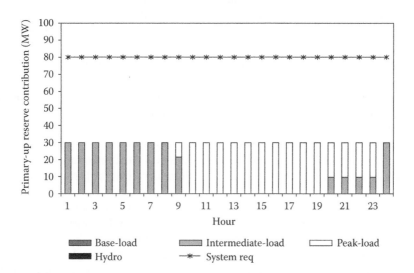

FIGURE 8.19 Primary up reserve contribution per unit type (case D.2).

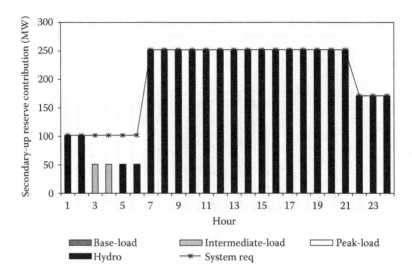

FIGURE 8.20   Secondary up reserve contribution per unit type (case D.2).

primary-up reserve as much as needed in order to lead the respective MCP to 0.02 €/MWh. In the same way, during high-demand hours, peak-load units operate below their technical maximum and, thus, they can contribute to primary up reserve in order to maintain the primary-up reserve MCP at 0.02 €/MWh. Secondary-up reserve is mainly contributed by hydro units (see Figure 8.20) and the respective clearing price is also equal to 0.02 €/MWh during the entire scheduling horizon. Similar results have been observed for the tertiary reserve market.

The explanation of the above scheduling strategy lies in the respective quota of the power company in the production and retail sectors. In this case study, the power company controls a significant percentage of the total production capacity and the entire (100%) retail sector (through its forward bilateral contracts). However, the Retailers cover all payments to the Producers for energy and reserve contribution.

Thus, the power company, being the only Retailer in the market, aims at minimizing the costs of the retail activities, in order to pay less to the competitors for their contribution in energy and reserves. When the quota of the power company in the retail sector drops below its percentage in the production sector, then the optimization leads the energy and reserve prices to the maximum allowed prices, as in this case the power company's income from the production sector is greater than its payments in the retail sector.

Table 8.8 presents the Producer's total daily profit in each of the first two cases studied. The profit is considerably lower in case D.2 than in case D.1 due to the remarkable reduction in the energy MCP (see Figure 8.18) and the fact that a portion of Producer revenues is now fixed at the forward contract price (100 €/MWh), which is significantly lower than the resulting constant MCP of case D.1 (150 €/MWh).

Table 8.9 shows the size of the resulting MILP model and the execution time for both cases, which includes generation of the model and the solution process. It should be noted that the optimization model results in proven optimal solution in both test cases.

TABLE 8.8   Daily Power Company Profits (Deterministic Case)

Cases	Total Profit (€)
Case D.1: Energy and reserves markets	15,917,641
Case D.2: Energy market, reserves markets, and forward bilateral contracts	10,535,364

TABLE 8.9   Size of the Mixed-Integer Linear Programming Model—Execution Times (Deterministic Case)

			Execution Time (s)	
Number of Equations	Number of Variables	Number of Integer (Binary) Variables	Case D.1	Case D.2
47,517	42,337	19,309	≈60	≈940

Also note that in case D.2, the solution algorithm reaches 0.1% of the optimal solution relatively quickly (in less than 120 s), whereas it takes much longer (≈940 s) to locate the proven optimal solution.

8.5.1.2.3 Variable Head Modeling

In the third test case, the Producer participates in the day-ahead energy and all reserves markets, whereas the effect of the variable head modeling on the self-scheduling program and profits of several hydroplants are examined. As already mentioned, the head variation mainly affects the daily operation of hydroplants with small reservoirs. In the hydro subsystem presented in Table 8.6, only hydroplants #2, #3, #6, and #9 may present a notable variation on their reservoir volume in a 24-h scheduling period due to their small size. Given that the incorporation of the variable head modeling formulation in the Producer optimization problem for all hydroplants leads to a considerable increase in the computing time due to the additional binary variables and constraints required, for the sake of simplicity the implementation of the head-dependent hydro operation is limited only in the above hydroplants/reservoirs, for which the full hydro model is adopted.

For this purpose, the main operating data of these hydroplants are modified to follow as consistently as possible the real hydroplant operation, as illustrated in Figures 8.10 and 8.11. The reservoirs of these hydroplants are considered to be discretized in three reservoir volume zones ($K = 3$), each with a distinct specific consumption, volume limits, and power output limits (under AGC also). Table 8.10 presents the modified technical data mainly based on those given in Reference [8] for the four small reservoirs (#2, #3, #6, and #9) of the hydro system. The remaining hydroplants with large reservoirs are considered to operate according to the hydraulic coupling.

To better illustrate the results from the implementation of the variable head modeling, hydroplants #2 and #3 are selected. Similar results are obtained for the remaining small hydroplants.

At first, we consider that all hydroplants operate under the hydraulic coupling model, as already discussed (case D.3.1). Figure 8.21 illustrates the trajectory of the water content and the associated power output of hydroplant #3. As already mentioned, in this case, no matter how high the head/water content of the associated reservoiris, the

TABLE 8.10   Modified Hydroplant Data (Deterministic Case—Variable Head Modeling)

			Hydroplants/(River) Subject to Variable Head Modeling			
			#2 (A)	#3 (A)	#6 (B)	#9 (C)
		V^{ini} (Hm³)	7.0	5.0	6.5	1.8
		V^{fin} (Hm³)	7.3	5.4	5.0	1.6
		Q^{min} (m³/s)	30	30	30	30
		Q^{max} (m³/s)	556	286	462	298
	1	sc^1 [(m³/s)/MW]	1.939	3.181	3.253	21.000
		$P^{min,1}$ (MW)	15.5	9.4	9.2	1.4
		$P^{max,1}$ (MW)	287.0	89.9	142.0	14.2
		$V^{max,1}$ (Hm³)	5.33	3.33	4.0	1.4
Reservoir Volume Zone	2	sc^2 [(m³/s)/MW]	1.851	2.914	3.168	14.875
		$P^{min,2}$ (MW)	16.2	10.3	9.5	2.0
		$P^{max,2}$ (MW)	300.0	98.2	145.8	20.0
		$V^{max,2}$ (Hm³)	10.67	6.67	7.9	2.7
	3	sc^3 [(m³/s)/MW]	1.764	2.647	3.083	8.750
		$P^{min,3}$ (MW)	17.0	11.3	9.7	3.4
		$P^{max,3}$ (MW)	315.2	108.0	149.8	34.1
		$V^{max,3}$ (Hm³)	16	10	11.9	4.1

specific consumption and the power output limits of hydroplant #3 remain unchanged for their whole operating range, as given in Table 8.6. Therefore, the operation of hydroplant #3 does not seem to follow any specific rule regarding the relationship between the stored water in the reservoir and the production schedule.

On the contrary, when the variable head modeling is considered for all small hydroplants (case D.3.2), a distinct production strategy is followed. Figure 8.22 shows the strong preference of hydroplant #3 to operate during the hours that the level of the water content is either in the medium or in the high reservoir volume zone (zones 2 and 3), so

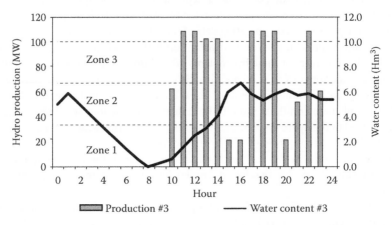

FIGURE 8.21   Power output of hydroplant #3 and water content of associated reservoir (case D.3.1: hydraulic coupling model).

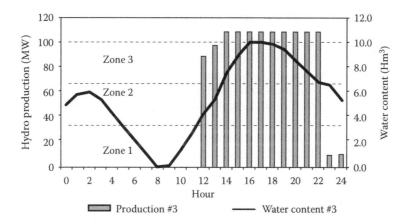

FIGURE 8.22 Power output of hydroplant #3 and water content of associated reservoir (case D.3.2: full hydro model).

that performance curve 2 or 3 is used. As a result, hydroplant #3 takes advantage of the lower specific consumption and the higher power output limits (see Table 8.10) to increase its total production.

It should be noted that, in both cases, the continuous decrease in the stored water of hydroplant #3 during the early morning hours (hours 2–8) is due to the pumping operation of the immediately upstream hydroplant #2, whereas in case D.3.2 the steep increase in the water volume of reservoir #3 during hours 10–16 is due to the intense water discharge and associated energy production of the upstream hydroplant #2.

Table 8.11 presents the total daily energy production of both hydroplants as well as the total daily Producer profit for each of the two cases studied in this paragraph. Although the adoption of the complex full hydro model instead of the simplified one (hydraulic coupling model) for hydroplant #2 and #3 results in a notable increase in their own energy production (and profits also, as the resulting MCP remains constant at 150 €/MWh for the whole scheduling period in both cases), the total daily Producer profit increases only slightly (by 0.05%). This is due to the fact that the total energy production of these hydroplants comprises a small portion of the total daily energy production of the Producer.

In case D.3.1, the optimization model results in proven optimal solution in similar time with that of case D.1, whereas in case D.3.2 the optimality gap is set to 0.1% and it is reached in about 55 min of computing time.

TABLE 8.11 Variable Head Modeling Effect (Deterministic Case)

		Hydraulic Coupling Model	Full Hydro Model	Difference (%)
Production (MWh)	Hydroplant #2	1,939.0	1,992.6	2.76
	Hydroplant #3	1,091.1	1,180.1	8.16
Total daily producer profit (€)		16,068,660	16,077,706	0.05

8.5.2 Stochastic Problem Application

In this section, the proposed two-stage stochastic programming model for the solution of the hydrothermal Producer self-scheduling problem has been implemented for the daily self-scheduling of a hypothetical price-maker hydrothermal Producer. Subsequently, a methodology for constructing the offer curves for each of the Producer generating units participating in the day-ahead electricity market is presented. For the sake of simplicity, it is considered that this Producer participates in the day-ahead energy market only; no reserves markets are considered in this case study. Additionally, this Producer is considered to act also as a Retailer through forward bilateral contracts with the end-consumers.

Regarding the operating constraints of the generating units, the same assumptions with the respective deterministic model are made: the fuel limitations of thermal generating units are relaxed (see Constraint 8.43), whereas convex cost functions are considered for all thermal units (constraints 8.46 and 8.47 are disregarded). In addition, no prohibited operating zones of hydroplants are considered (see Constraints 8.60 and 8.61) and no time delay between the discharge of a hydroplant and the resulting inflows of the downstream reservoir is taken into account (see Equation 8.65). Finally, as regards the hydro model, the effect of the variable head modeling is neglected and the hydraulic coupling model is implemented.

8.5.2.1 System and Units Data

In this case study, the hypothetical price-maker Producer owns five thermal units and two hydroplants. The total installed thermal and hydro capacities are 1664 and 812 MW, respectively. This Producer is considered to belong to the Greek power system, which is a medium-scale power system with an average yearly load of about 6000 MW. Therefore, the Producer's market share allows it to act as a price-maker during particular time periods affecting the resulting MCPs to the Producer's own benefit.

Table 8.12 gives a summary of the Producer's thermal system, whereas Table 8.13 presents an overview of the Producer's hydro system with the main technoeconomic data of hydroplants. Hydroplants are identical to hydroplant #1 and #4 of the deterministic test case and are considered to be located in different river basins.

The system load demand is a typical curve of the Greek power system [53]. The net system load curve (total system load minus the renewable energy sources injection) in

TABLE 8.12 Thermal System Overview (Stochastic Case)

Unit Type	Number of Units	Installed Capacity (MW)	Marginal Cost Range (€/MWh)	Start-up Cost Range (€/MWh)
Base load (lignite-fired)	2	548	29.0–30.5	38,000–64,000
Intermediate load (CCGTs)	2	928	54.0–60.0	11,000–32,000
Peak load (OCGTs, oil)	1	188	70.0–75.0	1700

Note: CCGT, combine-cycle gas turbine; OCGT, open-cycle gas turbine.

TABLE 8.13 Hydro System Overview (Stochastic Case)

Hydroplant/ Reservoir	River	P^{max} (MW)	P^{min} (MW)	$SC^{av}[(m^3/s)/ MW]$	I (m³/s)	V^{max} (Hm³)	V^{ini} (Hm³)	V^{fin} (Hm³)
#1	A	375	20	0.854	11.44	1158.6	447.0	441.9
#2	B	437	20	1.231	40.58	3320.9	1584.9	1588.4

combination with the competitors' energy offers (which, in this case study, are scenario-dependent) is used for the formulation of the Producer's scenario-dependent hourly residual demand curves. It is assumed that the competitor units' availability and the quantity of their own energy offers are considered to be deterministically known. Therefore, only uncertainties in the competitors' offer prices are modeled through the definition of a number of scenarios. The competitors' energy offers are discretized in 20 steps. The sum of the competitors' energy offer quantities at each hour is considered to be constant and equal to 9600 MW, whereas the respective nominal competitors' offer prices vary from 25.0 to 100.0 €/MWh.

To consider uncertainty in the residual demand curves, 11 scenarios are formulated, each representing a series of 24-h residual demand curves. Each series results from an initial series of 24 residual demand curves, where the nominal competitors' offer prices are multiplied by the product of (a) a scaling factor ranging from 0.75 to 1.25 and (b) a uniform distribution function in the interval [0.9, 1.1], in order to create randomness. Table 8.14 presents the 11 scenarios with their associated probabilities of occurrence. The 20 residual demand curves that the Producer faces in hour 5 are shown in Figure 8.23.

TABLE 8.14 Residual Demand Curves Scenarios

Scenarios	1	2	3	4	5	6	7	8	9	10	11
Probability	0.08	0.09	0.09	0.09	0.1	0.1	0.1	0.09	0.09	0.09	0.08
Scaling factor	0.75	0.80	0.85	0.90	0.95	1	1.05	1.10	1.15	1.20	1.25

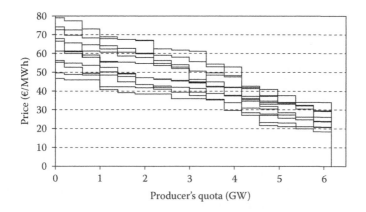

FIGURE 8.23 Residual demand curve realizations for all scenarios ($t = 5$).

TABLE 8.15   Forward Bilateral Contracts—Test Cases

Test Case	Contracted Quantity (MWh) (Percentage of the Hourly System Load Demand)	Contracted Price (€/MWh)
S.1	Without bilateral contracts	—
S.2	5%	100
S.3	10%	100
S.4	20%	100

Note that in this hour all residual demand curves consist of 16 steps, due to the significantly low net system load observed in that hour.

In this case study, four different test cases are examined, regarding the volume of forward bilateral contracts that the Producer has signed with the end-consumers (see Table 8.15) and how these contracts affect the Producer's day-ahead scheduling strategy under uncertainty. Test results referring to case S.4 (bilateral contracts covering 20% of the hourly total system load) are discussed in the following section, and comparative results from all test cases are also presented.

8.5.2.2 Test Results

As already discussed, the use of stochastic programming even under a limited number of scenarios usually leads to a dramatic increase in the size of the problem and may even lead to intractability. In the present formulation, where the Producer acts as a price-maker in the day-ahead energy market, the increase in the size of the stochastic model (as compared with the respective deterministic one) and the subsequent difficulty in solving it, is mainly due to the existence of the binary variables needed for the linearization of the residual demand curves, which are now scenario dependent (w_{bt}^{ω}), whereas the commitment decision variables (y_{it}, z_{it}, u_{it}, ...) remain scenario independent. However, the number of continuous variables also increases in proportion to the number of scenarios. In the stochastic price-taker Producer problem the number of the problem binary variables remains unchanged, irrespective of the number of scenarios examined.

The solution of the two-stage stochastic programming model yields the optimal commitment and dispatch schedule for each of the Producer's generating units and for

TABLE 8.16   Producer Unit Commitment Schedule (Case S.4)

												Hour												
Unit	0	1	2	3	4	5	6	7	8	9	10	11	12	13	14	15	16	17	18	19	20	21	22	23	24
#1	1	1	1	1	1	1	1	1	1	1	1	1	1	1	1	1	1	1	1	1	1	1	1	1	1
#2	1	1	1	1	1	1	1	1	1	1	1	1	1	1	1	1	1	1	1	1	1	1	1	1	1
#3	1	1	1	1	1	1	1	1	1	1	1	1	1	1	1	1	1	1	1	1	1	1	1	1	1
#4	1	1	1	1	1	1	1	1	1	1	1	1	1	1	1	1	1	1	1	1	1	1	1	1	1
#5	0	0	0	0	0	0	0	0	0	0	1	1	1	1	1	1	1	1	1	1	1	1	1	1	0
#6	0	0	0	0	0	0	0	0	0	0	1	1	1	1	1	1	1	1	1	0	1	0	0	0	0
#7	0	0	0	0	0	0	0	0	0	0	0	0	0	0	0	0	0	0	0	0	0	0	0	0	0

each scenario. Table 8.16 illustrates the optimal commitment schedule for all Producer units for case S.4, which is unique, as the unit commitment variables are first-stage variables and, thus, scenario-independent. Unit #1–5 correspond to the thermal units, whereas unit #6 and #7 are the hydroplants. Similarly with the deterministic case, the peak-load unit (unit #5) and hydroplant #6 are online during peak hours only, to increase the Producer's quota and increase profits. Note that hydroplant #7 does not start during the entire scheduling period to store water, as its target (final) reservoir volume is significantly higher than the initial one (see Table 8.13).

In this test case, the high volume of forward bilateral contracts incentivizes the Producer to reduce significantly the MCP to its own benefit, as already discussed in the deterministic model (see Section 8.5.1.2.2). Figure 8.24 illustrates the Producer's quota realizations for all scenarios, whereas the resulting MCP for each scenario is given in Figure 8.25. In this figure, the bold line denotes the average MCP over all scenarios. It is shown that during low-demand hours (hours 1–8) the total Producer's quota exhibits a considerable variability for the different scenarios, whereas during the remaining hours (hours 9–24) the Producer is rather certain about the total amount of energy to be sold. The explanation lies in the fact that during hours 1–8, due to the low system load demand, the resulting MCP lies in the range of the variable cost of the Producer's CCGTs (see Table 8.12). Therefore, the scheduling strategy of these units may seriously affect the MCP. On the contrary, during hours 9–24, the high system load demand leads the resulting MCP of almost all scenarios above the variable cost of the Producer's costliest generating unit (unit #5), making the Producer incapable of seriously affecting the MCP.

As already discussed, the 11 resulting MCPs, along with the 11 quantities decided by the Producer for each generating unit i, although corresponding to different market outcomes, constitute the offer curve that the generating unit should submit to the day-ahead market for each hour. Test results show that the decisions made by the Producer for each hour, for each unit, and for the 11 different realizations are not independent, but they constitute nondecreasing hourly offer curves, as required by constraints 8.99 through 8.102. For illustrative purposes, Figures 8.26 through 8.29 illustrate the optimal quantities

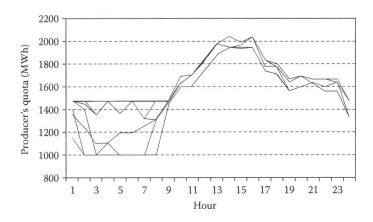

FIGURE 8.24 Producer's quota realizations for all scenarios (case S.4).

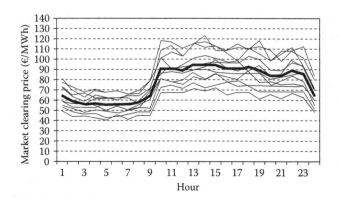

FIGURE 8.25 Market clearing price realizations for all scenarios (case S.4).

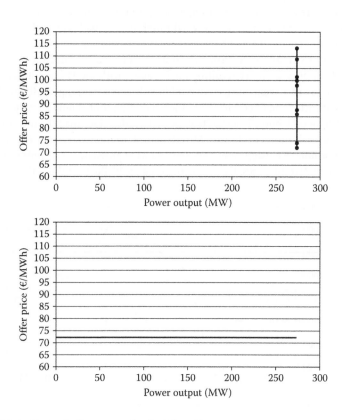

FIGURE 8.26 Optimal offer curves of unit #2 ($t = 15$).

decided by the Producer for some of the units and the resulting nondecreasing stepwise offer curves to be submitted to the day-ahead energy market. Note that the offer curve of lignite unit #2 consists of a single step, denoting that this unit is totally indifferent of the 11 scenario outcomes, as it is profitable for it to operate at its technical maximum, regardless of the resulting MCP.

Figure 8.30 illustrates the average Producer's quota over all scenarios for each test case studied. It is concluded that, as the volume of the contracted quantity increases, the Producer's quota also increases, to cause a significant reduction in the MCP and in the cost of energy supply (see also Section 8.5.1.2.2). Figure 8.31 shows how the average MCP decreases with an increase in the Producer's quota.

Table 8.17 presents the statistical properties of the stochastic model, that is, the expected profit and the CVaR for the four test cases studied. In this study, the CVaR is computed for $\beta = 0.5$ (moderate risk aversion) and $\alpha = 0.8$ (80% confidence level); that is, the CVaR denotes the expected value of the 20% scenarios with lowest profit. It is clear that, as the volume of the contracted quantity increases, both the expected profit and the CVaR increase, denoting that the Producer's risk exposure to the uncertain market conditions (i.e., competitors' energy offers) is significantly reduced in the presence of forward bilateral contracts.

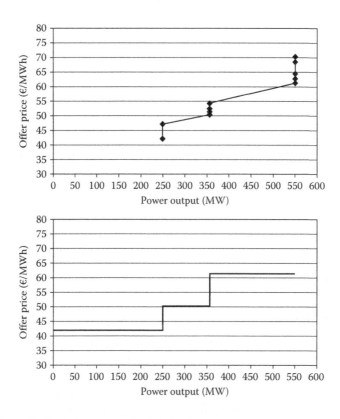

FIGURE 8.27 Optimal offer curves of unit #3 ($t = 4$).

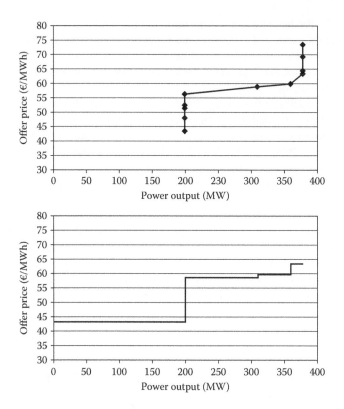

FIGURE 8.28 Optimal offer curves of unit #4 ($t = 2$).

The resulting stochastic MILP model includes 72,880 equations, 43,790 variables, and 10,040 binary variables. The present model results in proven optimal solution in all test cases and the computing time required to obtain this solution varies from 1.5 (case S.4) to 16 min (case S.1).

All simulation runs for the deterministic model were performed on a 2.8-GHz Intel Quad Core processor with 24 GB of RAM, while all runs for the stochastic model were performed on a 3.2-GHz Intel Quad Core processor with 24 GB of RAM. Both machines run 64-bit Windows, while the CPLEX 12.0 solver under GAMS 23.3 [37] was used for all simulation tests.

8.6 Conclusions

This chapter presented a detailed MILP formulation for the solution of the short-term self-scheduling problem of an electricity Producer with thermal and hydro generating units. Different optimization models regarding the role of the Producer in the electricity market as well as the thermal and hydro systems operation are analytically presented. Uncertainty in market conditions is also considered under a two-stage stochastic programming framework, whereas the CVaR metric accounting for risk management is also modeled.

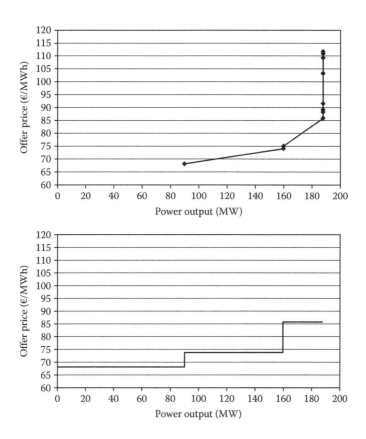

FIGURE 8.29 Optimal offer curves of unit #5 ($t = 18$).

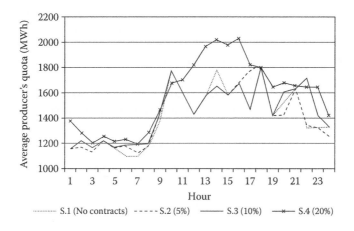

FIGURE 8.30 Average Producer's quota under uncertainty.

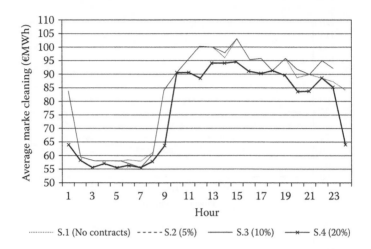

FIGURE 8.31 Average market clearing price under uncertainty.

Two different numerical applications are presented. First, the Producer is considered to have deterministic knowledge of the market conditions. In this case, the hydrothermal Producer acts as a price-maker in all day-ahead markets (energy and reserves), while also acting as a Retailer, through forward bilateral contracts with the end-consumers. Therefore, the Producer is considered to set up a power company with a dominant role in both the production and retail sectors of an electricity market. The proposed model provides a dominant power company with a useful tool to determine optimally its degree of involvement in each day-ahead market in order to maximize its total benefit. Numerical results show that in the case that the power company is involved in the production sector only, its dominant position permits the manipulation of the day-ahead energy and reserves markets, leading the day-ahead clearing prices for all market products to the respective price caps and maximizing its own profits. On the contrary, the power company's incentive to exercise market power and manipulate the short-term electricity markets is significantly reduced by the presence of forward bilateral contracts with the end-consumers and the procurement cost of reserves. The effect of the variable head modeling in the self-scheduling and profits of the Producer hydroplants is examined through an illustrative test case.

In the second case, the Producer (acting also as a Retailer) models the market uncertainty through stochastic residual demand curves for the day-ahead energy market only,

TABLE 8.17 Statistical Properties of the Stochastic Model

Test Case	Expected Profit (10^3 €)	CVaR ($\alpha = 0.8$, $\beta = 0.5$) (10^3 €)
S.1	1308.8	696.1
S.2	1451.0	1014.6
S.3	1600.6	1333.4
S.4	1974.8	1964.9

formulating a two-stage stochastic programming problem. The solution of the stochastic problem provides the optimal hourly offer curves for each Producer's generating unit to be submitted to the day-ahead energy market. Test results show that the Producer's risk exposure to the uncertain market conditions is significantly reduced in the presence of forward bilateral contracts.

8.7 List of Symbols

Sets

b (B^{pr})   Index (set) of steps of the residual demand curve of market product pr
f (F^i)   Index (set) of steps of the marginal cost function of thermal unit i
h (^{J})   Index (set) of pumped storage plants ($B \subseteq J$)
i (I)   Index (set) of thermal units and hydroplants
j (J)   Index (set) of hydroplants ($J \subseteq I$)
j^u (J^U)   Index (set) of hydroplants/reservoirs immediately upstream of hydroplant/reservoir j
k (K)   Index (set) of water volume zone of hydroplant/reservoir j
l (L)   Index (set) of unit start-up types $L = \{h, w, c\}$, where h represents hot, w warm, and c cold start-up
m (M)   Index (set) of reserve types $M = \{1+, 1-, 2+, 2-, 3\}$, where $m = 1+$ represents primary up, $m = 1-$ primary down, $m = 2+$ secondary up, $m = 2-$ secondary down, and $m = 3$ tertiary (spinning: 3S and nonspinning: 3NS)
n (N^h)   Index (set) of the units of pumped storage plant h
pr (PR)   Index (set) of market products, where E represents energy and m reserves
s (S^i)   Index (set) of steps of the soak process of thermal unit i
t (T)   Index (set) of hours of the scheduling horizon
T^- Planning horizon extended to the past (set)
T^+ Planning horizon extended to the future (set)
z (Z)   Index (set) of the permissible operating zones of hydroplant j
ω (Ω)   Index (set) of scenarios

Functions

$\pi_t^{\mathrm{pr}}(\xi_t^{\mathrm{pr}})$ Residual demand curve of market product pr in hour t, expressed as a stepwise monotonically nonincreasing function that represents the MCP as a function of the Producer quota ξ_t^{pr}
c_{it} (p_{it})   Total production cost of thermal unit i in hour t at level p_{it} (in €/h)

Parameters

λ_t^{pr} forecast clearing price of market product pr in hour t (in €/MWh)
π_{bt}^{pr} Price of step b of the residual demand curve of market product pr in hour t (in €/MWh)

C_{if} Marginal cost of step f of thermal unit i marginal cost function (in €/MWh)

CQ_t^{pr} Forward contract quantity of market product pr in hour t (in MWh)

CP_t^{pr} Forward contract price of market product pr in hour t (in €/MWh)

D_t System load demand in hour t (in MW)

D_{bt}^{pr} Size of step b of the residual demand curve of market product pr in hour t (in MW)

$D_{bt}^{pr,min}$ Summation of power blocks from step 1 to step $b-1$ of the residual demand curve of market product pr in hour t (in MW) ($D_{1t}^{pr,min} = 0, \quad \forall t \in \mathbf{T}$)

DT_i Minimum down time of unit i (in h)

E_{if} Size of step f of thermal unit i marginal cost function (in MW)

E_j^{max} Maximum energy production of hydroplant j during the scheduling period (in MWh)

E_j^{min} Minimum energy production of hydroplant j during the scheduling period (in MWh)

e_h^A Pumping efficiency of pumped storage plant h

$\bar{F}_i$ Maximum energy production of unit i during the scheduling period due to fuel limitations (in MWh)

I_{jt} Forecast net water inflows in reservoir j in hour t (in m³/s)

M Conversion factor equal to 0.0036 (in Hm³s/m³h)

NLC_i No-load cost of unit i (for 1 h operation) (in €/h)

$P_{A,h}^n$ Constant pumping load of unit n of pumped storage plant h (in MW)

P_i^{max} Maximum power output of thermal unit/hydroplant i (in MW)

$P_i^{min,AGC}$ Maximum power output of thermal unit/hydroplant i while operating under AGC (in MW)

P_i^{min} Minimum power output of thermal unit/hydroplant i (in MW)

$P_i^{min,AGC}$ Minimum power output of thermal unit/hydroplant i while operating under AGC (in MW)

$P_{i,s}^{soak,\ell}$ Power output of unit i corresponding to the sth interval of the soak phase under type ℓ start-up (in MW)

$\bar{P}_i^{soak}$ Fixed power output of unit i while in the soak phase (in MW)

P_i^{syn} Synchronization load of unit i (in MW)

$P_j^{max,k}$ Maximum power output of hydroplant j while operating in reservoir volume zone k (in MW)

$P_j^{max,AGC,k}$ Maximum power output of hydroplant j while operating under AGC in reservoir volume zone k (in MW)

$P_j^{min,k}$ Minimum power output of hydroplant j while operating in reservoir volume zone k (in MW)

$P_j^{min,AGC,k}$ Minimum power output of hydroplant j while operating under AGC in reservoir volume zone k (in MW)

$P_j^{max,z}$ Upper bound of the zth permissible operating zone of hydroplant j (in MW)

$P_j^{min,z}$ Lower bound of the zth permissible operating zone of hydroplant j (in MW)

$prob_\omega$ Probability of scenario ω

Q_j^{max} Maximum water discharge of hydroplant j (in m³/s)

$Q_j^{\max,AGC}$	Maximum water discharge of hydroplant j while operating under AGC (in m³/s)
$Q_j^{\min}$	Minimum water discharge of hydroplant j (in m³/s)
$Q_j^{\min,AGC}$	Minimum water discharge of hydroplant j while operating under AGC (in m³/s)
R_i^m	Maximum contribution of thermal unit/hydroplant i in reserve type m (in MW)
RD_i	Ramp-down rate of thermal unit/hydroplant i (in MW/min)
RD_i^{AGC}	Ramp-down rate of thermal unit/hydroplant i while operating under AGC (in MW/min)
RR_t^m	System requirement in reserve type m during hour t (in MW)
RU_i	Ramp-up rate of thermal unit/hydroplant i (in MW/min)
RU_i^{AGC}	ramp-up rate of thermal unit/hydroplant i while operating under AGC (in MW/min)
sc_h^A	Average specific consumption of pumped storage plant h while operating in pump mode [in (m³/s)/MW]
sc_j^{av}	Average specific consumption of hydroplant j [in (m³/s)/MW]
sc_j^k	Specific consumption of hydroplant j while operating in reservoir volume zone k [in (m³/s)/MW]
SDC_i	Shutdown cost of thermal unit i (in €)
SUC_i^ℓ	Start-up cost of thermal unit i from type ℓ standby until load with synchronization (in €)
T_i^ℓ	Time off-load before going into longer standby conditions ($\ell = w$: hot to warm, $\ell = c$: hot to cold) of unit i (in h)
T_i^{des}	Time from technical minimum power output to desynchronization of thermal unit i (in h)
$T_i^{syn,\ell}$	Time to synchronize unit i under type ℓ start-up (in h)
$T_i^{soak,\ell}$	Soak time of unit i under type ℓ start-up (in h)
UT_i	Minimum up time of unit i (in h)
V_j^{ini}	Initial volume of water stored in reservoir j at the beginning of the scheduling horizon (in Hm³)
V_j^{fin}	Final volume of water stored in reservoir j at the end of the scheduling horizon (in Hm³)
$V_j^{\max}$	Maximum volume of water stored in reservoir j (in Hm³)
$V_j^{\max,k}$	Maximum volume of water stored for zone k of reservoir j (in Hm³)

Variables

$\mathbf{x}_{it}$	Vector of the continuous and the binary decision variables associated with the operation of unit i during hour t

Continuous Variables

ξ_t^{pr}	Producer quota of market product pr (summation of power output or contribution in reserves of the Producer thermal units/hydroplants) in hour t (in MW)

d_{bt}^{pr}	portion of step b of the residual demand curve of market product pr that corresponds to the quantity accepted by the ISO in hour t (in MW)
e_{ift}	Portion of step f of the ith unit's marginal cost function loaded in hour t (in MW)
p_{it}	Power output of unit i accepted by the ISO in hour t in the day-ahead energy auction (in MW)
p_{it}^{des}	Power output of thermal unit i during the desynchronization phase in hour t (in MW)
p_{it}^{soak}	Power output of thermal unit i during the soak phase in hour t (in MW)
p_{ht}^{A}	Pumping load of pumped storage plant h in hour t (in MW)
q_{jt}	Total water discharge of hydroplant j in hour t (in m³/s)
q_{jt}^{δ}	Portion of water discharge in excess of the minimum water discharge of hydroplant j in hour t (in m³/s)
q_{ht}^{A}	Water flow from the lower to the upper reservoir of a pumped storage plant h when operating in pump mode in hour t (in m³/s)
r_{it}^{m}	Contribution of thermal unit/hydroplant i in reserve type m during hour t (in MW)
sp_{jt}	Spillage over reservoir j during hour t (in m³/s)
v_{jt}	Volume of water stored in reservoir j at the end of hour t (in Hm³)

Binary Variables

g_{ht}^{n}	Binary variable that is equal to 1 if unit n of pumped storage plant h is in pump mode in hour t
u_{it}	Binary variable that is equal to 1 if thermal unit/hydroplant i is committed during hour t
u_{it}^{AGC}	Binary variable that is equal to 1 if thermal unit/hydroplant i operates under AGC and provides secondary reserve during hour t
u_{it}^{n}	Binary variable that is equal to 1 if unit i is in operating phase n during hour t, where $n = syn$: synchronization, $n = soak$: soak, $n = disp$: dispatchable, $n = des$: desynchronization
u_{jt}^{z}	Binary variable that is equal to 1 if hydroplant j operates in permissible zone z during hour t
w_{bt}^{pr}	Binary variable that is equal to 1 if step b is the last step needed to obtain Producer quota ξ_{t}^{pr} or market product pr in hour t
x_{jt}^{k}	Binary variable used for the discretization of the performance curves of hydroplant j during hour t
y_{it}^{ℓ}	Binary variable that is equal to 1 if a type ℓ start-up of thermal unit i is initiated during hour t
y_{it}	Binary variable that is equal to 1 if thermal unit/hydroplant i is started during hour t
z_{it}	Binary variable that is equal to 1 if thermal unit/hydroplant i is shut down during hour t

All continuous and binary variables, if augmented with superscript "ω," represent their realization in scenario ω.

Appendix A8.1: Modeling of Intertemporal Constraints

Intertemporal constraints, 8.11, 8.13, 8.15, 8.18, 8.23, and 8.24, invoking the binary variables related to the unit commitment status (e.g., y_{it}, z_{it}, u_{it}) are "backward looking," that is, the binary variables within the summations that appear on one side of the constraints refer to the past of the time interval where the binary variable on the other side of the constraint is calculated. The specification of the initial conditions requires the extension of the planning horizon to the past (negative time). The scheduling horizon extended to the past is as follows:

$$T^- = T^{\text{ext}-} \cup T = \{-T^-,...,0\} \cup \{1,...,T\}$$

where $T = 24$ h for daily scheduling, and T^- is a large number of hours in the past that satisfy the condition

$$T^- > \max_i \{T_i^c\}$$

In our simulations, $T^- = 100$ h, which is on the safe side (typical reservation time for cold start of lignite-fired units is 3 days or 72 h).

The extension of the planning horizon to the negative time axis provides a natural and easy way to introduce initial conditions, without increasing the number of problem binary variables, as all problem variables are known for $t \in T^{\text{ext}-}$. By allowing some of the binary variables that refer to the past to be determined through the problem constraints, the specification of the initial conditions may be further simplified at the expense of a small increase in the number of variables.

Suppose, for example, that, at $t = 0$, lignite unit #1 has already been online for 3 h, following a warm start-up. The only initial conditions that are required are

$$\{y_{it}^w = 1 \text{ for } t = -2\} \quad \text{and} \quad \{y_{it} = 1 \text{ for } t = -2\}$$

Constraints 8.13 through 8.27 will then ensure that the unit will begin the day at the first step of its soak sequence and will remain online at least until hour 5, when the unit's minimum up time constraint is satisfied, according to the unit's data in Table 8.2 (the unit's time to synchronize after a warm start-up is 3 h, soak duration is 3 h, and minimum up time is 8 h). Additional initial conditions, such as $\{u_{it}^{\text{syn},w} = 1 \text{ for} - 2 \le t \le 0\}$ or $\{u_{it} = 1 \text{ for} - 2 \le t \le 5\}$, may also be specified, but are not necessary.

Intertemporal constraints related to unit desynchronization, 8.21, 8.22, and 8.40, are forward-looking and require the extension of the planning horizon to the future, as time index "τ" related to the summations in constraints 8.21 and 8.22 as well as time index

"$t + T_i^{des}$" in constraint 8.40 may refer to hours beyond the T planning horizon. The planning horizon extended to the future is defined as follows:

$$T^+ = T \cup T^{ext+} = \{1,...,T\} \cup \{T+1,...,T+T^+\}$$

where $T^+ = \max_i\{T_i^{des}\}$ and $T = 24$ h for daily scheduling.

The extension of the planning horizon to the future is a natural and easy way to introduce final conditions.

From constraints 8.21, 8.22, and 8.40, it is noted that future values of only the unit shutdown binary variables, z_{it}, are needed in the model. In the model implementation, the time domain of definition of the unit shutdown variables, z_{it}, is extended to include future time intervals ($t \in T^{ext+}$). As the time to desynchronize, T_i^{des}, of thermal units is small and the extension of the time horizon to the future depends on the maximum time to desynchronize, only a few additional binary variables, z_{it}, $i \in I$, $t \in T^{ext+}$, are added to the model. The future values of the shutdown variables (z_{it}, $i \in I$, $t \in T^{ext+}$) are determined by the model solution.

References

1. A. G. Bakirtzis, N. P. Ziogos, A. C. Tellidou, and G. A. Bakirtzis, Electricity producer offering strategies in day-ahead energy market with step-wise offers, *IEEE Transactions on Power Systems*, 22(4), 1804–1818, 2007.

2. A. Baillo, M. Ventosa, M. Rivier, and A. Ramos, Optimal offering strategies for generation companies operating in electricity spot markets, *IEEE Transactions on Power Systems*, 19(2), 745–753, 2004.

3. J. Yang and N. Chen, Short term hydrothermal coordination using multi-pass dynamic programming, *IEEE Transactions on Power Systems*, 4(3), 1050–1056, 1989.

4. H. Z. Yan, P. B. Luh, X. H. Guan, and P. M. Rogan, Scheduling of hydrothermal power systems, *IEEE Transactions on Power Systems*, 8(13), 1358–1365, 1993.

5. M. S. Salam, K. M. Nor, and A. R. Hamdan, Hydrothermal scheduling based Lagrangian relaxation approach to hydrothermal coordination, *IEEE Transactions on Power Systems*, 13(1), 226–235, 1998.

6. C. Li, P. J. Jap, and D. L. Streiffert, Implementation of network flow programming to the hydrothermal coordination in an energy management system, *IEEE Transactions on Power Systems*, 8(3), 1045–1053, 1993.

7. O. Nilsson and D. Sjelvgren, Mixed-integer programming applied to short-term planning of a hydro-thermal system, *IEEE Transactions on Power Systems*, 11(1), 281–286, 1996.

8. C. E. Zoumas, A. G. Bakirtzis, J. B. Theocharis, and V. Petridis, A genetic algorithm solution approach to the hydrothermal coordination problem, *IEEE Transactions on Power Systems*, 19(2), 1356–1364, 2004.

9. D. N. Simopoulos, S. D. Kavatza, and C. D. Vournas, An enhanced peak shaving method for short term hydrothermal scheduling, *Energy Conversion and Management*, 48(11), 3018–3024, 2007.

10. H. Y. Yamin, Review on methods of generation scheduling in electric power systems, *Electric Power Systems Research*, 69(2–3), 227–248, 2004.

11. J. M. Arroyo and A. J. Conejo, Optimal response of a thermal unit to an electricity spot market, *IEEE Transactions on Power Systems*, 15(3), 1098–1104, 2000.

12. J. M. Arroyo and A. J. Conejo, Optimal response of a power generator to energy, AGC, and reserve pool-based markets, *IEEE Transactions on Power Systems*, 17(2), 404–410, 2002.

13. J. M. Arroyo and A. J. Conejo, Modeling of start-up and shutdown power trajectories of thermal units, *IEEE Transactions on Power Systems*, 19(3), 1562–1568, 2004.

14. G. W. Chang, Y. D. Tsai, C. Y. Lai, and J. S. Chung, A practical mixed integer linear programming based approach for unit commitment, in *Proceedings IEEE Power Engineering Society General Meeting*, June 2004, Vol. 1, pp. 221–225.

15. J. Garcia-Gonzalez and J. Barquin, Self-unit commitment of thermal units in a competitive electricity market, in *Proceedings 2000 IEEE Power Engineering Society Summer Meeting*, 2000, Vol. 4, pp. 2278–2283.

16. S. de la Torre, J. M. Arroyo, A. J. Conejo, and J. Contreras, Price maker self-scheduling in a pool-based electricity market: A mixed-integer LP approach, *IEEE Transactions on Power Systems*, 17(4), 1037–1042, 2002.

17. G. Chang, M. Aganagic, J Waight, J. Medina, T. Burton, S. Reeves, and M. Christoforidis, Experiences with mixed-integer linear programming based approaches on short-term hydro scheduling, *IEEE Transactions on Power Systems*, 16(4), 743–749, 2001.

18. A. J. Conejo, J. M. Arroyo, J. Contreras, and F. A. Villamor, Self-scheduling of a hydro producer in a pool-based electricity market, *IEEE Transactions on Power Systems*, 17(4), 1265–1272, 2002.

19. A. Borghetti, C. D'Ambrosio, A. Lodi, and S. Martello, An MILP approach for short-term hydro scheduling and unit commitment with head-dependent reservoir, *IEEE Transactions on Power Systems*, 23(3), 1115–1124, 2008.

20. J. P. S. Catalão, S. J. P. S. Mariano, V. M. F. Fendes, and L. A. F. M. Ferreira, Scheduling of head-sensitive cascaded hydro systems: A nonlinear approach, *IEEE Transactions on Power Systems*, 24(1), 337–346, 2009.

21. J. P. S. Catalão, H. M. I. Pousinho, and V. M. F. Fendes, Scheduling of head-dependent cascaded hydro systems: Mixed-integer quadratic programming approach, *Energy Conversion and Management*, 51(3), 524–530, 2010.

22. A. Diniz and M. E. P. Maceira, A four-dimensional model of hydro generation for the short-term hydrothermal dispatch problem considering head and spillage effects, *IEEE Transactions on Power Systems*, 23(3), 1298–1308, 2008.

23. T. Li and M. Shahidehpour, Price-based unit commitment: A case of Lagrangian relaxation versus mixed-integer programming, *IEEE Transactions on Power Systems*, 20(4), 2015–2025, 2005.

24. S. Bisanovic, M. Hajro, and M. Dlakic, Hydrothermal self-scheduling in a day-ahead electricity market, *Electric Power Systems Research*, 78(9), 1579–1596, 2008.

25. M. A. Plazas, A. J. Conejo, and F. J. Prieto, Multimarket optimal bidding for a power producer, *IEEE Transactions on Power Systems*, 20(4), 2041–2050, 2005.

26. C. Triki, P. Beraldi, and G. Gross, Optimal capacity allocation in multi-auction electricity markets under uncertainty, *Computers & Operations Research*, 32(2), 201–217, 2005.

27. G. B. Shrestha, S. Kai, and L. Goel, An efficient stochastic self-scheduling technique for power producers in the deregulated power market, *Electric Power Systems Research*, 71(1), 91–98, 2004.

28. S.-E. Fleten and T. K. Kristoffersen, Short-term hydropower production planning by stochastic programming, *Computers & Operations Research*, 35(8), 2656–2671, 2008.

29. C. De Ladurantaye, M. Gendreau, and J.-Y. Potvin, Strategic bidding for price-taker hydroelectricity producers, *IEEE Transactions on Power Systems*, 22(4), 2187–2203, 2007.

30. A. J. Conejo, F. J. Nogales, J. M. Arroyo, and R. Garcia-Bertrand, Risk-constrained self-scheduling of a thermal power producer, *IEEE Transactions on Power Systems*, 19(3), 1569–1574, 2004.

31. R. A. Jabr, Robust self-scheduling under price uncertainty using conditional value-at-risk, *IEEE Transactions on Power Systems*, 20(4), 1852–1858, 2005.

32. T. Li, M. Shahidehpour, and Z. Li, Risk-constrained bidding strategy with stochastic unit commitment, *IEEE Transactions on Power Systems*, 22(1), 449–458, 2007.

33. H. Haghighat, H. Seifi, and A. R. Kian, On the self-scheduling of a power producer in uncertain trading environments, *Electric Power Systems Research*, 78(3), 311–317, 2008.

34. P. Beraldi, D. Conforti, and A. Violi, A two-stage stochastic programming model for electric energy producers, *Computers & Operations Research*, 35(10), 3360–3370, 2008.

35. L. P. Garces and A. J. Conejo, Weekly self-scheduling, forward contracting and offering strategy for a producer, *IEEE Transactions on Power Systems*, 25(2), 657–666, 2010.

36. R. Dahlgren, C.-C. Liu, and J. Lawarree, Risk assessment in energy trading, *IEEE Transactions on Power Systems*, 18(2), 503–511, 2003.

37. A. Brooke, D. Kendrick, and A. Meeraus, *GAMS User's Guide*. Redwood City, CA: The Scientific Press, 1990. Available at: http://www.gams.com/docs/document.htm

38. B. R. Szkuta, L. A. Sanabria, and T. S. Dillon, Electricity price short-term forecasting using artificial neural networks, *IEEE Transactions on Power Systems*, 14(3), 851–857, 1999.

39. F. J. Nogales, J. Contreras, A. J. Conejo, and R. Espinola, Forecasting next-day electricity prices by time series models, *IEEE Transactions on Power Systems*, 17(2), 342–348, 2002.

40. A. Mateo Gonzalez, A. M. San Roque, and J. Garcia-Gonzalez, Modeling and forecasting electricity prices with input/output hidden markov models, *IEEE Transactions on Power Systems*, 20(1), 13–24, 2005.

41. N. Amjady, Day-ahead price forecasting of electricity markets by a new fuzzy neural network, *IEEE Transactions on Power Systems*, 21(2), 887–896, 2006.

42. G. Li, C-C. Liu, C. Mattson, and J. Lawarree, Day-ahead electricity price forecasting in a grid environment, *IEEE Transactions on Power Systems*, 22(1), 266–274, 2007.

43. C. M. Ruibal and M. Mazumdar, Forecasting the mean and the variance of electricity prices in deregulated markets, *IEEE Transactions on Power Systems*, 23(1), 25–32, 2008.
44. D. Rajan and S. Takriti, Minimum up/down polytopes of the unit commitment problem with start-up costs, *IBM Research Report*, RC 23628, June 2005.
45. M. Carrion and J. M. Arroyo, A computationally efficient mixed-integer linear formulation for the thermal unit commitment problem, *IEEE Transactions on Power Systems*, 21(3), 1371–1378, 2006.
46. M. T. Nowak and W. Römisch, Stochastic Lagrangian relaxation applied to power scheduling in a hydro-thermal system under uncertainty, *Annals of Operations Research*, 100(1–4), 251–272, 2000.
47. X.-F. Wang, Y. Song, and M. Irving, *Modern Power Systems Analysis*. New York: Springer Science+Business Media, LLC, 2008, p. 382.d
48. F. N. Lee and A. M. Breipohl, Reserve constrained economic dispatch with prohibited operating zones, *IEEE Transactions on Power Systems*, 8(1), 246–254, 1993.
49. J. R. Birge and F. Louveaux. *Introduction to Stochastic Programming*. New York: Springer-Verlag, 1997.
50. P. Kall and S. W. Wallace. *Stochastic Programming*. Chichester, UK: John Wiley & Sons, 1994.
51. V. P. Gountis and A.G. Bakirtzis, Bidding strategies for electricity producers in a competitive electricity marketplace, *IEEE Transactions on Power Systems*, 19(1), 356–365, 2004.
52. M. V. Pereira, S. Granville, M. C. Fampa, R. Dix, and L. A. Barroso, Strategic bidding under uncertainty: a binary expansion approach, *IEEE Transactions on Power Systems*, 20(1), 180–188, 2005.
53. *The Hellenic Transmission System Operator* [Online]. Available at: http://www.desmie.gr.

9

Unit Commitment and Economic Dispatch for Operations Planning of Power Systems with Significant Installed Wind Power Capacity

Barry G. Rawn

Madeleine Gibescu

Bart C. Ummels

Engbert Pelgram

Wil L. Kling

9.1 Introduction

Committing generation units ahead of time is necessary to manage the risk associated with operating the power system to serve load under uncertain future conditions. Optimization of unit commitment (UC) decisions is, however, not an easy task as it must take into account all possible reasons for bringing units online or shutting them down [1]. After UC has been decided upon, an economic dispatch (ED) is then performed, distributing expected load between the committed units such that overall operating costs are lowest while still providing necessary reserves in addition to demand. These alterations can be applied to controllable generation and resources, and in some cases controllable loads. A forecast of the demand, which is periodically updated but still contains some uncertainty, is central to the activity of UC–ED. UC–ED algorithms are used by operators of power plants and power systems market operators.

In this chapter, wind generation is viewed as an unscheduled resource with insignificant operating costs that acts to lower electricity demand. A forecast of wind generation, therefore, also becomes necessary once wind offsets a significant portion of demand in power systems. If wind power reduces demand significantly, the "minimum load" problem is encountered: some conventional generating units have minimum generation levels that cannot be violated without complication and/or unacceptable loss of efficiency. For example, some units generate electricity as a coproduct while supplying heat for industry or residential use. Nuclear generators require a long time to safely shutdown or start, and are usually operated at a constant output as "base loads."

Unit commitment decisions are currently reassessed typically once or twice a day, whereas generation dispatch is carried out throughout the day. With the reasonable predictability of system load, intraday calculations for unit commitment are, in principle, only necessary when unexpected, significant changes occur in generation (e.g., plant outages) or demand. This changes when large amounts of wind power must be taken into account, as its variations are more difficult to predict.

The emergence of international markets and the growth of wind power have complicated the optimization of UC–ED in the sense that more variables and uncertainties (i.e., market prices, wind power forecasts) must be taken into account. In liberalized markets, the generation owners are responsible for supplying their own customers (i.e., long-term contracts and short-term trading agreements) and for providing certain ancillary services such as reserve and regulating power to the transmission system operator. Each individual owner therefore optimizes the UC–ED of the generation units under its control, taking the market price into account. For existing systems, ideal markets, in principle, lead to the same outcome regarding the scheduling of generation as would have been the case with central optimization. It is, therefore, still highly relevant to formulate and solve the central optimization problem [2]. The issue of market effects will be revisited in Section 9.4, when stating the central assumptions made in this chapter's approach to solving the problem.

To realistically explore issues related to the integration of wind power, a UC–ED problem formulation should include models for wind power, interconnecting capacity to neighboring power systems, and energy storage facilities. For some systems, it can also be important to include appropriate models of combined heat and power (CHP) units, which constrain the power system by coupling it with residential and industrial heat demand. It is also important to acknowledge that the commitment of units must cope in reality with generation unit outages, which may be either planned or accidental. Sections 9.2 and 9.3 are dedicated to discussing the features of realistic generation, load, and international trade models.

The approach to unit commitment discussed in this chapter involves some simplified assumptions and emphasizes the effects of wind power. Simulations are performed using a commercial software package (PwrSym4) that can represent multiple areas, works from a database of fuel costs and emissions by technology, and accepts input data for load and heat demand. It is a chronological simulation that conducts a "rolling window" optimization, making adjustments over several time scales, and accounting for ramp rates and minimum up and down times. It can include the effects of forced outages using a Monte Carlo randomization. A schematic overview of this process is the main subject of Section 9.5. With the knowledge of the previous sections in hand, in Section 9.6 the outcomes of a case study on large wind power penetration scenarios in the Netherlands and Germany and their effect on the Northwest European system are examined.

9.2 Modeling Input Data

9.2.1 Load Time Series

System load is the most crucial input component of a unit commitment study. The difference between its peaks and values along with its minimum and maximum values provide a condensed signature of any power system, and hold a clue to the composition of its current generation mix. The character of daily and weekly load patterns relative to that of renewable resources must be retained to properly produce realistic unit commitment schedules. For unit commitment, aggregate loads are being considered, and its forecast can be viewed as being perfect. Cooperation with a transmission system operator to obtain data is ideal; such entities have historically and to date usually still log the output of major generators supplying loads. For studies of future years, it is often assumed that the load pattern can be scaled to reflect a chosen rate of growth, even though this assumption does not take into account possible, but uncertain developments like growth of the use of air conditioning during summer, energy savings, or the future use of electric cars.

If only the patterns of demand are considered, the period of a year is widely accepted as necessary to exhibit the full range of extreme load levels. When large amounts of wind power that alter the pattern of demand are involved, interyear variations in this net demand can be significant, and it may be necessary to consider a number of years to capture high or low wind power scenarios and extreme situations such as shutdown due to storms.

9.2.2 Wind Power Time Series

Meteorological stations provide a dataset of wind speeds that can be transformed into wind power time series on the basis of a set of assumptions discussed here. In the example in this chapter, station data from the Netherlands (see Figure 9.1) is used to produce estimates for desired locations. The locations of onshore wind parks are determined by extrapolating the present distribution of onshore wind turbines in the Netherlands to larger wind power capacity levels, taking into account provincial targets [3]. Locations of offshore wind parks are based on a selection of locations proposed by the Dutch Ministry of Economic Affairs [4] and under consideration by Noordzeeloket [5]. The minimum number of sites required

FIGURE 9.1   Wind measurement sites.

for reproduction of the relevant patterns and geographical diversity is difficulty to general-ize. For the case of the Netherlands, most possible wind park locations are found in an area of approximately 80–100 km², to the west of Hoek van Holland to Den Helder, and a small section of the Wadden Sea, north of Groningen, as can be seen in Figure 9.2.

In addition to a geographically representative cover of measurements for the region being studied, data or assumptions regarding the area and machine type of the wind

FIGURE 9.2   Interpolated wind farm locations.

farms in question are used in the conversion of wind speed time series to wind farm power output, as is now elaborated following the method described in Reference [6]. For a deeper discussion of the method, the reader is directed toward Reference [8].

9.2.2.1 From Meteorological Observations to Wind Farm Locations

First, wind speed time series from measurements or mesoscale model output are used to determine periodic patterns at each location. The use of time series guarantees that correlations of wind speeds (variations over space and time) are automatically taken into account. The wind speed data are transferred from measurement height to wind turbine hub-height. The data are then transferred from the measurement sites to existing and foreseen locations of wind parks by linear interpolation, taking into account the spatial correlation between the sites.

Analysis of the wind speed measurement data reveals that the sample variance of the wind speed increases with the average wind speed [6]. To suppress this so-called heteroscedasticity, a variance stabilizing transformation [7] is applied and the logarithm of the wind speed is used instead of wind speed itself. To arrive at a suitable wind speed time series model, any periodic effects in the wind patterns must be investigated first. For the sites onshore and offshore in the Netherlands, the daily pattern was determined (see Reference [8]) to be more significant than any seasonal pattern. In Figure 9.3, the average wind pattern is plotted for each of the wind speed measurement locations for one day. The lower curves correspond to locations onshore and the upper ones to

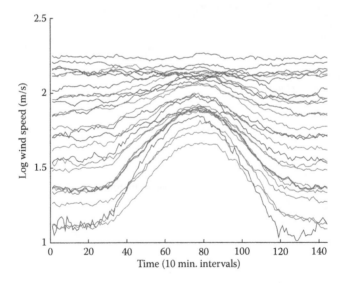

FIGURE 9.3 Deterministic component $\mu(\mathbf{x}, t)$ of wind speed at different locations: A daily wind speed pattern for observed (dark) and interpolated (light) signals. (Adapted from A. J. Brand, M. Gibescu, and W. W. de Boer. in *Wind Power*, S M Muyeen (ed.). Croatia: InTech, 2010. ISBN: 978-953-7619-81-7; Available at: http://www. intechopen.com/articles/show/title/variability-and-predictability-of-large-scale-wind-energy-in-the-netherlands.)

offshore. It can be observed that onshore measurement locations have a typical maximum occurring around midday, offshore locations have a rather flat daily profile with a higher average, and coastal locations fall somewhere in between.

This daily pattern comes from large-scale geographical features and local effects, and the pattern varies smoothly but significantly with location. This pattern is discerned by averaging in bins for each hour of the day over all days in the series. The remaining stochastic component of the signal, which at times overpowers the daily pattern, can then be characterized through statistical analysis. The log wind speed field $w(\mathbf{x},t)$ is modeled as

$$w(\mathbf{x},t) = \mu(\mathbf{x},t) + \varepsilon(\mathbf{x},t) \tag{9.1}$$

where $\mathbf{x}$ is a vector holding the location coordinates, t is time, $\mu(\mathbf{x},t)$ is a deterministic variable representing the daily wind pattern, and $\varepsilon(\mathbf{x},t)$ is a zero-mean random process variable representing shorter-term variations around the daily mean. The covariance structure of $\varepsilon(\mathbf{x},t)$ must take into account the geographical correlations between different locations, especially for smaller areas such as the Netherlands. It is assumed that the wind is a Markov process and that only the lag 1 autocorrelation of the signal should, therefore, be captured. The full development concerning the synthesis of new data from old is detailed in Reference [8], and requires determination of means, spatial covariances, and autocorrelation. In this text, we examine the results for the latter two characteristics.

For the estimation of the random component $\varepsilon(\mathbf{x},t)$, the covariance $\text{cov}(\varepsilon(x_i,t), \varepsilon(x_j,t))$ between two locations x_i and x_j is calculated. In Figure 9.4, wind speed covariances are plotted versus the distance between measurement locations. Assuming that covariance reaches zero at large distances, it is modeled through an exponential decay:

$$\text{cov}(\varepsilon(x_i), \varepsilon(x_j)) = a_0 \exp\left(-\alpha_0 \|x_i - x_j\|\right) \tag{9.2}$$

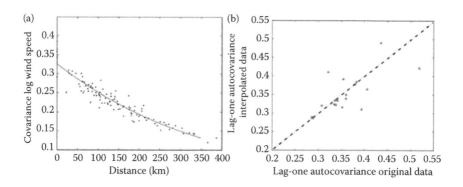

FIGURE 9.4 Space and time properties of stochastic component $\varepsilon(\mathbf{x}, t)$ of wind field. Spatial correlation as a function of distance is shown with exponential fit (solid line) in (a), while the autocovariance in time of the measured and synthetic data is compared in (b). (Adapted from A. J. Brand, M. Gibescu, and W. W. de Boer. in *Wind Power*, S M Muyeen (ed.). Croatia: InTech, 2010. ISBN: 978-953-7619-81-7;   Available at: http://www.intechopen.com/articles/show/title/variability-and-predictability-of-large-scalewind-energy-in-the-netherlands.)

where parameters a_0 and α_0 are estimated using a least-squares fit, also shown in Figure 9.4, with $1/\alpha_0$ as the characteristic distance or decay parameter. Translation of this decay fit from log wind speed to wind speed gives a characteristic distance of 610 km, a value in line with those reported elsewhere [9,10]. A similar functional form has been used to characterize temporal dependence in the following covariance between samples:

$$\mathrm{cov}\Big(\varepsilon(\boldsymbol{x}_i(t)),\varepsilon(\boldsymbol{x}_j(t-1))\Big) = a_1 \exp\Big(-\alpha_1 \big\|\boldsymbol{x}_i - \boldsymbol{x}_j\big\|\Big)$$

(9.3)

where parameters a_1 and α_1 are estimated using a least-squares fit. The lag 1 autocovariance of the measurements has been compared in Figure 9.4 with that of the synthesized data for multiple sites. Although not perfect, there does not appear to be any structural bias [6].

The measurement height for meteorological stations does not correspond to the hub height of modern wind turbines (70–120 m). A height transformation can be applied to the measurements based on logarithmic approximation describing how flowing air is slowed close to the surface of the earth due to drag and other ground effects. A detailed exposition of can be found in the literature [6] or texts [8].

The wind speeds at the farm locations marked in Figure 9.2 are estimated by interpolation of the wind speed data at 18 measurement locations. In order to obtain the wind speeds at other locations, a linear spatial interpolation can be applied for all locations within the boundary of the measurement locations. For locations outside this boundary, nearest-neighbor interpolation is applied.

The linear interpolation takes into account the spatial correlations among multiple sites to arrive at wind speed time series for existing and foreseen wind power locations. The results have been cross-validated by removing one location from the n-site measurement set at a time and using the remaining $n-1$ measurement sites to estimate it.

9.2.2.2 From Wind Speed to Farm Power

Wind speed and wind power are governed by a third-order relationship. The actual relationship between wind speed and the wind power output of a wind turbine is defined by the wind turbine power curve, defining the amount of power generated by the wind turbine P_{wt} at wind speed v [11]:

$$P_{wt} = \rho C_p(\lambda,\theta)A_r v^3$$

(9.4)

where ρ is the density of air (kg/m³), C_p the power coefficient of the wind turbine, λ the tip speed ratio between the turbine blade tip speed v_t (m/s) and the wind speed upstream from the rotor v (m/s), θ the blade pitch angle (degree), and A_r the swept area of the turbine rotor blades (m²). Wind turbines control their λ and θ and thereby C_p in order to maintain rated electric power generation at higher wind speeds and to prevent mechanical overloading of the turbine's moving components and structure. The maximum power coefficient C_p of an ideal wind turbine rotor is 16/27, which is known as the Lanchester–Betz–Joukowsky limit [12]. As it is only possible to maximize C_p for a limited range of wind speeds, the design and control of C_p and the wind turbine are such that the conversion efficiency is highest at the wind speed range where most energy can be captured. Wind speeds can be converted to wind power using the wind turbine's wind speed–power curves, which is based on the fundamental relationship shown in Equation 9.4, but accounts for the effect of controls. Apart from the power coefficient, which is specifically designed for different wind classes, wind speed–power curves are also determined by wind turbine technology and type.

Throughout the development of wind power in the past decades, different wind turbine technology concepts have been used, each with different power curves. These concepts can be categorized by generator type. The four most commonly used machine types for wind turbines are the fixed speed wind turbine with induction generator (type A), variable speed with variable rotor resistance (type B), variable speed with doubly fed induction generator (type C), and direct drive turbine with permanent magnet generator (type D) [13,14].

Since the late 1990s, most wind turbine manufacturers have changed to variable speed for power levels from about 1.5 MW and above. The different turbine concepts have different power curves. Wind speed–power curves for each type of wind turbine are shown in Figure 9.5 [15], assuming an air density of 1.225 kg/m³ and no noise constraints for turbine operation. Modern types C and D involve turbines with larger capacities, whereas type A does not produce a flat power curve at wind speeds exceeding 15 m/s (rated wind speed). Regions of the power curves of special interest for power system integration are in the range of 5–15 m/s, where changes in wind speed correspond to relatively large changes in electrical power output, and the cut-out speed (20 m/s for NEG-MICRON NM48, 25 m/s for Vestas V52 and V90), where the power output of the wind turbine changes from full and no power.

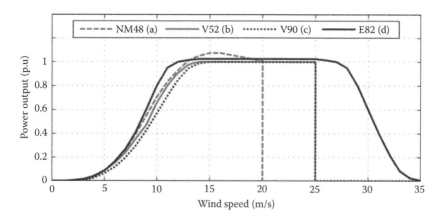

FIGURE 9.5 Wind turbine power curves for the NEG-MICON NM48 (a), Vestas V52 (b), Vestas V90 (c), and Enercon E82 with storm control (d). (Adapted from B. C. Ummels. Wind integration: Power system operation with large-scale wind power in liberalized environments. PhD thesis, Technische Universiteit Delft, the Netherlands, 2009.)

Manufacturer power curves for a single turbine for the relation of wind speed to wind power are commonly used for transformation of measured and synthetic wind speed data series to an input for unit commitment or power system studies [16,17]. Applying such curves to predict the output of an entire farm will significantly overestimate power variations near cut-out wind speeds, because different machines experience different wind speeds, especially offshore where distances between wind turbines are significant. A methodology for the development of wide-area, aggregated wind park power curves is presented in Reference [18], enabling the development of wind power data based on locational shares of the total capacity spread out over large geographical areas. The approach makes use of the distance between wind farms to develop a Gaussian filter (normal distribution function) that characterizes wind speed deviations around the park's average wind speed at a given wind park location. The filter is convolved with the power curve of a single turbine to yield a new smoothed power curve. This methodology was intended for the production of country-scale equivalents for wind power, but its underlying principles can also be applied to develop power output data for farm-sized areas. The width of an appropriate Gaussian filter can be derived by taking into account regional variations of wind speeds based on exponential decay [18,6] and the layout of farms [8].

For each wind park, a location-dependent power curve is thus developed based on the local standard deviation of wind speed and the geographical size of the wind park. Such an approach ignores park effects such as wind turbine wakes, which can slightly alter annual yields [19] but are much less significant than the differences in wind resource from year to year. More significant sources of error are the assumptions made regarding wind park locations, distances of turbines, and wind park layout.

Figure 9.6 shows an example of a multiturbine curve developed for an offshore wind park. The influence of the filter is most visible around the cut-out wind speed (25 m/s), where the wind turbine alternates between full power and no power. This illustrates how

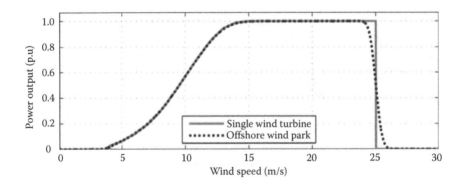

FIGURE 9.6   Example of a multiturbine power curve for an offshore wind park. (Adapted from A. J. Brand, M. Gibescu, and W. W. de Boer. in *Wind Power*, S M Muyeen (ed.). Croatia: InTech, 2010. ISBN: 978-953-7619-81-7; Available at: http://www.intechopen.com/articles/show/title/variability-and-predictability-of-large-scale-wind-energy-in-the-netherlands.)

the wind speed deviations from the wind park's average wind speed at individual turbine locations result in a smoothing of the wind park's power curve. The wind speed data developed above for each wind park location can be multiplied with the parks' multiturbine power curve in order to develop wind power data.

The last step for the development of the wind power time series for each wind park is the incorporation of the unavailability of wind turbines. This can be done in different ways, with a full Monte Carlo outage approach delivering the most accurate results. The large number of wind turbines (<1000 for the smallest penetration level) and the aggregation of wind power output at system level makes the added value of this approach limited compared to an averaged availability rate. For the example in this chapter, wind turbine availability rates are assumed to be constant at 98% for onshore wind turbines and 95% for offshore turbines, the latter due to more difficult access. It can be noted that availability rates for wind turbines placed offshore may well be lower than this figure [20]. For system integration studies, assuming a higher rate builds some conservatism into the results. For a consideration of portfolio, it would be wise to select several values for availability or conduct a sensitivity analysis.

9.2.3 Wind Power Forecasts

In Chapter 5, the state-of-the-art wind power forecasting was introduced. Wind farm owners and utilities may well obtain these sophisticated forecasts from third parties. The basic time series and numerical weather forecasts introduced in Sections 5.2 and 5.3 in Chapter 5 are sufficient for most unit commitment studies.

For the example in this chapter, numerical weather forecasts were augmented using a physically motivated [21,22] statistics module that takes into account the local influences of roughness, obstacles, and stability on wind speeds at the specified height. The forecasts are based on underlying runs of the atmospheric high-resolution limited area model (HIRLAM), which is a numerical weather prediction (NWP) model. HIRLAM numerically approximates the physical state of the atmosphere at 6-h intervals with initial conditions taken from recent observations. The statistics module was developed by the Energy Research Centre of the Netherlands to create an energy supply forecast for renewable energy sources (Aanbodvoorspeller Duurzame Energie or ADVE). The wind speeds approximated by HIRLAM are postprocessed by AVDE into 15-min averaged wind speed at hub-height for two onshore and five offshore measurement locations. These approximated wind speeds are then compared to the measured wind speeds to obtain wind speed forecast errors. Using the same method applied for interpolating wind speed data, the wind speed forecast errors are interpolated to the foreseen locations, and finally added to the interpolated wind speeds to develop forecast wind speed time series at the locations of interest. As the time dependency of wind speeds is taken into account in the wind speed interpolation method, time dependence is automatically included in the forecast time series as well.

The minimum lead time of the NWP model feeding the forecasting method that underlies the augmented forecast is 6 h. Therefore, the first 0–6 h are filled in by a persistence-based forecasting method fed by real-time measurement data. For this, the

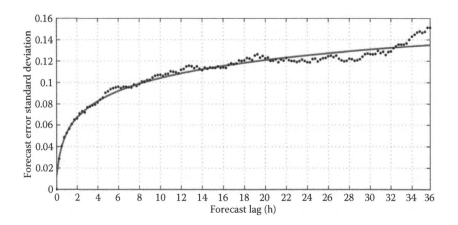

FIGURE 9.7 Normalized standard deviation of wind power forecast error for 12 GW wind power.

12–36-h-ahead aggregated wind power forecast errors are modeled as a first-order autoregressive moving average (ARMA) process:

$$\varphi(t) = a\varphi(t-1) + b\gamma(t-1) + \sigma\gamma(t) \tag{9.5}$$

where $\gamma(t) \sim N(0,\sigma)$ is a zero mean, normally distributed noise term of standard deviation σ. The a, b and σ parameters of the ARMA process are estimated by the maximum likelihood estimator method [7]. The resulting forecast error standard deviation from these two sources is plotted as a function of prediction horizon in Figure 9.7.

One 1–36-h-ahead wind power forecast is developed for each hour of the year, and incorporated into the unit commitment algorithm as a forecast matrix (see Figure 9.15 later in this chapter). The wind power time series generated using the methods described in Section 9.2.2.2 determines the actual net load. As the UC–ED algorithm proceeds to process this net load, it uses best forecast information on a rolling basis for intrahourly optimization and shifted forecast information to simulate international markets with closures (discussed in Section 9.5.3).

9.3 Modeling of Conventional Units

In this chapter, we label as conventional units all those types of generation that currently dominate generation mixes in most power systems and are dispatchable by virtue of a controllable fuel source.

9.3.1 Modeling of Thermal Units

Conventional thermal units derive electricity from steam turbines, where the steam can be produced from input water flows by reacting fossil fuels or nuclear fuels. In this

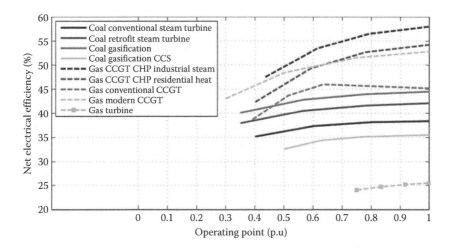

FIGURE 9.8 Thermal fossil unit efficiency at different operating points. CCGT denotes "combined-cycle gas turbine" and CCS denotes carbon capture and storage. (Adapted from B. C. Ummels. PhD thesis, Technische Universiteit Delft, the Netherlands, 2009.)

chapter, we omit other possible forms of thermal units such as geothermal units. Many types of fossil units exist; most are fueled with either coal or natural gas.

Figure 9.8 demonstrates the operational range and efficiency of several types of fossil-fueled forms of power generation. The acronym CCGT denotes "combined cycle gas turbine." It is notable that each type of generation has different maximum and minimum production levels when operating. Also, the efficiency of generators depends on the level of production. These factors complicate the UC–ED problem.

The operating cost of thermal generation units include fixed and variable operating costs such as start-up, fuel, and emission cost. The fuel-related costs has the general form [23]

$$C(P,H) = (p_f + p_e)(c_0 + c_1 P + c_2 P^2 + c_3 H + c_4 H^2 + c_5 PH) \qquad (9.6)$$

where the prices of fuel and emissions p_f and p_e are based on the fuel consumed, which is modeled by the bracketed quantity. Fuel consumed has a constant component c_0 associated with operational losses, costs having a linear and quadratic dependence on the power P and heat H being produced (indicated by coefficients c_1, c_2, c_3, and c_4), and a component related to, in some cases, the product of P and H, expressed by the coefficient c_5 [23]. Each technology has a different set of coefficients, some of which may be zero. The cost of starting a unit can be given a time-dependent formulation depending on time since shutdown and on fixed start costs, but detailed information is not always available for all units. It is practical to assume a constant cost for all startups,

$$SC = c_6 N_{starts} \qquad (9.7)$$

where c_6 is the start-up cost and N_{starts} is the number of start-ups. The two costs C and SC together give the costs of operating a generating unit.

The most important emissions resulting from the use of fossil fuels in power generation units are carbon dioxide (CO_2), sulfur dioxide (SO_2), and nitrogen oxides (NO_x). The emissions released into the air are, in principle, defined by the elements involved in the chemical reactions between air and fuel. This is mostly methane (CH_4) for natural gas, and carbon (C), metal compounds, and several other compounds involving hydrogen (H), sulfur (S), and nitrogen (N) for coal. Emission levels of CO_2 and SO_2 are connected only to the fuel composition and the operating efficiency of the unit. For NO_x, the emission levels are also significantly dependent upon a range of factors in the thermodynamic processes in the plant [24]. Through the price p_e in Equation 9.6 they become an intrinsic cost of operation.

Some power systems include CHP units, which serve a local demand for heat by providing steam, and also generate electricity. The benefits of CHP include a very high overall fuel efficiency (electricity plus heat), up to 87% at the best operating point. However, the operation of CHP units is dominated by the local demand for heat or steam. Therefore, they introduce the influence of another varying input. Additionally, CHP units have operating constraints associated with the technical operational area (power P and heat H) and with their operational status due to heat demand. The operation area of each CHP unit can be described as a set of n linear inequality constraints of the type:

$$d_iP + e_iH \geq f_i \tag{9.8}$$

where P and H denote electrical power and heat generation, respectively, and d_i, e_i, and f_i some positive coefficients. These inequalities each define a side of a polygonal region with a form like that shown in Figure 9.9, which has the optimal operating point in the top-right corner.

There are two important limitations introduced by this constraint that are relevant to unit commitment. At high heat production, the flexibility of the unit decreases, and at times of low load, CHP units serving industrial loads have a "must-run" status considering the needs of the steam supply. This means they cannot so easily accommodate schedule changes necessary to meet the load. For units serving residential loads, heat boilers and buffers may offer some operational flexibility for the CHP (i.e., temporal

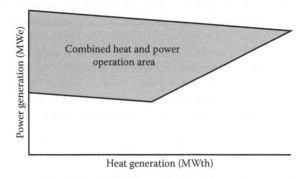

FIGURE 9.9 Operation area of combined heat and power (CHP) units.

decommitment based on economics). For new CHP units, sets of linear inequality constraints are developed on the basis of the unit's operating efficiency.

The fuel consumption for the heat production by CHP units is included in their overall heat rate levels, which can be part of the optimal dispatch procedure. CHP units are assigned to certain heat districts, which may comprise one or more CHP units, heat boilers and heat buffers. Local demand for heat or steam takes priority in the scheduling of CHP units, with boilers and buffers, if available, standing by for peak-demand situations.

Because of their possible limiting effects on the deployment of wind power, possibilities have been investigated for using heat storage or the use of heat boilers [25,26]. This is easiest to do in residential areas. Additional operational flexibility of industrial CHP units would be technically possible but is often considered as unfeasible due to the risks associated with starting and stopping. This consideration is likely to change with the integration of large-scale wind power, as the impact of wind power on market prices can be significant, especially during periods of low load. A must-run status of CHP, allowing operation only between minimum and maximum output, becomes less profitable with the development of large-scale wind power.

Introducing possibilities for storing heat allows a decoupling of the generation of heat and power and thereby brings additional operational flexibility of the CHP units. A higher flexibility of CHP units reduces the amount of must-run capacity in the system and it can be viewed as an alternative to the development of energy storage to alleviate problems during periods of high wind power and low load. This prospect is tested in the chapter case study.

9.3.2 Hydro and Storage Units

The absence of combustion processes, fossil fuels, or emissions simplifies the modeling of hydropower units. However, the cascading of different hydro units, precipitation, and natural variations of the reservoir inputs introduce significant complexity into the scheduling and optimization of these units. For the purposes of this chapter, a relatively simple treatment of hydropower is given. The interested reader is directed toward one of a number of excellent references on this subject. Three different types of hydro units can be identified: run-of-river hydro, reservoir hydro, and pumped hydro. The latter can be modeled such that it also represents different energy storage technologies [15], as are studied in Section 9.6.

Run-of-river hydro comprises hydropower units with a generating capacity depending on the availability of the primary energy source, water. In case water is available, the unit must produce power or the water must be spilled, resulting in an opportunity loss. Run-of-river hydro usually involves smaller generation units located in water streams. Run-of-river models include specifications for minimum output level (MW), maximum output level, and the unit's water inflow. The water inflow can be simplified as a constant inflow of gross energy on a weekly or an hourly basis. In first pass of the unit commitment process, this is used as a baseline level for the small hydro plants.

Reservoir hydro consists of a unit connected to a hydro reservoir, allowing decoupling of water inflow and electricity generation. The reservoir size (GWh) and losses (MWh/h) are defined additional to the minimum and maximum output levels (MWh) and the reservoir inflow, which is modeled as net energy (GWh/week) available for dispatch. The minimum power level of the reservoir hydro reflects flow constraints and any run-of-river elements of this type of hydropower. As an alternative to the explicit modeling of run-of-river hydro units, the minimum level of a large reservoir hydro unit can be adjusted to account for such units as a single aggregate unit. Because of its large storage capacity, the operation of reservoir hydro can be dispatched to minimize cost over a week-long horizon.

Pumped hydro is capable of storing energy by converting electricity into potential energy. Pumped hydro is modeled as a reservoir hydro unit with a pumping facility between two reservoirs. Using the pumps for storing energy in the upper reservoir increases the head level H_{res} (m) by

$$H_{res} = \sqrt{(R+B)/A} \tag{9.9}$$

where R is the reservoir's energy content (GWh) and A (GWh/m^2) and B (GWh) are constants depending on the physical size of the reservoir. The head level is important especially for pumped hydro units with a small height difference between the reservoirs, as conversion efficiency is partly dependent on the head level in this case. Pumped hydro operation includes the constraint that R must be the same at the beginning and at the end of each week. A ramp rate may be included in the model in order to take into account any technical limitations to the operational flexibility or by setting the minimum time needed for changing from pump to generator operation, if applicable.

Wind power and hydropower and/or energy storage form a natural combination. Therefore, wind power and (hydro) energy storage have been considered in a back-to-back configuration [27] or as a hybrid system to provide firm power [28]. Wind power may be used to fill up storage reservoirs during high wind periods, either by pumping up or by saving water, and the stored energy may be used for electricity generation during low wind periods. When low prices coincide with high wind availability, storage units operating independently in the market will play such a role.

Hydropower and energy storage can be optimized by use of the value-of-energy method based on marginal cost. Time-related constraints such as generation cost, operational aspects of thermal generation units and storage reservoir size are major determinants in the dispatch of energy storage. Thus, energy storage generates if the marginal cost is higher than the generating value of energy and stores if the marginal cost is lower than the pumping value of energy, and remains idle otherwise.

9.4 UC-ED Problem Formulation

In principle, UC–ED is a problem of optimization. The power production of the ith generator in area n is a function of time $P_i^n(t)$, and each generator's production must be chosen to minimize overall operating cost:

$$\min \left\{ \sum_{n \in Ni} \sum_{\in G_n} \left(SC_i^n + \sum_{t \in T} C_i^n \left(P_i^n(t), H_i^n(t) \right) \right) \right\} \tag{9.10}$$

where N is the set of area indices, G_n is the set of generator/load bus indices in area n, T is the set of time indices in the full optimization horizon of discrete time steps, and SC and C are as defined in Equations 9.6 and 9.7, but indexed by generator i and area n. The condition of actually serving load by dispatching generation within its safe limits can be expressed as a number of constraints:

$$\sum_{i \in G_n} P_i^n(t) - L_i^n(t) + \sum_{m \in M} X_m(t) = 0, \quad \forall n \in N \tag{9.11}$$

$$\sum_{i \in G_n} H_i^n(t) - D_i^n(t) = 0, \quad \forall n \in N \tag{9.12}$$

$$X_m(t) \le \overline{X}_m, \quad \forall\, t \in T,\ \forall\, m \in M \tag{9.13}$$

$$\underline{P}_i^n \le P_i^n(t) \le \overline{P}_i^n, \quad \forall\, i \in G_n,\ \forall n \in N,\ \forall\, t \in T \tag{9.14}$$

$$\max_t \left| \frac{P_i^n(t+1) - P_i^n(t)}{\Delta t} \right| \le R_i^n(t), \ \forall\, i \in G_n,\ \forall n \in N \tag{9.15}$$

$$\sum_{kn_{wk}}^{(k+1)n_{wk}} P_i^n(t) \le E_i^n(k), \ \forall\, i \in G_n,\ \forall n \in N,\ \forall\, k \in K \tag{9.16}$$

$$\sum_{i \in G_n} \overline{P}_i^n - P_i^n(t) \ge SR_n, \ \forall n \in N,\ \forall t \in T \tag{9.17}$$

$$P_i^n(t) = 0, \quad \forall t : A_i^n(t) = 0 \tag{9.18}$$

where M is the set of interconnection indices, $L^n(t)$ is electricity demand, and k is the week number. The overbar indicates an upper limit on a variable, whereas the underbar indicates a lower limit, and additional symbols are defined as follows. Constraint 9.11 sums the interarea electric power transfers $X_m(t)$ with generation and load to state that electricity production and import must equal electricity demand in each area n. Equality of heat production H and heat demand D is imposed by constraint 9.12. The maximum capacity of transmission between areas must be observed, as given by constraint 9.13. The next three relations express constraints on the capability of the generating units. Each unit has a minimum and a maximum level of generation as denoted by underbar and overbar notations, a maximum ramp rate R_i^n, and, in the case of hydro generators, each week k a finite energy $E(k)$ in its reservoir. This $E(k)$ value would be derived from a separate long-term optimization of hydro capacity. Relation 9.17 expresses that each area must maintain a spinning reserve SR_n, and relation 9.18 expresses the fact that unavailable units have a zero output. The vector $A_i^n(t)$ indicating a unit's availability is set, as will be discussed in Section 9.5, to reflect scheduled outage, forced outage, and decommitment decisions.

The following assumptions have been applied:

1. The transmission system within each area n has no losses and sufficient capacity (copper plate assumption).
2. Demand is inelastic.
3. Adjustments are not made by operators to optimize generator dispatch level between day-ahead energy markets and intraday balancing markets.

9.5 Solution Approach: Optimization Heuristics and Horizons

The unit commitment problem as formulated is a mixed-integer optimization problem. A practical solution of the minimization problems 9.10 through 9.18 will be pursued in this chapter by applying several simplifying assumptions, and breaking it into three subproblems, each with its own time horizon: annually, weekly, and a short-term operational time step (e.g., hourly). An annual horizon is used to realistically address unit outages and generate the initial availability indicator functions $A_i^n(t)$ to reflect maintenance schedules and forced outages; a weekly horizon is used for the optimal scheduling of hydro and energy storage units; and a short-term horizon is used for the simulation and optimization of thermal unit operation and exchanges. Heuristic solution methods are used that may not be globally optimal, but have a common sense interpretation and reflect utility practices.

Thus, the following further assumptions have been applied:

1. Weeks may be optimized independently of each other.
2. Hydropower always has the lowest variable cost.
3. Wind power has negligible variable cost.

9.5.1 Annual Horizon: Determining Unit Availability

Outages comprise all events leading to a partial unavailability of generation units in the system. A differentiation can be made between unforced or planned outages, such as due to scheduled maintenance, and forced outages, such as due to unexpected technical failures of the unit. Planned outages are usually scheduled with the objective of minimizing opportunity losses or minimizing reliability risk, for instance through the loss-of-load probability (LOLP) calculation, a widely used reliability measure in generation planning. This can be done by scheduling maintenance when prices are assumed to be low (low load periods) or by distributing maintenance between different

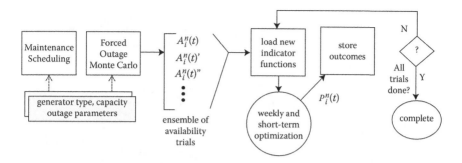

FIGURE 9.10 An ensemble of availability vectors derived from an optimized maintenance schedule and forced outages comprise the input to the unit commitment–economic dispatch optimization algorithm.

units over different periods, resulting in a rather constant generation adequacy during the year.

For the example in this chapter, the LOLP reliability index is computed in hours per year using the cumulant method [29]. After calculation of the LOLP, load carrying ability and capacity surplus/deficit are calculated relative to a specified reliability index. This index is used for an annual optimization of the maintenance schedules and the production of availability indicator functions $A_i^n(t)$ for all units.

An outage model further adjusts the availability indicator using a random number generator for a specified number of trials (i.e., Monte Carlo draws). Each trial is saved and used as input for a weekly simulation. The expected unit availability for individual iterations and across all iterations can be computed for later evaluation.

An ensemble of these functions is created where each trial has further unavailability imposed on it based on a full Monte Carlo assessment of forced outages. The resulting indicator functions can then form the input to runs of the UC–ED optimization algorithm, as shown in Figure 9.10.

9.5.2 Weekly and Intrahourly Horizon: Determining Dispatch

The unit commitment is optimized initially by heuristics based on the load prediction and wind power forecast. Subsequently, it is improved in three iterative processes for hydro, thermal units and exchange, and energy storage units. The general flow is as shown in Figure 9.11. The first and final decision boxes come into play when allowing the possibility of international exchange markets with advance gate-closure times. As detailed in a Section 9.5.3, this involves two successive applications of the UC–ED algorithm.

In both cases, the steps to optimize dispatch over a week are as follows. First, hydropower stations are scheduled using a price leveling algorithm based on their weekly energy constraint, load prediction, and wind power forecast. The hydro schedules are subject to hourly minimum and maximum generation levels and ramp-rate limitations.

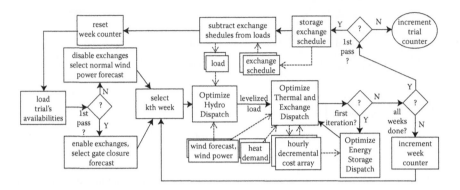

FIGURE 9.11 Sequence and dependence of optimizations carried out over weekly and hourly horizons.

Second, the model then uses local heat demands, system load, wind power, and wind power forecasts for the scheduling of the thermal generation units on the short-term operational time step, considering both generator and area technical constraints such as spinning reserve. In this step, international exchanges are also set. The starting condition of the short-term simulation is defined by the output of the weekly simulation and the generation units' states at the end of the previous week, taking into account outages and minimum up and down times. The heuristic optimization process is done using the equal incremental cost method [23], which is illustrated in Figure 9.12 and comprises the following steps:

1. Set all available generation units at maximum power level, taking into account operating constraints.
2. Calculate decremental cost arrays for each short-term time step.
3. Find the largest unit decremental cost.
4. Decommit or ramp down most expensive unit(s).

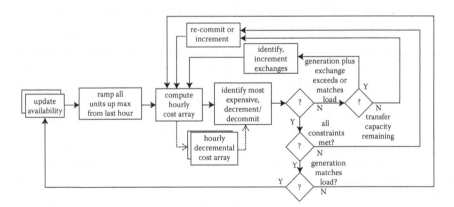

FIGURE 9.12 Sequence and dependence of intrahourly optimization.

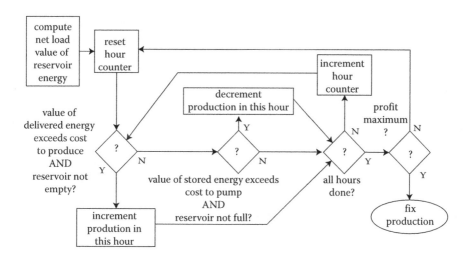

FIGURE 9.13 Procedure to dispatch energy storage to maximize profit on the basis of marginal cost information for each hour.

The construction of the decremental cost arrays takes into account system load, heat dispatch, and decommitment cost while satisfying minimum up and down times and ramp rates. The costs of generating heat by CHP units are taken into account in the calculations of the marginal cost of these units. In case the system comprises different areas, decremental cost arrays are constructed for each area taking into account transarea transmission constraints (i.e., net transfer capacity between areas), if applicable. After completion of the decrement procedure, system costs are calculated and these steps are repeated until load, heat demand, and reserve requirements are balanced. The repetitive process provides marginal cost for all areas and all units. It can be noted that the must-run status of many base-load units (coal, industrial CHP) in the Dutch system studied here reduce the complexity of unit commitment optimization.

Third, based on the hourly decremental cost array obtained through the first thermal optimization, energy storage is scheduled such that total operating costs over the week are minimized. This is equivalent to maximizing the profit of the energy storage operator. The general operating strategy, as depicted in Figure 9.13, is to generate when marginal costs are high, to store energy when marginal costs (and prices) are low, and to do nothing otherwise.

9.5.3 Multiarea Representation and International Gate Closure

As mentioned in the assumptions, basic unit commitment analysis usually neglects the electric network, assuming it can connect any generator to any load without limit within a certain area. However, the ability to model some degree of trade between different power markets or market areas is required to account for the possibilities of international trade and to assess the potential for new transmission lines. An example of a set

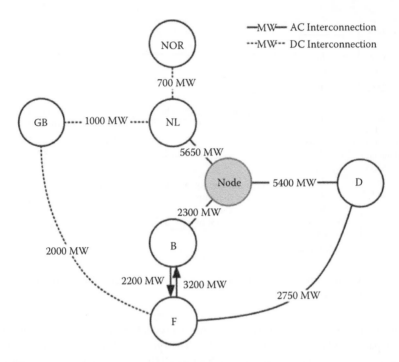

FIGURE 9.14   Multiple area model used in this chapter's example. Acronyms denote Northwest European countries.

of areas is shown in Figure 9.14; these areas are used in the case study found at the end of this chapter.

A complication posed by international markets is that they have a "gate-closure time"; the transfer amounts are settled some period of time in advance of their delivery. Gate-closure times are commonly a day ahead or earlier. In principle, it can be assumed that an international market optimizes dispatch in the same way a regular electricity market does, and both approach the solution given by an economic dispatch analysis algorithm. The difference is that international markets must use advance information that is necessarily more uncertain and likely less accurate.

A useful approach to acknowledge the differing information quality associated with trade is to solve the unit commitment problem using two passes [15]. On the first pass, transfers between different areas up to a predefined available capacity are allowed during the dispatch phase. However, for wind power, the information at a given time is drawn from the prediction made from some time ago in the past. Three intervals are used, to correspond to different gate-closure times considered: day-ahead, 3-h ahead, and 1-h ahead. This concept is illustrated in Figure 9.15, with the help of a "forecast matrix" that depicts the forecast available (a row) at a successive hours.

The first column contains the actual power at a given time, and can be viewed as a perfect forecast. Other columns contain a slice of the forecast matrix that can be used to simulate 1 or 3-h gate closure. For example, at time $h = 1$, decisions about 3 h in the

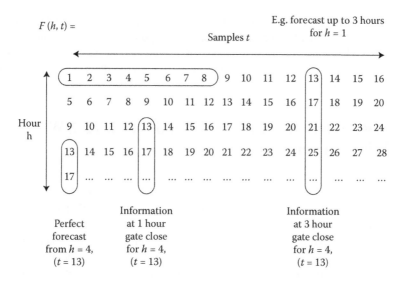

FIGURE 9.15   Forecast matrix; for each hour h, a forecast of 15-min values with index t (rows) is available, where the first element of every row is the true wind power. A rolling sample of a previous hour's forecast (subcolumns) is used to simulate information available for different international gate-closure times for a given hour.

future can only be made using this hour's forecast about sample $t = 13$. Therefore, even a rolling forecast can only use information as good as is contained in the right-most vertical column in the figure. This situation would be simulated by starting at hour 4, and applying the unit commitment algorithm (which computes the ideal actions of this market) based on the right-most subcolumn as the expected wind power output.

The exchanges that occur from a simulation of the shifted subcolumns model how the ahead market dispatch decisions would be made, and these exchanges are the only information kept from the first pass. They are applied to the loads of all areas to create a new set of loads for the second pass. On the second pass, the UC–ED algorithm is run again, but it can be seen as the action of the market after gate closure, where trade is only possible internally.

To summarize, the methodology consists of the following steps:

1. Selection of the wind power forecast matrix.
2. Creation of wind power time series consisting of the selected forecast data.
3. First simulation of the UC–ED and calculation of international exchange, with the wind power time series as input.
4. Adaption of the original area load files with the international exchange levels as settled in the first simulation run.
5. Second simulation of the UC–ED.

It uses the actual wind power as input, with no extra international exchanges allowed.

For the market design with 1-h-ahead gate closure, it is assumed that no wind power forecast errors are present. For this market gate closure, a single simulation run is performed using wind power data as input and assuming that international exchange is always possible. In order to specifically consider wind power in the Dutch power system, it is assumed that wind power in Germany can be predicted very well; that is, there are no wind power forecast errors present at market closure, regardless of the international market gate closure.

9.6 Case Study for the Netherlands with 12 GW Wind Power

This case study shows how the commitment and dispatch logic and realistic inputs as discussed in this chapter can be combined and applied to quantify the technical, economic, and environmental impacts of wind power scenarios for the Netherlands. System variants that are expected to shape these outcomes are explored in numerous simulations. The base simulation variants consider seven levels for wind power capacity installed in the Netherlands, four designs of international markets and three wind power forecast methods. Three solutions for increasing wind power integration are also examined.

9.6.1 Details and Discussion of System Variants

For all base variants, it is assumed that wind power is integrated into the system by taking into account wind power in the optimization of the UC–ED of conventional generation capacity. In case no international exchange market is available, only the Dutch conventional generation units are used. All base simulations were carried out for six wind power penetration levels and for a 0-MW wind power variant to be used as a reference.

9.6.1.1 International Exchange

International exchange is modeled and simulated for four market designs:

- No international exchange
- International market gate-closure time day ahead
- International market gate-closure time 3 h ahead
- International market gate-closure time 1 h ahead

In the base case, the Netherlands is seen as an isolated power system. International exchange with Belgium, France, Germany, Norway, and Great Britain is assumed to be zero at all times. This variant serves as a reference to consider the integration of wind power in the Dutch power system using the technical capacities available in the Netherlands only.

The other market designs all comprise international exchange possibilities between the Netherlands and its neighbors, but using different gate-closure times. This means that the imports and exports of the Netherlands are optimized using the wind power forecast available at market gate closure, using the first pass of the UC–ED algorithm. After market gate closure, the international exchange schedules become fixed and are executed as scheduled (second pass UC–ED). For the day-ahead (12–36 h) market closure, wind power forecast errors are significant. This will result in a suboptimal scheduling of imports and exports from a wind power integration point of view. Forecast errors will have decreased by about 50% if market gate closure is delayed up to 3 h ahead of operation, and no forecast errors are present for the market design with near-real-time operation (1 h ahead), which allows an optimal scheduling of international exchange considering wind power.

9.6.1.2 Wind Power Forecasts

The following wind power forecasts are considered as base-case variants:

- 0-MW wind power forecast (fuel saver approach [30])
- Best available forecast
- Perfect forecast

For the 0-MW wind power forecast, the UC–ED is optimized without incorporating wind power capacity in the planning stage, although actual wind power output is taken into account in the operational stage. This forecast leads to an over-commitment of conventional generating capacity and serves as a worst-case planning situation. The best available forecast comprises an hourly update of the wind power forecast ("rolling forecast"), on the basis of updated wind power forecasts using the current row of the forecast matrix, and a subsequent recalculation of the UC–ED taking these into account. For a perfect forecast, the actual wind power levels are exactly known in all stages of the UC–ED. It is important to note that, for all wind power forecasts, the real-time wind power output level is assumed to be exactly known and used as an input for economic dispatch in the following hours. Furthermore, it is assumed that UC–ED is continuously optimized up until the hour of operation (1 h ahead).

9.6.1.3 Flexibility, Thermal and Electrical Storage

Technical limits to the system integration of wind power in the power system may be revealed by a UC–ED analysis. Different alternatives for overcoming these integration limits have been evaluated as system variants of the case study. More flexible base-load generation capacity may provide additional technical space for wind power during low-load, high-wind situations. For this alternative, the commitment status of selected industrial CHP units was changed from must-run to economic. The existing heat boilers could, in principle, take over the generation of steam for the industrial processes at times when the CHP is shut down to allow a further integration of wind power. Industrial-scale energy storage, often proposed to play the same role, exists in several forms including surface and underground pumped accumulation energy storage, and compressed air energy storage. The former two units can be readily modeled, as was explained in Section 9.3.2. The latter type of unit is modeled as a high-efficiency CCGT [31]. Finally, creating additional interconnection capacity between the Netherlands and the hydropower-dominated system of Norway is another way of gaining technical space. In this study, the option of extra interconnection capacity is regarded as a storage option. No specific attempt was made to optimize the design of the energy storage: the energy storage capacities and reservoir sizes applied in Reference [32] have been adopted here.

9.6.1.4 Summary

Simulations are performed for seven wind power penetrations from 0 to 12 GW in increments of 2 GW each, four international market designs (no international exchange, international market gate closure at 24, 3, and 1 h ahead) and three cases for wind power forecasts (best available/rolling forecast, perfect prediction of wind power, 0-MW wind power forecast). Furthermore, solutions for wind power integration are explored: flexible CHP units, surface and underground pumped accumulation energy storage, compressed air energy storage, and a dedicated transmission connection to Norway.

9.6.2 Assumptions and Simulation Setup

The UC–ED is calculated using the equal marginal cost method elaborated in Section 9.5, in which the objective function is the total cost for generating both heat and power given in Equation 9.10. A calculation of unit dispatch is performed hourly using the given load profile and an estimation of the wind power production levels. This recalculation of the UC–ED is in fact much more frequent than presently applied by Dutch generation portfolio owners, who usually do this two to four times a day. It can also be noted that a central optimization of the UC–ED does not take into account the behavior of individual market parties or generation clusters on the electricity market (i.e., fuel contracts, individual reserve power considerations). The availability of a better wind power forecast may be very beneficial for the market operating strategy of traders and leads to significant revenues. Here, only the opportunities of wind power forecasts for the maximization of the system integration of wind power are considered.

Fuel and emission costs have been determined on the basis of price forecasts stated in Reference [33] for the year 2015. The prices for coal, lignite, gas, oil, and uranium, and CO_2 used in the example are 2.00, 1.36, 5.00, 10.50, and 1.00 €/GJ, and 25.00 €/ton, respectively. Emission costs are included in the calculation of the marginal operating cost of each thermal generation unit. The sensitivity of the simulation results to these assumptions are very small regarding technical limits for wind power integration, but are considerable for operating cost and emissions. The simulation results for these two are, therefore, only a first-order estimate, and are reported only relative to the base case.

9.6.3 Simulation Outcomes

9.6.3.1 Load, Wind Power, and Conventional Generation Dispatch

Figure 9.16 provides an overview of the dispatch of units in the Netherlands during 1 week for the scenario with 12 GW wind power. The graph shows generation levels for distributed generation, thermal units, integrated wind power, and the amount of wasted wind energy.

Wasted wind means wind power that could not be integrated due to technical constraints.

Total generation by conventional thermal generation units follows the system load, distributed generation, and wind power. In this particular week, wind power is ramped down at moments of high wind power and low load (all nights, except Sunday when

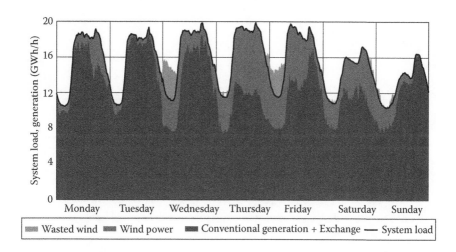

FIGURE 9.16   Example of a unit commitment–economic dispatch for 1 week in the Netherlands for a 12-GW installed wind power capacity and no international exchange.

there is little wind power available) to prevent minimum load problems. A good example of the use of thermal generation for balancing the combined variations of load and wind power can be seen on early Sunday morning (when thermal generation ramp up and wind power is decreasing).

In Figure 9.17, the change in annual electricity output between different generation technologies is shown for the Netherlands (no international exchange) with increasing wind power capacity. Nuclear power, being a full-load must-run technology, is not affected by wind power integration. Wind power does reduce the full-load hour equivalents of coal-fired units, CCGT CHP and CCGT. Importantly, the profits of these units also decrease during the hours that they are in operation, as wind power always replaces the most expensive unit in operation (as far as technically feasible). Because of the large share of coal-fired generation in the Dutch generation park modeled in this example, the electricity generation (TWh/y) of coal is reduced most.

Notably, the technical flexibility of coal, CCGT CHP, and CCGT does not require additional operating hours of peak-load gas turbines for wind power integration. DG (in this case gas engines found in greenhouses) decreases its operation hours only very slightly: the must-run part is fixed, and the flexible units produce heat and power during other periods, with the heat being stored.

On a relative scale, the output of CCGT is affected most by the integration of wind power: CCGT operates only during medium- and peak-load hours, during which it is often the marginal technology and, therefore, the first to be replaced by wind power. As coal and CCGT CHP have a part-load must-run status, the integration of wind power reduces their output only to a certain extent.

9.6.3.2 Scenarios for International Exchange

In case international exchange is possible, the integration of wind power in the Netherlands influences, in principle, the exchanges between all countries. In Figure 9.18,

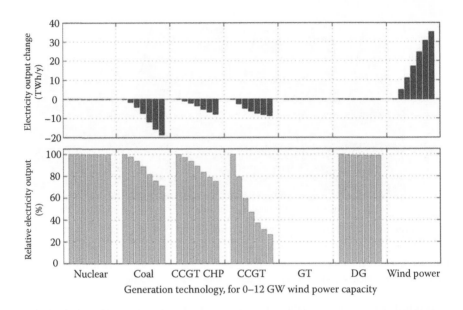

FIGURE 9.17 Absolute electricity production change and relative output per technology in the Netherlands for different wind power penetration scenarios in increments of 2 GW, with no international exchange. Relative electricity output for wind is omitted due to relatively large percentages above 100.

imports and exports are shown for each of the areas originally introduced in Figure 9.14, with each bar representing a wind power penetration scenario. Clearly, the Netherlands increases its annual exports and decreases its imports in case more wind power is installed. This influences mainly imports and exports of Germany and Great Britain, and Belgium to a limited extent. Due to the presence of large interconnection capacity between Germany and the Netherlands, the Dutch wind power decreases not only the full-load

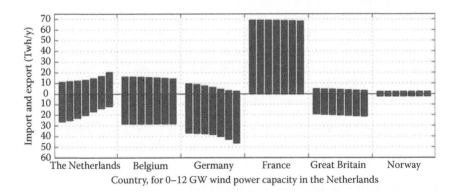

FIGURE 9.18 International exchange in Northwest Europe for 0- to 12-GW installed wind power capacity in the Netherlands, in increments of 2 GW per bar.

hours of base-load coal and lignite in Germany, but also the hours of some CCGT. Wind power furthermore reduces the exports of base-load coal power from Belgium and to a lesser extent from France during periods of low load (nights and weekends). Germany reduces its imports from France at times of high wind in the Netherlands. Exchanges with Norway stay constant in volume since it is modeled as such, although the moments of exports and imports are increasingly determined by wind power as its installed capacity in the Netherlands increases. These results clearly show the importance of the larger German system for the integration of wind power in the Dutch system.

9.6.3.3 Operating Cost Savings

Figure 9.19 shows the annual savings in operating cost due to wind power, for the Netherlands with and without international exchange. The outcome for a 1-h-ahead market gate-closure time for the Northwest European system as a whole is shown as a solid line, and for the Dutch system as a dotted line. The figure shows that operating cost savings by wind power increase with the amount of wind power installed. For the fuel and operating costs assumed here, the overall annual operating cost savings by wind power are estimated to be in the order of 1.5 billion euros annually for a 12-GW wind power capacity. The slight differences in total cost savings between an isolated Dutch system and a Northwest European system are due to a different generation mix in which wind power is integrated. Thus, the base-case with 0 MW wind power is already different with respect to marginal costs. Differences in total cost savings with and without international exchange at high wind power penetrations are due to the wasting of wind in an isolated Dutch system (additional fuel cost and emissions).

In case no international exchange is possible for exports of excess wind power, the relative cost savings gained from wind power start to decrease from 8 GW installed capacity onwards.

Limits in the operational flexibility of conventional plants lead to suboptimal dispatch, reduced operating efficiencies, and, ultimately, increased wasting of available wind resources. Some improvement is possible for the isolated system when different

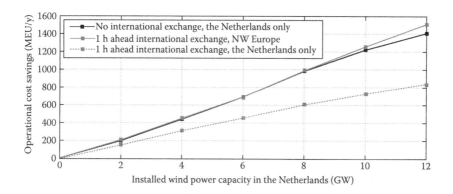

FIGURE 9.19 Annual operating cost savings by wind power for different wind power levels.

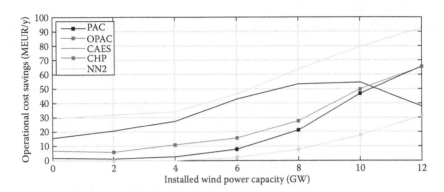

FIGURE 9.20 Annual operating cost savings with no international exchange, but with flexibility solutions (three energy storage types, combined heat and power boilers, interconnection to Norway) implemented.

technologies are introduced, as shown in Figure 9.20. However, such savings are but small improvements over the base cost savings from wind power.

In case the Netherlands is part of an international Northwest European market, the technical integration limits for wind power lie further away (highest line, Figure 9.19). Because more wind power can be integrated in a Northwest European system, the operating cost savings are greater than for an isolated Dutch system, where wind power must instead be wasted. In an international environment, slightly over one half of the total economic benefit of wind power is realized in the Netherlands, the rest is realized in neighboring areas.

Operating costs are also influenced by wind power forecasts. However, savings in operating cost as a result of the use of wind power predictions differ between weeks and are in the order of 0.2% of the total operating cost per year. Application of the best available wind power forecast, however, does not save operating costs for each operation hour. Over-predictions of wind power may lead to an under-commitment of base-load units: when the wind power falls short compared to the forecast, extra units must be committed at a higher cost. Therefore, at times choosing the 0-MW wind power forecast would be cheapest, and the choice is related to the quality of the forecast. The extra improvement that could come from applying 0 MW and the best available forecast methods at different times throughout the week (i.e., from choosing among options of an "ensemble forecast") is an additional integration of 1 TWh or 2.4% of available wind energy for a 12-GW installed wind power capacity, compared to using the best available forecast only. However, this would be difficult in practice. Summarizing, improved wind power forecasts have some benefit for system operation (wasted wind, operating cost) but little influence on the total amount of wasted wind energy. From a technical point of view, a frequent update of the UC–ED using real-time information on wind power together with the application of updated wind power forecasts is sufficient. It can be noted that the benefits of improved wind power forecasts for trading on markets are not considered here. Such benefits may be significant for the individual market parties for the formulation of their market trading strategy [34].

9.6.3.4 Emissions Reduction

The simulation results clearly show that wind power leads to a saving of significant amounts of CO_2 emissions. In Figure 9.21, the annual emission savings are shown for the Netherlands without international exchange, and for the Northwest European system as a whole (international exchange is possible in this case), with the Dutch part of that as a dotted line. Emission savings are estimated to lie around 19 Mton annually for 12 GW wind power, with higher savings for the isolated Dutch system due to the large share of coal in the generation mix. In case international exchange is possible, more expensive, less efficient units (mainly CCGT) have already been pushed out of the market at 0 MW wind power. Wind power will also replace more expensive CCGT generation during the day rather than coal-fired units in the isolated Dutch system, resulting in lower emission savings for the cases with international exchange. It can be noted that emission savings also positively impact operating costs, as CO_2 emission savings are part of the total operating cost. The change in steepness of the curves at 2- and 6-GW installed wind power capacity is due to the higher capacity factor of offshore wind power. For the isolated Dutch system, there is a change at 8 GW wind power due to the increasing amounts of wasted wind energy. The results for emission savings for SO_2 and NO_x show similar trends as CO_2. Total annual emissions show an estimated decrease of 11 Mton and 20 kton for SO_2 and NO_x, respectively, for an isolated Dutch system.

The results also show that the emissions reduction benefits of the three technological solutions are very limited. Figure 9.22 shows that, for all energy storage scenarios, emissions savings resulting from these technologies are much smaller than the total emissions, or the differences between international exchange scenarios. Also, for levels of wind power below 8 GW, emissions are actually increased. The additional emission of CO_2 can be explained by two factors. First, it must be understood that energy storage is operated to minimize system operating cost, within the technical constraints of the system. For cost optimization, the storage reservoirs are filled when prices are low, to be emptied for generating electricity when prices are high. In the Dutch system, energy

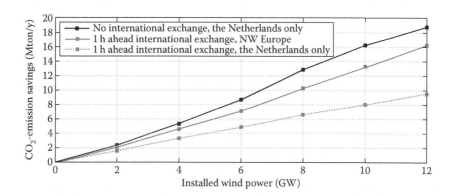

FIGURE 9.21   Carbon dioxide emission savings by wind power, with and without international exchange.

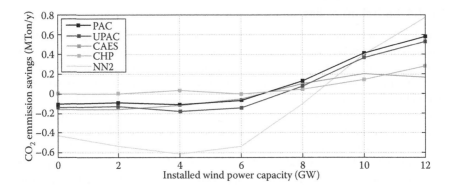

FIGURE 9.22 Carbon dioxide emission savings with no international exchange, but with flexibility solutions (three energy storage types, combined heat and power boilers, interconnection to Norway) implemented.

storage in fact substitutes peak-load gas-fired production by base-load coal-fired production. Since coal emits about twice as much CO_2, on a megawatt-hour basis, as gas, the net coal-for-gas substitution by energy storage increases the overall amount of CO_2 emitted by the Dutch system. Second, energy storage brings about conversion losses that must be compensated by additional generation from thermal units, which again increases CO_2 emissions, especially as this is also done by coal-fired units, being the cheapest option. It follows that for the system and assumptions applied in this research, from a CO_2 perspective, energy storage is an environmentally friendly option only for very high wind penetration levels, when energy storage prevents wasted wind. The same is the case for an extra interconnector to Norway operated as assumed here.

Notably, the use of heat boilers not only saves operating costs but also CO_2 emissions. Since the use of heat boilers at CHP locations specifically tackles the minimum load problem as a result of CHP unit operating constraints, heat boilers reduce the amount of wasted wind. As the CO_2 emissions of boilers and wind power are lower than CO_2 emissions of CHP units, boilers reduce the overall amount of CO_2 emitted by the system as well.

9.6.3.5 Wasted Wind

In Figure 9.23, wind energy integrated into the Dutch power system is shown for different wind power penetrations and different market designs. Wasted wind energy becomes significant in the range of 6- to 8-GW installed wind power capacity for the Dutch power system, in the market design without international exchange. The slight change in steepness of the available wind energy curve at 2- and 6-GW installed wind power capacity is due to an increased capacity factor of wind power (offshore vs. onshore). The use of international exchange provides significant additional space for the integration of wind power (middle-gray area is additionally integrated wind energy). The light gray area representing wasted wind in a 1-h-ahead market gate closure is very small and not visible in this figure.

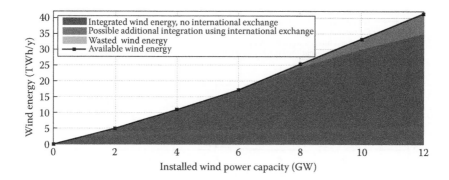

FIGURE 9.23 Integrated and wasted wind energy in the Netherlands.

In Figure 9.24, the amount of wasted wind energy for all five integration improvement options can be observed. Clearly, all options considered here reduce the amount of wind wasted in the Netherlands due to minimum-load problems. Energy storage and heat boilers increase the flexibility of the Dutch system and thereby enable larger amounts of wind energy to be integrated. An extra interconnection to Norway creates a virtual energy storage with the same effects. When considering an isolated Dutch system, pumped accumulation energy storage has the highest potential of all storage options for reducing the amount of wasted wind. An extra interconnector to Norway would provide a similar potential for this, if it could be used as assumed here. However, none of the energy storage options is sufficient to prevent wasted wind energy altogether. In case international exchange is possible with a 1-h-ahead market gate closure, wasted wind energy is reduced much more than using any of the integration improvement options investigated here.

Figure 9.25 focuses further on a comparison of the amount of wasted wind energy for different market designs. Only wind power forecast errors in the Netherlands are considered here. In case no interconnection capacity is available for balancing purposes, an

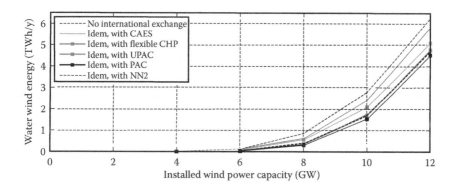

FIGURE 9.24 Wasted wind with flexible combined heat and power units, energy storage options and extra interconnection to Norway, with no international exchange.

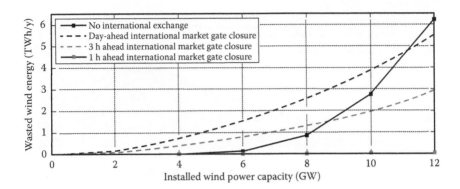

FIGURE 9.25 Wasted wind energy for different international market designs and best available wind power forecast.

estimated amount of 6.2 TWh/y or 15% of available wind energy in the Netherlands cannot be integrated into the system. In case international exchanges can be used for exports at high wind power levels, additional wind power can be integrated, with only 0.05 TWh or 0.1% of available wind energy being wasted for the 1-h-ahead market gate closure.

Interestingly, a day-ahead or 3-h-ahead international market gate-closure time results in larger amounts of wasted wind power at smaller wind power capacities. This is the result of the methodology applied for the optimization of international exchange at market gate closure, which is based on the assumption that all feasible international transactions are being made. In case a significant wind power forecast error is present at the moment that these transactions become fixed, scheduled imports may prevent the integration of unpredicted surpluses of wind power, leading to larger amounts of wasted wind energy. For large wind power penetrations, however, the benefits of international exchange capacity outweigh the disadvantage of forecast errors. Clearly, a more conservative scheduling of international exchanges (less imports) will result in less wasted wind energy. This result illustrates the benefits of postponed international market gate-closure times for integrating wind power.

In case interconnection capacity is available and the market design allows an adjustment of international exchange up until the moment of operation (1-h-ahead international market gate closure), the potential for additionally integrated wind energy is high. Still, even the most flexible international market design cannot prevent a small amount of wasted wind energy, starting from 8- to 10-GW installed wind power capacity in the Netherlands (bottom of Figure 9.25, not visible in Figure 9.23). The reason for this is that, even though international transmission capacity may be sufficient, this capacity is not always fully available for exports. This applies to Germany in particular. Germany has a significant must-run conventional generation capacity and a large amount of wind power (32 GW in the year 2014) that is highly correlated (0.73) to wind power in the Netherlands. Both factors reduce the possibilities for export of wind power from the Netherlands, especially during critical periods.

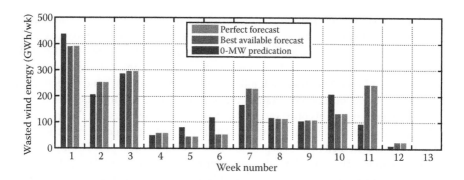

FIGURE 9.26 Weekly wasted wind energy for different forecast methods, with 12 GW wind power and no international exchange.

The different types of wind power forecasts only have a minor influence on the amount of wasted wind energy on an annual basis. Figure 9.26 shows the amount of wasted wind energy for several weeks for the 12-GW wind power variant, with no international exchange allowed. Interestingly, some differences are present in the amount of wasted wind between 0 MW and the best available forecast method. The simulation results show that an over-commitment of conventional generation capacity may be beneficial for wind power integration during some hours of the week. Generally, however, the differences in wasted wind between the 0-MW wind power forecast, the best available wind power forecast, and the perfect forecast are small (<<5% of wasted wind energy). This can be explained by the frequent recalculation of the UC–ED that is applied to all simulations, which allows the inclusion of the real-time wind power output and of regularly updated wind power forecasts. As actual wind power levels are accurately known and wind power output generally does not change significantly between 15-min intervals, the conventional generation units in operation will typically be adequate for the next time interval as well, explaining the relatively good performance of the 0-MW wind power forecast.

9.7 Conclusions

Analysis of UC–ED is an important tool for assessing the effect of wind power in large power systems. Many factors that determine the cost and emissions of a power system are properly combined in such an approach. This is crucial to obtaining realistic results. From the results shown in this chapter, it is undeniable that absolute emissions of CO_2 are significantly reduced with the installation of wind power. Operating costs of the system are also reduced due to fuel and emissions pricing.

In this chapter, a simple but effective way of representing international trade in a UC–ED analysis by using a wind power forecast matrix was described. International and interregional markets are becoming more flexible, and this means the approach of considering a continuously updated forecast will only become more relevant in the future. The analysis for different penetrations of wind power has indicated that last-minute gate-closure times will allow the integration of more wind energy by reducing the impact of imperfect wind forecasts. However, the effect of forecasts on wasted winds and operation costs was relatively small for the case where the Dutch power system was isolated. In this example, international exchange clearly affected the technical limits on integrating wind power in the Netherlands. It follows that when significant wind generation is deployed in a neighboring power system, the technical limits to wind integrations can shift in both countries. This observation is a crucial one for settings like Europe where nuclear phase-out and wind farm construction are proceeding in some jurisdictions.

For the system studied in this chapter, limits to integration were posed by high wind, low load situations. This underlines the importance of having accurate daily patterns for wind power information. From a cost and emissions perspective, it was interesting to note that international exchange was a much more effective technological solution for increasing the amount of wind power than was storage, and that among the storage options the use of heat boilers was most effective for reducing emissions.

References

1. P. P. J. van den Bosch. Short-term optimisation of thermal power systems. PhD thesis, Technische Hogeschool Delft, the Netherlands, 1983.
2. M. Carrion and J. M. Arroyo. A computationally efficient mixed-integer linear formulation for the thermal unit commitment problem. *IEEE Transactions on Power Systems*, 21(3): 1371–1378, 2006.
3. Wind Energy News. Available at: http://windenergie-nieuws.nl/ [Last accessed: October 2011].
4. Dutch Ministry of Economic Affairs. Connect 6.000 MW, Eindrapportage [in Dutch]. *Technical Report*. The Haag, the Netherlands: Dutch Ministry of Economic Affairs, 2004. 44pp.
5. Noordzeeloket. Overzicht van startnotities windparken op zee, 2008. Available online at: http://www.noordzeeloket.nl/ [Last accessed: October 2008].
6. M. Gibescu, A. J. Brand, and W. L. Kling. Estimation of variability and predictability of large-scale wind energy in the Netherlands. *Wind Energy*, 12(3): 241–260, 2009.
7. P. J. Brockwell and R. A. Davis. *Time Series: Theory and Models*, 2nd ed.. New York: Springer-Verlag, 1991.
8. A. J. Brand, M. Gibescu, and W. W. de Boer. Variability and predictability of large-scale wind energy in the Netherlands, in *Wind Power*, S M Muyeen (ed.). Croatia: InTech, 2010. ISBN: 978-953-7619-81-7; Available at: http://www.intechopen.com/articles/show/title/variability-and-predictability-of-large-scale-wind-energy-in-the-netherlands
9. G. Giebel. On the benefits of distributed generation of wind energy in Europe. PhD thesis, Carl von Ossietzky Universitaüt Oldenburg, Germany, 2001.
10. H. Holttinen. Hourly wind power variations in the Nordic countries. *Wind Energy*, 8: 173–195, 2005.
11. J. F. Walker and N. Jenkins. *Wind Energy Technology*. New York: Wiley, 1997.
12. G. A. M. van Kuik. The Lanchester–Betz–Joukowsky limit. *Wind Energy*, 10: 289–291, 2007.

13. A. D. Hansen and L. H. Hansen. Wind turbine concept market penetration over 10 years (1995–2004). *Wind Energy*, 10: 81–97, 2006.

14. H. Polinder, F. F. A. van der Pijl, G.-J. de Vilder, and P. J. Tavner. Comparison of direct-drive and geared generator concepts for wind turbines. *IEEE Transactions on Energy Conversion*, 21(3): 725–733, 2006.

15. B. C. Ummels. Wind integration: Power system operation with large-scale wind power in liberalized environments. PhD thesis, Technische Universiteit Delft, the Netherlands, 2009.

16. R. Doherty and M. J. O'Malley. A new approach to quantify reserve demand in systems with significant installed wind capacity. *IEEE Transactions on Power Systems*, 20(2): 587–595, 2005.

17. A. J. M. van Wijk. Wind energy and electricity production. PhD thesis, Universiteit Utrecht, The Netherlands, 1990.

18. P. Nørgavrd and H. Holttinen. A multi-turbine power curve approach. In *Proceedings of Nordic Wind Power Conference*, Gothenburg, Sweden, March 1–2, 2004, 5pp.

19. F. W. Koch, M. Gresch, F. Shewarega, I. Erlich, and U. Bachmann. Consideration of wind farm wake effect in power system dynamic simulation. In *Proceedings of IEEE PowerTech Conference*, St. Petersburg, Russia, June 27–30, 2005, 7pp.

20. G. J. W. van Bussel and M. B. Zaaijer. Reliability, availability and maintenance aspects of large-scale offshore wind farms, a concepts study. *Proceedings of MAREC Marine Renewable Energies Conference*, 113(1): 119–126, 2001.

21. A. J. Brand and J. K. Kok. Aanbodvoorspeller Duurzame Energie Deel 2: Kortetermijn prognose van windvermogen [in Dutch]. *Technical Report, Energy Research Centre of the Netherlands (ECN)*, ECN–C–03–049, 2003.

22. M. Lange and U. Focken. *Physical Approach to Short-Term Wind Power Prediction*. Berlin and Heidelberg: Springer-Verlag, 2005.

23. Operation Simulation Associates, Inc. *PowrSym3 User's Manual v361d2*, Technical Report, 2008.

24. E. Denny and M. J. O'Malley. Quantifying the total net benefits of grid integrated wind. *IEEE Transactions on Power Systems*, 22(2): 605–615, 2007.

25. H. Lund. Large-scale integration of wind power into different energy systems. *Energy*, 30: 2402–2412, 2005.

26. P. Meibom, J. Kiviluoma, R. Barth, H. Brand, C. Weber, and H. V. Larsen. Value of electric heat boilers and heat pumps for wind power integration. *Wind Energy*, 10: 321–337, 2007.

27. E. D. Castronuovo and J. A. Peças Lopes. Optimal operation and hydro storage sizing of a wind-hydro power plant. *Electrical Power and Energy Systems*, 26: 771–778, 2004.

28. O. A. Jaramillo, M. A. Borja, and J. M. Huacuz. Using hydropower to complement wind energy: A hybrid system to provide firm power. *Renewable Energy*, 29: 1887–1909, 2004.

29. J. P. Stremel, R. T. Jenkins, R. A. Babb, and W. D. Bayless. Production costing using the cumulant method of representing the equivalent load curve. *IEEE Transactions on Power Apparatus and Systems*, 99(5): 1947–1956, 1980.

30. B. Fox, D. Flynn, L. Bryans, N. Jenkins, D. Milborrow, M. O'Malley, R. Watson, and O. Anaya-Lara. *Wind Power Integration: Connection and System Operational Aspects*, IET Power and Energy Series. London: Institution of Engineering and Technology, 2007.

31. B. C. Ummels, P. Pelgrum, W. L. Kling, Integration of large-scale wind power and use of energy storage in the Netherlands' electricity supply *Renewable Power Generation*, 2(1): 34–36, 2008.

32. Transition Platform Sustainable Electricity Supply. Onderzoek naar de Toegevoegde Waarde van Grootschalige Elektriciteitsopslag in Nederland [in Dutch]. *Technical report*, 2008.

33. International Energy Agency. World energy outlook 2008. Available at: http://www.worldenergyoutlook.org/ [Last accessed: October 2008].

34. M. Gibescu, E. W. van Zwet, W. L. Kling, and R. D. Christie. Optimal bidding strategy for mixed-portfolio producers in a dual imbalance pricing system. In *Proceedings of Power Systems Computational Conference*, Glasgow, UK, July 14–18, 2008, 7pp.

10

Operational Optimization of Multigeneration Systems

Pierluigi Mancarella

Gianfranco Chicco

10.1 Introduction

The evolution of power systems is being deeply influenced by the growing need for enhancing energy saving and cutting carbon dioxide (CO_2) emissions from energy generation in order to meet challenging environmental targets set by governments worldwide. In this context, the perspective of increasing energy efficiency and environmental performance on the generation side is boosting the interest toward deploying integrated energy systems and combined production of different energy vectors. In this respect, multigeneration (MG) can be defined as the production of multiple energy vectors such as electricity, heat, cooling, and so on, from a unique source of fuel (Mancarella and Chicco, 2009a). Such an approach can bring energy, environmental, and economic benefits with respect to a classical approach where energy vectors are produced in separate production.

As the "base" and more traditional case of MG, cogeneration or combined heat and power (CHP) plants allow more efficient fuel energy input utilization with respect to classical separate production whereby electricity is generated in centralized power plants and heat in traditional boilers (Horlock, 1997; Martens, 1998). Depending on the characteristics of the CHP plant and of the separate production references, CHP higher overall efficiency can thus bring energy savings along with CO_2 emission reduction, also depending on the fuel carbon content and on the emission intensity of the displaced sources (Horlock, 1997; Chicco and Mancarella, 2008b).

CHP system operability can be significantly affected by low thermal loads in the summertime, when the need for space heating is generally not present and only domestic hot water makes up the thermal demand. Hence, the CHP unit, sized on the basis of the winter thermal demand, could operate at partial load and often be switched off below a certain loading threshold, losing all or at least part of the benefits from cogeneration production. Further decisions on operating the CHP system could depend on economic aspects (energy prices set up on the basis of contracts or within energy markets), for instance leading to switch the CHP unit off during the night because of low electricity prices making it unprofitable to sell electricity to the grid.

The presence of cooling requirements paves the way to adopt combined cooling, heat, and power (CCHP) or trigeneration (Borbely and Kreider, 2001; Heteu and Bolle, 2002; Hernandez-Santoyo and Sanchez-Cifuentes, 2003; Chicco and Mancarella, 2006; Ziher and Poredos, 2006). In this case, a typical application is to exploit cogenerated heat for cooling production by means of a water absorption/adsorption refrigerator group (WARG) (Maidment and Prosser, 2000; Minciuc et al., 2003). Hence, in a CCHP plant, the CHP prime mover can be operated at high loading level also in the summertime, contributing to cover an air-conditioning demand that is steadily rising even in mild climate.

A more general approach to trigeneration has been recently put forward as well (Chicco and Mancarella, 2009a; Mancarella and Chicco, 2009a), where a number of different solutions for electricity and heat can be coupled to the CHP system, including classical compression electric refrigerator group (CERG), electric heat pump (EHP), engine-driven chiller (EDC), engine-driven heat pump (EDHP), and gas absorption heat pump (GAHP). Hence, such CCHP plants can, in turn, be seen as a particular case of the more general category of distributed multigeneration (DMG) systems (Chicco and Mancarella, 2009a) enabling the dispatch of different types of energy and the conversion from one type of energy to another through suitably sized components, with other possible external networks for further exploitation of the energy products.

The authors have illustrated and discussed DMG concepts and applications in recent references, following a research line developed to highlight the perspectives and assess the potential of DMG applications, in terms of both energy efficiency improvement (Chicco and Mancarella, 2007a,b) and environmental impact reduction (Chicco and Mancarella, 2008b; Mancarella and Chicco, 2008, 2010), up to the formulation of a unified approach to define structured indicators to quantify the technical and environmental performance of MG systems (Chicco and Mancarella, 2008a). Relatively decentralized solutions for generation of different energy vectors can be effectively integrated within a number of contexts. Above all, the applications in urban areas are of particular interest, owing to wide availability of adequate loads in case aggregated through district networks (Calì and Borchiellini, 2004; Danny Harvey, 2006), as well as of incumbent large-scale energy networks such as for gas and electricity. In addition, new energy vectors could play a key role for the development of future energy systems, such as hydrogen owing to its characteristics of being transported over long distances (like electricity) and of being stored (in a relatively easier way than electricity).

While development of MG solutions can represent a fundamental milestone in the evolution of future energy systems, assessment of such systems requires powerful tools and methodologies from different viewpoints, particularly economic and environmental ones. In addition, different timescales and purposes can be considered when optimizing MG systems.

Analysis and optimization of MG systems can be addressed with reference to different time frames. Considering the time frames in descending order of their duration, it is possible to synthetically identify:

- *Long-term* time frame, with multiyear duration, including strategic problems (e.g., referring to investments or sustainable energy development) or *design* problems linked to the choice of the most convenient technological solution among a set of predefined planning alternatives.
- *Short-term* time frame, which can be further partitioned into an *operation* timescale, indicatively ranging from 1 min to 1 week, serving as the basis for the formulation of operational problems that can span up to an annual time horizon, and a *real-time* scale, indicatively ranging from seconds to minutes to represent the dynamics of various nature, mainly related to the electrical side including the action of control systems, but also to atmospheric (sun, wind), chemical (e.g., for battery storage and fuel cells), and economic (real-time pricing) aspects.

In this context, this chapter deals with optimization of MG systems in the short-term *operation* timescale. The formulation and application of various approaches presented in the literature are summarized to provide a synthetic view of how optimization is addressed (with different objective functions and constraints) and solved (with different computational approaches). In particular, operational optimization can be seen as a specific problem, as well as part of a more comprehensive combined optimization of

MG system planning and operation; for instance, to simultaneously perform the selection of the most convenient planning alternative together with its operational schedule, as for instance in Burer et al. (2003), Cardona et al. (2006a), Kavvadias et al. (2010). This chapter assumes that the technological choices have already been made by the decision-maker, and focuses only on detailed operational aspects that can indicatively range from minutes to a week, serving as the basis for the formulation of operational problems that can span up to an annual time horizon.

Besides energy efficiency, the optimization aspects analyzed include both economic and environmental issues, and are thus naturally orientated toward multiobjective optimization. In this respect, classical economic objective functions such as for cost minimization or profit maximization can be more or less conflicting with environmental objectives of local dimension (for instance, for pollution control) or of global dimension such as primary energy resources conservation or greenhouse gas (GHG) emission control. Suitable formulations of objective functions and constraints, as well as of performance criteria, are outlined in this work.

The contents of this chapter are organized as follows. Section 10.2 recalls the structures and characteristics of MG components. Section 10.3 addresses the different types of objective functions for single-objective and multiple-objective optimizations. Section 10.4 illustrates various constraints that can be encountered in the formulation of MG operational optimization problems. Section 10.5 summarizes the solution techniques used in various applications. Section 10.6 presents an illustrative example of operational optimization of a trigeneration system.

10.2 MG Components and Structures

10.2.1 MG Components and Operational Characteristics

Representation of MG components for formulation of optimization problems is a challenging task requiring a suitable modeling framework. A convenient formulation is based on a *black-box approach* (Horlock, 1997; Mancarella, 2006; Mancarella and Chicco, 2009a) whereby the input–output characteristics of individual pieces of equipment are represented through synthetic efficiency models, without describing, for instance, thermodynamic details of the plant. This approach is consistent with classical representation of electricity-only power plants, where the fuel input is typically expressed as a polynomial function of the electricity output. In addition, such a representation is synthetic and effective at the same time, limiting the number of variables used to describe the energy flows of individual components and of the overall plant, while holding sufficient information for technoeconomic characterization. As illustrated below, sets of components can be aggregated into equivalent black boxes and the whole plant can, in turn, be seen as a black box for network interaction (Geidl and Andersson, 2007a; Chicco and Mancarella, 2009b).

Focusing on the most typical end-use energy vectors, namely, electricity, heat, and cooling (trigeneration systems), an MG plant can be composed of a number of different units, ranging from CHP systems to heat-fired generators (for both heating and cooling) as well as electrothermal technologies.

Regarding CHP prime movers, microgenerators at the household level typically include internal combustion engines and Stirling engines (Onovwiona and Ugursal, 2006; Kuhn et al., 2008), whereas larger prime movers, up to relatively large-scale CHP for district energy networks, also include microturbines and gas turbines. Natural gas is the most widespread fuel, although biomasses and biogas represent an important option for clean development. Cluster operation, for instance for microturbines, could ensure higher operational flexibility, but raises the issue of taking into account the partial-load microturbine characteristics, in terms of energy efficiency reductions and pollutant emissions increase at partial load (Boicea et al., 2009).

Let us adopt a black-box representation of CHP systems, in which the terms W, Q, and F denote electricity, heat, and fuel thermal energy, respectively, whereas the superscript y points out cogeneration entries. For the purposes of the models presented here, the energy vectors appearing in the definition of performance indicators can indicate both power (i.e., average power in a given time interval) and energy. The CHP system is characterized by its energy performance indicators (first law efficiencies) such as the electrical efficiency η_W and the thermal efficiency η_Q (output-to-input energy ratios). Furthermore, an important cogeneration characteristic linking electricity and heat is the *heat-to-electricity* (or heat-to-power) *cogeneration ratio* λ^y (Horlock, 1997). These quantities are defined as

$$\eta_W = \frac{W^y}{F^y}, \quad \eta_Q = \frac{Q^y}{F^y}, \quad \lambda^y = \frac{Q^y}{W^y} = \frac{\eta_Q}{\eta_W} \tag{10.1}$$

The specific models of the performance indicators can be developed with different levels of detail, from constant values to part load-dependent linear or polynomial approximations, depending on the type and purpose of the study. For the sake of exemplification, constant values will be adopted here.

The most widespread technologies for heat/cooling generation include:

- Auxiliary boiler (AB), providing thermal back up to the CHP system and/or used to top up the heat production to cover thermal peak loads.
- CERG (Voorspools and D'haeseleer, 2003; Danny Harvey, 2006), where the input is electricity W to power an electrical compressor and the output is cooling R, typically in the final form of chilled water or cooled air. Some options are available where the condensing temperature can be raised to allow heat to be recovered at a temperature useful for other purposes, although this usually implies a drop in efficiency in the cooling production.
- Water absorption/adsorption refrigeration group (WARG) (Tozer and James, 1998; Kreider, 2001; Danny Harvey, 2006; Deng et al., 2011), where heat Q at different enthalpy levels (for instance, in the form of hot water, steam, or exhausts, specifically depending on the equipment) is used as input to a so-called thermochemical compressor to generate cooling energy R. In particular, adsorption chillers are emerging as viable alternatives to absorption chillers, above all for exploiting low-temperature waste heat, for instance, from solar energy, for residential applications. Again, heat recovery options may be available.

- Gas absorption refrigeration group (GARG) (Kreider, 2001; Wu and Wang, 2006), direct-fired by gas whose thermal content F is transformed into cooling R again through an equivalent thermochemical compression.
- Engine-driven chiller (EDC) (Danny Harvey, 2006; Lazzarin and Noro, 2006), for which a conventional compression chiller is driven by a mechanical compressor directly connected to the shaft of an engine; again, the energy input is represented by fuel thermal content F, and the output is cooling energy R. Heat can be recovered at a particularly high value, by extracting energy from the exhausts of the driving engine.
- EHP of different types (e.g., ground source, water source, air source, and so on), where electricity W is input to extract "free" heat available in the environment and provide end-use heat Q at a higher temperature than the environment. EHPs can also be used to enhance the enthalpy level of waste heat, for instance, recovered from WARG (Chicco and Mancarella, 2007b) or CERG, and are usually reversible machines that can be used as chillers.
- Engine driven heat pump (EDHP) and absorption heat pump (AHP), where again the output is heat Q and the input can be fuel thermal energy F in a combustion engine driving a mechanical heat pump (EDHP), or heat Q into the AHP (Danny Harvey, 2006; Lazzarin and Noro, 2006; Costa et al., 2007).

In addition, other technologies can be used within MG systems for various purposes. For instance, dedicated systems can be coupled to the CHP plant for desalination, desiccant, or dehumidification effects, as well as chemical products used in specific processes (Liu et al., 2004; Uche et al., 2004; Wang and Lior, 2007; Badami and Portoraro, 2009). Also, extending the scope beyond conventional trigeneration, hydrogen is envisaged to play an important role in the future, with several solutions such as local production for instance from natural gas or from electrolysis, storage, further distribution to other plants, direct use for specific processes or for transportation, and so forth (Hemmes et al., 2007).

In terms of black-box modeling, as for CHP systems, the above cooling/heating components can be modeled as input–output energy performance indicators. For instance, for cooling generation equipment, the most used performance indicator is the coefficient of performance (COP), generally defined as output (cooling energy R) to the relevant input depending on the specific component; for instance, electrical energy W for a CERG, thermal energy Q for a WARG, fuel thermal energy F from an EDC, and so forth (Mancarella and Chicco, 2009a):

$$\mathrm{COP^{CERG}} = \frac{R}{W}; \quad \mathrm{COP^{WARG}} = \frac{R}{Q}; \quad \mathrm{COP^{EDC}} = \frac{R}{F} \tag{10.2}$$

Other components can be modeled similarly, while heat recovery from a chiller can also be modeled through specifically built indicators. Hence, it is clear how multiple energy vectors actually need to be modeled at an individual component level, besides the overall MG plant. An effective way to formalize the black-box approach to consider multiple input and output energy vectors at a component level is by introducing a matrix notation. More specifically, considering the set Y containing equipments or components belonging to the MG system, the black box of each component $Y \in \mathbf{Y}$ can be described by an *efficiency matrix* with the relevant performance indicators in positions to opportunely model the flow connections between input and output. Let us denote the general form of an efficiency matrix for an individual component Y as $\mathbf{H}^Y$. The corresponding output-to-input relationship is written as (Chicco and Mancarella, 2008c, 2009b):

$$\mathbf{v}_o^Y = \mathbf{H}^Y \cdot \mathbf{v}_i^Y \tag{10.3}$$

where $\mathbf{v}_o^Y$ is the array containing the relevant output vectors and $\mathbf{v}_i^Y$ the relevant inputs for the component Y. For instance, if we consider four energy vectors (with the same types of *ordered entries* for both inputs and outputs), namely, F, W, Q, and R, in a trigeneration plant, and we identify the entries of $\mathbf{H}^Y$ by means of a two-letter subscript, with the first letter referring to the output energy vector and the second one to the input energy vector, we can write

$$\begin{pmatrix} F_o^Y \\ W_o^Y \\ Q_o^Y \\ R_o^Y \end{pmatrix} = \begin{pmatrix} \eta_{FF}^Y & \eta_{FW}^Y & \eta_{FQ}^Y & \eta_{FR}^Y \\ \eta_{WF}^Y & \eta_{WW}^Y & \eta_{WQ}^Y & \eta_{WR}^Y \\ \eta_{QF}^Y & \eta_{QW}^Y & \eta_{QQ}^Y & \eta_{QR}^Y \\ \eta_{RF}^Y & \eta_{RW}^Y & \eta_{RQ}^Y & \eta_{RR}^Y \end{pmatrix} \cdot \begin{pmatrix} F_i^Y \\ W_i^Y \\ Q_i^Y \\ R_i^Y \end{pmatrix} \tag{10.4}$$

An extension to the overall MG plant is described in Section 10.2.2.2.

10.2.1.1 Energy Loads

An MG system is purposely built to supply different loads, the most common types of which are electricity, heat, and cooling. In general, different time resolutions and analysis window could be applied depending on the specific study. The steady-state types of analysis object of this work assume that the real-time dynamics have already reached their equilibrium. Hence, hourly or half-hourly resolution is typically considered sufficient to capture the variations involved in economic (energy prices) and environmental (emission factors, see Section 10.3.2.2) parameters. In terms of load modeling and data availability, usually there is much more information available for electrical loads (even with granularities of seconds, sometimes) than for thermal loads (for which information is sometimes available only on an integral energy basis with daily or even longer resolutions). To decrease the computational burden, characteristic days can be simulated to represent the typical seasonality levels that occur in terms of thermal and electrical demand, by simulating for instance weekdays and weekend days in winter, summer, and intermediate periods.

10.2.1.2 Local Generators and Control Strategies

At the individual component level, in the case of equipment with single input and output, control strategies are, in general, not an issue, and the typical operation is to follow the demand of the output energy vector. Hence, heat/cooling generation equipment are operated under heat/cooling following mode. For instance, if heat needs to be provided by an EHP, the heat requirements are satisfied by supplying the needed electricity.

This corresponds to having direct transformation from the output (end-use) energy vectors into input energy vectors, with energy shifting properties (Mancarella, 2009) according to the relevant performance indicator model. On the other hand, the presence of MG at an individual component level or for the overall plant introduces significant additional complexity to the above reasoning. The most important example is, of course, that of a CHP system producing both electricity and heat and which can therefore be operated under a heat-following or electricity-following mode.

More specifically, under *thermal load tracking* (TLT) mode (also called thermal load following or heat tracking mode) the CHP generator follows the thermal load, and electricity is produced accordingly (regardless of the actual local load), as resulting from the characteristics cogeneration ratio in Equation 10.1. This is assumed to be the "classical" strategy for CHP, which maximizes the environmental effectiveness of cogeneration with no heat waste, whereas positive or negative electricity local generation/load imbalances are made up through grid connection. Thus, it is important to highlight that, from the electrical outlook, CHP in TLT mode may be considered as *uncontrolled* generation.

On the other hand, under *electrical load tracking* (ELT) mode (also called electrical load following or electricity tracking mode) the CHP generator follows the local electrical load, and heat is produced accordingly (regardless of the actual thermal load), again as resulting from the characteristics cogeneration ratio. Such a strategy may typically be used to minimize the energy to be sold back to the grid when this is not economically convenient, or to minimize the network impact from "uncontrolled" generation in TLT mode. In this case, an AB is assumed to be put in operation to supply the thermal load if the thermal production is not sufficient. On the other hand, if the cogenerated heat is higher than the actual local thermal demand, heat is wasted. Therefore, either because additional boiler production is needed, or because heat production is wasted, the ELT strategy brings smaller environmental benefits relative to thermal load following. Hence, also for modeling, it is important to highlight that the actual *useful* heat Q^u may be different from the overall cogenerated heat Q^y defined above, as also further discussed in Section 10.2.2.2. Thermal energy storage can be an effective option to minimize heat waste.

Besides classical output following strategies or their hybrid versions (Kavvadias et al., 2010; Mago et al., 2010), optimal operational strategies can be put forward whereby a given objective is pursued. The definition of two types of operational strategies (primary energy reduction or CO_2 emission reduction) for the CHP in a trigeneration system that supplies a building is addressed in Fumo et al. (2009). The CHP system operates when its primary energy (or CO_2 emissions, respectively) is lower than that in separate production. In the primary energy reduction strategy, further criteria are defined to run the CHP also in other cases.

Specific strategies can be formulated to minimize the plant operational cost or maximize its revenues. In this case, optimization based on spark spread (SS) figures can be quite simple to represent and solve, as discussed in Section 10.3.2.3. However, such optimization schemes require a certain level of flexibility enabled in the CHP plant by additional components such as AB or thermal storage. Similarly, the economic optimization of an overall MG system can become significantly more complex when taking into account coproduction from other equipment, as shown in Porteiro et al. (2004) and Cardona et al. (2006a) in the presence of a CHP–EHP combined scheme.

10.2.1.3 Network Interfaces

Interconnection among various MG plants allows optimal management of portfolios of resources. On the other hand, interconnections with energy distribution networks, such as gas distribution system (GDS), electrical distribution system (EDS), district heating/cooling networks, and so forth, introduce further components to be managed in the study of optimal operational strategies. These mainly refer to taking optimal decision for buying/selling gas, electricity, heat, and so on, according to the rules for energy provision (depending on the tariff system or market structure referring to the different energy vectors). A situation that is envisaged to occur more and more frequently is that of coordinated MG plants belonging to the same energy district (in case of private property) and managed by the same energy company. In this case, specific objectives can be pursued by optimizing the MG operation within the local system.

An alternative, more likely and more challenging situation is that of a set of MG plants managed in a coordinated way and interconnected through public (or different owners') networks. In terms of modeling, while the individual components can be modeled as black boxes, the whole MG plant itself can be modeled in a similar way to link the end user's energy flows to the plant energy inputs. In the same fashion, it is possible to model the interactions among different MG plants, each of which is seen as a black box through input–output equivalent energy efficiencies (see Section 10.2.2.2). This representation is also consistent with the *energy hub* approach (Geidl et al., 2007; Geidl and Andersson, 2007b; Krause et al., 2011), as an MG plant can be interpreted as an energy hub with multiple energy interactions at input and output.

10.2.1.4 Storage

MG system operation can potentially be improved by the possibility of storing energy under different forms, also depending on the economics of storing different types of energy. Currently, the cheapest option is to store thermal energy, particularly in the form of heat (for instance, hot water), whereas storing electricity is quite expensive. Hence, thermal storage is widespread in MG applications, particularly to decouple the generation of electricity and heat from the CHP unit from the energy requirements. In these cases, the thermal storage allows the creation of an energy buffer to be profitably deployed for load shifting, for instance, providing preheating to smooth the morning heat generation ramp and limit its peak value. On the other hand, electricity can be transformed and stored under different forms if more convenient, for instance, as heat produced through EHPs or hydrogen produced through electrolysis. These options will be more and more relevant for both economic and environmental purposes in the presence of cheap and clean electricity produced by renewable energy sources such as wind or photovoltaic power. In addition, storage systems in MG plants can support firming up of renewable energy sources that are often highly variable and unpredictable (Geidl and Andersson, 2007b; Koeppel and Korpås, 2008).

From a modeling point of view, in analogy to the above component matrices it is possible to introduce the storage coupling matrix **S**, which generally describes how changes of the stored energies affect the MG output flows (Geidl, 2007).

10.2.2 MG Plant Structures

10.2.2.1 Connection Schemes

The different components forming an MG system can be connected according to different schemes. Focusing again on CCHP systems, it is possible to conceive two basic schemes with reference to the utilization of the energy produced by the CHP side, namely (Mancarella, 2006; Mancarella and Chicco, 2009a):

- *Parallel generation*: Cogenerated heat and electricity are used to *directly* supply the user's needs, and are not used to fire other devices for further production. Therefore, cooling can be generated for instance through a GARG or an EDC.
- *Bottoming generation*: Cogenerated heat and/or electricity are used to supply cascaded equipment for further production; therefore, heat can be used for instance to supply a WARG for cooling production, and electricity to generate cooling through a CERG or heat through an EHP. An example of bottoming schemes (Chicco and Mancarella, 2008d) is reported in Figure 10.1.

For modeling purposes, plant component connections and relevant energy flows among components internal to the plant need to be conveniently described. An effective approach to describe the connections among the component black boxes and with the external energy networks is to introduce a connectivity matrix (Geidl et al., 2007; Geidl and Andersson, 2007b; Chicco and Mancarella, 2009b) that accounts for the plant topology. In addition, for optimization and definition of control strategies, time-domain energy flows can be modeled through properly defined *dispatch factors* or splitting ratios (Valero and Lozano, 1997; Shivakumar and Narasimhan, 2002; Grekas and Frangopoulos, 2007) that are usually part of the decision variables set. A dispatch factor can be defined as the relative amount of an energy vector at flow splitting points

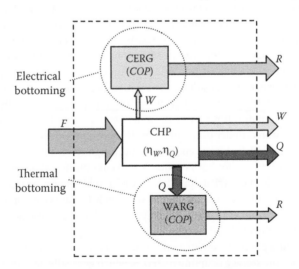

FIGURE 10.1 Combined cooling, heat, and power system with bottoming schemes.

(bifurcations) to supply different components. Hence, for energy conservation, the following properties hold:

i. The number of dispatch factors originated from the same energy vector at a bifurcation is equal to the number of split flows from that vector minus 1; hence, the sum of the dispatch factors originated from the same energy vector must be equal to unity.
ii. Each dispatch factor is limited between 0 and 1.

10.2.2.2 Black-Box Representations

According to the above discussion, MG plant components can individually be represented through black boxes characterized by performance indicators suitably arranged in matrix form (*component efficiency matrix*) to take into account the multiplicity of energy vectors that can be relevant to MG. In addition, a connectivity matrix and dispatch factors can be used to model the topological aspects of the plant and the energy flow interaction internal and external to the plant. Within this framework, and through geometric rules and algorithms (Chicco and Mancarella, 2009b), the overall plant representation can be reduced to a black box characterized by a *plant efficiency matrix* that synthesizes plant interconnection and energy flow description. As mentioned earlier for the single components, this *output-to-input black-box* approach is synthetic and effective at the same time, limiting the number of variables used to describe the plant energy flows, but holding sufficient information to have a clear picture of the plant.

Focusing again on a trigeneration system to exemplify the approach, let us consider the CCHP system shown in Figure 10.2, where the CHP prime mover is sided by an AB and the bottoming cooling plant is composed of both a WARG and a CERG. Input energies are fuel F_i from the GDS and electricity W_i from the EDS. Dispatch factors are

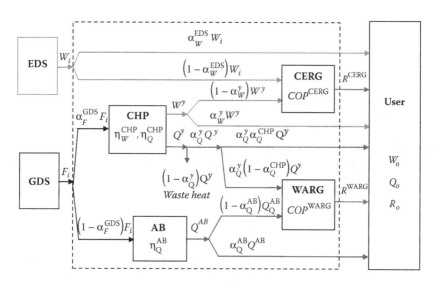

FIGURE 10.2 Black-box representation of a combined cooling, heat, and power system.

indicated with α, with superscripts indicating the origin of the relevant energy flow and subscripts indicating the type of energy output on which the dispatch factors are defined; this is consistent with the matrix representation of the individual components.

The useful heat Q^u is not directly represented among the input or output entries, as it appears explicitly only once in the mathematical model, so that it can be calculated *a posteriori* from the system solution as $Q^u = \alpha_Q^y Q^y$.

By putting together the component matrices, taking into account their connections, the overall MG plant can be represented as

$$\mathbf{v}_o = \mathbf{H} \cdot \mathbf{v}_i \tag{10.5}$$

where $\mathbf{v}_o$ is the array of the ordered output energy vectors (F_o, W_o, Q_o, and R_o) and $\mathbf{v}_i$ is the array of the ordered input energy vectors (F_i, W_i, Q_i, and R_i). Clearly, null entries are used if the corresponding energy vector does not appear in the system representation.

By direct inspection of all the MG plant energy flows in Figure 10.2, and taking into account the efficiency characteristics of the individual plant components (namely, η_W and η_Q for the CHP, η_t for the AB, and COP^{WARG} and COP^{CERG} for the chillers), the output-to-input efficiency matrix for the overall MG plant can be expressed as

$$
\begin{pmatrix} 0 \\ W_o \\ Q_o \\ R_o \end{pmatrix} =
\begin{pmatrix}
0 & 0 & 0 & 0 \\
\alpha_W^{CHP}\eta_W\alpha_F^{GDS} & \alpha_W^{EDS} & 0 & 0 \\
\alpha_Q^{AB}\eta_t\left(1-\alpha_F^{GDS}\right)+\alpha_Q^{CHP}\eta_Q\alpha_F^{GDS} & 0 & 0 & 0 \\
\eta_{RF} & COP^{CERG}\left(1-\alpha_W^{EDS}\right) & 0 & 0
\end{pmatrix}
\begin{pmatrix} F_i \\ W_i \\ 0 \\ 0 \end{pmatrix}
\tag{10.6}
$$

with

$$
\eta_{RF} = COP^{CERG}\left(1-\alpha_W^y\right)\alpha_F^{GDS}\eta_W + COP^{WARG}
$$
$$
\times\left[\left(1-\alpha_Q^{AB}\right)\left(1-\alpha_F^{GDS}\right)\eta_t + \alpha_F^{GDS}\alpha_Q^y\alpha_Q^{CHP}\eta_Q\right]
\tag{10.7}
$$

From Equation 10.6, it can be appreciated that availability of manifold components and interconnections allows a certain degree of freedom in the way that demand can be satisfied, which poses the basis for developing and solving an optimization problem. More specifically, if we consider the following set of dispatch factors

$$\alpha = \left[\alpha_F^{GDS}\alpha_W^y\alpha_Q^{CHP}\alpha_Q^{AB}\alpha_Q^y\alpha_W^{EDS}\right]^T \tag{10.8}$$

they can be selected as representative of the energy flow in the plant, and constitute the decision variables of an optimization problem together with the input electricity W_i and fuel F_i. This will be extensively dealt with in the following sections.

10.3 Objective Functions for MG Optimization

10.3.1 Formulations of the Operational Optimization Problems

Operational optimization of an MG system is run by considering a given set of equipment with assigned size, to identify the possible control strategy according to a predefined objective function (or multiple objective functions). Operational optimization can be used as an inner loop within a design or planning procedure. In this case, the type and size of the equipment are chosen among predefined alternatives in an outer loop, and the inner loop returns the operational conditions in which each alternative exhibits its best performance.

Different classes of problems can be identified on the basis of the nature of the phenomena to be analyzed. Two specific aspects pertain to the time domain and to the information domain.

The first aspect is linked to the typical partitioning of the total *time interval* of analysis (e.g., 1 year) into subintervals of specific duration (e.g., 1 h). A basic classification, with a heavy impact on the problem formulation and on the choice of the solution methods, can be made through the distinction between decoupled-in-time (indicated here as DiT) and coupled-in-time (indicated here as CiT) problems. More specifically:

i. In *DiT problems*, it is considered that the events analyzed in one subinterval do not affect any other subinterval. In this case, the various subintervals are studied independently of each other. Furthermore, it is possible to group together a number of subintervals regardless of the fact that they can be nonsuccessive. This leads to the exploitation of the *duration curves*, in which the values of the quantity under analysis in the various time intervals are simply sorted in descending order. The resulting duration curve can be further simplified by reducing the number of steps through the grouping of similar values of the quantity under analysis into an equivalent average value, as for instance in the equivalent load approach illustrated in Mancarella and Chicco (2009b). If the integration of the quantity in the time period is meaningful from the energy viewpoint, the equivalent average value is generally defined in such a way as to reproduce the same area of the grouped quantities. For MG systems, the representation of the energy quantities through their duration curves has the advantage of summarizing in a synthetic way the overall energy need of each energy quantity. However, it has to be taken into account that in the duration curve representation the *simultaneity* among the various quantities is lost, and as such no direct comparison can be made through superposition of the duration curves of different quantities. A dedicated discussion on the duration curve characteristics for MG systems is provided in Piacentino and Cardona (2008b).

ii. In *CiT problems*, the events analyzed in one subinterval may affect the other subintervals. In this case, the subintervals must be analyzed by maintaining their consecutiveness. The coupling-in-time of the subintervals may be, in general, due to integral quantities or constraints, or to operational constraints linked to the time evolution of some variables. For instance, it is possible to identify the following aspects:

1. Limits on *storage* systems: Different types of storage, such as hydro, batteries, plug-in hybrid electric vehicles, heat and cooling storage systems, and so on, have a limited capacity.

2. *Integral limits* on the units: Transients in the thermal units are relatively long (e.g., with respect to electrical transients); the design conditions of the thermal units are set up for continuous operation and include minimum start-up and shutdown times, directly reflecting on the operational schedules; for systems allowing faster switch on/off operations, a maximum number of these operations can be introduced for maintenance reasons.

3. *Rate-of-change* limits: Ramp-rate limits depending on the physical nature of the energy provision can be introduced both for thermal generation systems and for some storage systems (e.g., battery discharge rate).

4. Cumulative constraints on *emissions*: For global or local emissions of substances or pollutants that can remain in the atmosphere for relatively long time periods (compared with the timescales of the operational problems analyzed), the constraint can be set on the cumulative emissions on a given period of time; cumulative emissions can also be of interest to address emission trading schemes (Rong and Lahdelma, 2007; Tsikalakis and Hatziargyriou, 2007; Kockar et al., 2009).

5. Provision of *reserves*: Time coupling can be relevant to avoid simultaneous unavailability of many units (for instance, due to maintenance reasons), calling for the introduction of reserve supply (Kockar et al., 2009).

6. Exploiting *electricity market* prices as a driver: The production schedules from different units can be adjusted to follow the time evolution of electricity prices, with complex implications that can be handled through specific models (Makkonen et al., 2003; Deng and Jiang, 2005; Carrión et al., 2007).

The second aspect refers to the nature of the *data* available for the study. In this respect, it is possible to consider deterministic or stochastic optimization, namely:

iii. In *deterministic optimization*, each variable is introduced in a deterministic form (or as a predefined probabilistic distribution feature such as average value, median, mode, or a given percentile), and the probabilistic distribution of the results is not of interest as an output.

iv. In *stochastic optimization*, one or more variables are introduced in probabilistic form, usually because the probability distribution of the results is relevant; probabilistic inputs are typical of atmospheric variables (from sun or wind) and of some energy loads, for instance, residential electricity loads, heavily depending on consumers' lifestyle (Herman and Kritzinger, 1993; Capasso et al., 1994; Carpaneto and Chicco, 2008).

For the classes of problems indicated above, optimization is conceptually different when a *single objective* function is used or a *multiobjective* problem is formulated (Salgado and Pedrero, 2008), as discussed below.

10.3.1.1 Single-Objective Optimization

Let us define a vector **x** containing the decision variables of the optimization problem. The decision variables may include energy input components, dispatch factors, and in the case of storage also stored energy vectors in different time intervals.

The general formulation of the single-objective optimization problem with objective function f(**x**) to be minimized under the equality constraints represented by the vector function **h**(**x**) and the inequality constraints represented by the vector function **ψ**(**x**) is

$$\left(\tilde{\mathbf{x}}, f(\tilde{\mathbf{x}})\right) = \min_{\mathbf{x}}\left\{f(\mathbf{x})\right\} \tag{10.9}$$

subject to

$$\mathbf{h}(\mathbf{x}) = \mathbf{0} \tag{10.10}$$

$$\mathbf{\psi}(\mathbf{x}) \leq \mathbf{0} \tag{10.11}$$

where $\tilde{\mathbf{x}}$ is the vector containing the values of the decision variables corresponding to the optimal objective function $f(\tilde{\mathbf{x}})$.

A single objective function may be formulated for a problem in which only one objective is of interest, or for a multiobjective problem in which the various objectives are merged into a single objective. A single objective function can be set up also when multiple objectives are prioritized with predefined hierarchical rules, introducing the terms related to the other objectives within associated boundary constraints (Cardona et al., 2006b).

In the *operational* time frame, a typical partitioning of the overall period of analysis (for instance, 1 year) can be found by considering $s = 1, \ldots S$ seasons (e.g., winter, summer, and intermediate) and $d = 1, \ldots D$ day types (e.g., weekdays, Saturdays, and Sundays), as for instance done in Beihong and Weiding (2006), Arcuri et al. (2007), and Carvalho et al. (2011). Within each day, $h = 1, \ldots, H$ time intervals are considered, with duration Δt_h for the hth time interval, not necessarily corresponding to regular partitioning of the time intervals along the day.

Considering n_{sd} days of type d in season s and indicating as $\mathcal{F}(\mathbf{x}_{sdh})$ a function of the decision variables to be set up at season s, for day type d and at time interval h, on an annual basis the optimization problem can be formulated as the minimization

$$\left(\tilde{\mathbf{x}}, f(\tilde{\mathbf{x}})\right) = \min_{\mathbf{x}}\left\{f(\mathbf{x}) = \sum_{s=1}^{S}\sum_{d=1}^{D}\sum_{h=1}^{H}\mathcal{F}(\mathbf{x}_{sdh})n_{sd}\Delta t_h\right\} \tag{10.12}$$

In a planning problem, a decision variable can contain a single value to be identified for the whole period (e.g., the size of an MG unit). In operational problems, a decision variable could contain a set of values to be established for each season, day type and time interval, to identify its optimal trajectory (of variation in time) in combination with the trajectories of the other decision variables.

The individual objectives can be of different nature, typically arising from technical, environmental, or economic concepts. Some examples are illustrated in Section 10.3.2.

10.3.1.2 Multiobjective Optimization

In the general formulation of the multiobjective optimization problem with N objectives and decision variables vector $\mathbf{x}$, the objective function to be minimized becomes a vector type, that is, $\mathbf{f}(\mathbf{x}) = \{f_1(x), f_2(x), \ldots, f_N(x)\}$, whereas the equality constraints are still represented by the vector function $h(x)$ and the inequality constraints by the vector function $\psi(\mathbf{x})$. The multiobjective optimization can end up with more than one solution point, each of which is characterized by a different combination of the decision variables. Considering M solution points, the formulation can be expressed as follows:

$$\left(\left\{\tilde{\mathbf{x}}_1,\ldots,\tilde{\mathbf{x}}_M\right\},\left\{\mathbf{f}\left(\tilde{\mathbf{x}}_1\right),\ldots,\mathbf{f}\left(\tilde{\mathbf{x}}_M\right)\right\}\right) = \min_{\mathbf{x}}\left\{\mathbf{f}(\mathbf{x})\right\} \tag{10.13}$$

subject to

$$h(\mathbf{x}) = \mathbf{0} \tag{10.14}$$

$$\psi(\mathbf{x}) \leq \mathbf{0} \tag{10.15}$$

where $\left\{\tilde{\mathbf{x}}_1,\ldots,\tilde{\mathbf{x}}_M\right\}$ is the set of decision variables vectors corresponding to the optimal objective functions $\left\{\mathbf{f}\left(\tilde{\mathbf{x}}_1\right),\ldots,\mathbf{f}\left(\tilde{\mathbf{x}}_M\right)\right\}$.

In MG system operational optimization, the various objectives generally do not lead to the same solutions. Indeed, it is very likely to find formulations with *conflicting objectives*, that is, optimizing one objective leads to worse solutions for other objectives, as for instance the reduction of the operational costs and the reduction of the local emissions. The occurrence of conflicting objectives calls for modifying the approach to operational optimization. Some decision-making could be needed, for instance, according to the following categorization (Mavrotas et al., 2007):

- *A priori* decision-making, in which priorities or preferences among the objectives or specific goals are established beforehand.
- *Interactive* decision-making, in which successive phases of calculation and decision are alternated in order to drive the search toward the most preferred solution; however, the decisions affect the trajectory of evolution of the solutions and may reduce the breadth of the solution space analyzed.
- *A posteriori* decision-making, in which a number of *compromise* solutions are calculated before identifying the preferred solution.

On these bases, multiobjective formulations of two different kinds can be identified:

1. Transformation of the multiple objectives into a *single objective*: The transformation is typically formulated by resorting to a weighted sum of the objectives, assuming a set of weights that carry the twofold goals of making the numerical

values of the objectives more uniform (also providing compensation of the effects of using different measuring units) and of establishing preferences among the objectives.

$$\mathcal{F}(\mathbf{x}_{sdh}) = \sum_{j=1}^{N} w_j f_j(\mathbf{x}_{sdh}) \tag{10.16}$$

with positive weighting factors $w_1, \ldots, w_N$, whose sum can be in case constrained to unity.

If the multiple objectives are not conflicting, the choice of the weights is not relevant, as the optimal solution will be the same for all the objectives. In case of conflicting objectives, the transformation into a single objective leads to renouncing to exploit the variety of possible solutions with respect to the other kind of formulation. As a particular case (Mago and Chamra, 2009), the weights can be set to the inverse of the quantity to minimize evaluation in conventional conditions. A further approach (Tsay, 2003) is based on considering

$$\mathcal{F}(\mathbf{x}_{sdh}) = \sqrt{\sum_{j=1}^{N} w_j \left(\frac{f_j(\mathbf{x}_{sdh}) - \tilde{f}_j}{\tilde{f}_j - \check{f}_j} \right)^2} \tag{10.17}$$

where $\tilde{f}_j$ is the best solution found for objective $j = 1, \ldots, N$ from the solution of a single-objective problem with the same objective j, and $\check{f}_j$ is the worst solution obtained for objective j in the optimal solution of a single-objective problem referring to any (other) individual objective. In Equation 10.17, the weighting factors $w_1, \ldots, w_N$ are positive and their sum is constrained to unity.

Moreover, goal programming (Chang and Fu, 1998) can be used to transform a multiobjective problem into a single-objective one.

2. Handling the multiple objectives *simultaneously*, with the aim of seeking not only the optimal values of the individual objectives but rather identifying an extended class of *compromise solutions* appearing in hyperspace formed by representing the individual objectives on orthogonal directions: The compromise solutions are interesting as feasible alternatives to assist the decision-making process. The conceptual framework for obtaining the compromise solutions is based on the definition of a *nondominated* solution as a solution for which it is not possible to find other solutions with better performance for *all* the individual objective functions. The *Pareto front* is thus defined as the set of all possible nondominated solutions.

10.3.2 Single Objectives and Performance Indicators

In the formulation of a single-objective function, the focus is generally set on the definition of suitable indicators that can be conveniently used in energy system operation studies. These indicators consider that the relevant quantities represent constant values or average values in the time interval under analysis, for any duration of the time interval. As such, for the sake of simplicity, the subscript referring to the time interval is not explicitly represented in the illustration of the indicators provided below, reporting the links to the general formulation 10.12 at the end of each section.

10.3.2.1 Technical Performance Indicators

As an example of MG, let us consider a combined trigeneration system, in which energy efficiency can be assessed by resorting to synthetic indicators of the overall plant performance. By using the black-box approach indicated in Section 10.2.2.2, the only energy flows to take into account are the ones visible from the *outside*, no matter what happens *inside* the plant.

For this purpose, let us consider the trigeneration primary energy saving (TPES) indicator (Chicco and Mancarella, 2007a) that quantifies the primary energy saving from a generic trigeneration plant with respect to the conventional separate production serving the *same net useful energy outputs* (electricity W_o, heat Q_o, and cooling R_o) to the user. Let us further consider the overall fuel thermal input F_i to the trigeneration system (for CHP prime mover and boiler) and the total fuel thermal energy input F^{SP} required for the *separate production* of electricity W_o through a power plant with conventional reference efficiency η_e^{SP}, heat Q_o through boilers with conventional reference efficiency η_t^{SP}, and cooling R_o through a CERG with conventional reference efficiency COP^{SP}.

The TPES indicator is expressed as

$$\text{TPES} = \frac{F^{SP} - F_i}{F^{SP}} = 1 - \frac{F_i}{\dfrac{W_o}{\eta_e^{SP}} + \dfrac{Q_o}{\eta_t^{SP}} + \dfrac{R_o}{\eta_e^{SP} COP^{SP}}} \tag{10.18}$$

More generally, in Chicco and Mancarella (2008a), the TPES indicator has been extended to a polygeneration system supplying different types of energy vectors to the users. For this purpose, let us consider the pairs (X,x), where X is the energy vector and x is the corresponding energy that characterizes the useful net output (for instance, with X corresponding to heat Q and x corresponding to thermal energy t). By introducing the pairs (X,x) in the set D, the resulting polygeneration primary energy saving (PPES) indicator is written as

$$\text{PPES} = \frac{F^{SP} - F_i}{F^{SP}} = 1 - \frac{F_i}{\displaystyle\sum_{(X,x)\in D} (X_o/\eta_x^{SP})} \tag{10.19}$$

Each term η_x^{SP} represents the *conventional* separate production efficiency for the energy vector X. The numerical values to assign to these efficiencies have to be specified by the relevant regulatory bodies (Cardona and Piacentino, 2005; Mancarella and Chicco, 2009a).

The TPES or PPES indicator is positive when the trigeneration or polygeneration system requires an amount of fuel F_i lower than the amount F^{SP} needed for running the separate production system to serve the same user's load. Hence, *positive* TPES or PPES values (with upper value conceptually limited to unity) indicate convenience in energy efficiency terms of exploiting trigeneration to serve the composite energy load. Negative TPES or PPES values indicate lack of convenience of trigeneration or polygeneration, and their absolute value is conceptually not limited.

Elaborations of the PPES concept can be made by exploiting the entries of Equation 10.19. For instance, Cardona and Piacentino (2003) propose to adopt for short-term analyses (e.g., within 1 h) a management criterion for a CHP system that could be more effective than the usual TLT mode. This criterion is based on rewriting the condition $F \le F^{SP}$ by expressing F^{SP} as the sum of the contributions from the EDS and from boilers, yielding the positive energy saving criterion

$$F_i \le \frac{\eta_e^{SP} Q}{\eta_t^{SP} \left(\eta_e^{SP} - \eta_e^{CHP} \right)} \tag{10.20}$$

A similar rationale can be adopted to identify acceptable operating conditions by rewriting the energy system equations in specific MG cases.

In the minimization framework of Equation 10.12, the objective function to be evaluated is

$$\mathscr{F}(\mathbf{x}_{sdh}) = 1 - \text{PPES}(\mathbf{x}_{sdh}) \tag{10.21}$$

10.3.2.2 Environmental Performance Indicators

Environmental performance indicators may refer to *global* or *local* emissions of various types of pollutants that can create hazards to public health or to other receptors (Canova et al., 2008).

For *global* emissions, the mass of CO_2 is evaluated by resorting to the *emission factor model* (Cârdu and Baica, 2002; Meunier, 2002; Minciuc et al., 2003) to represent the reference-specific emissions $\mu_{CO_2}^{X,SP}$ for producing the energy vector X in separate production. The emission factor model is based on considering the mass $m_{CO_2}^X$ of CO_2 emitted to produce the useful energy output X, and expressing it as a function of the energy and of the *emission factor* $\mu_{CO_2}^X$ (or *specific mass emissions* per unity of X, in [g/kWh]), so that

$$m_{CO_2}^X = \mu_{CO_2}^X \cdot X \tag{10.22}$$

For a polygeneration system, Chicco and Mancarella (2008a) introduced an indicator consistent with respect to the PPES indicator used for energy efficiency assessment. This indicator is called polygeneration CO_2/carbon dioxide emission reduction (PCO2ER/PCDER), and is based on the relative difference between the mass $m_{CO_2}^F$ of CO_2 emitted by the polygeneration system and the mass $m_{CO_2}^{F,SP}$ emitted in separate production to serve the same amount of demand energy vectors (included in the set D):

$$\text{PCDER} = \frac{m_{CO_2}^{F,SP} - m_{CO_2}^{F}}{m_{CO_2}^{F,SP}} = 1 - \frac{\mu_{CO_2}^{F} F}{\sum\limits_{(X,x) \in D} \mu_{CO_2}^{X,SP} X} \tag{10.23}$$

Other GHGs seen as global pollutants can be handled by means of their global warming potential (GWP), expressed as multiple of CO_2 (for which $\text{GWP}_{CO_2} = 1$), thus introducing for a generic GHG g the *equivalent* emission factor (Chicco and Mancarella, 2008b; Mancarella and Chicco, 2010)

$$\mu_{CO_2eq,g}^{X} = \text{GWP}_g \cdot \mu_g^{X} \tag{10.24}$$

summing up the contributions of the various GHGs belonging to the set G emitted from the same source

$$\mu_{CO_2eq}^{X} = \sum\limits_{g \in G} \mu_{CO_2eq,g}^{X} \tag{10.25}$$

and replacing the emission factors related to CO_2 in Equation 10.23 with the corresponding equivalent ones $\mu_{CO_2eq}^{F}$ and $\mu_{CO_2,eq}^{X,SP}$.

By further introducing the CO_2 *emission equivalent efficiency* $\eta_{CO_2,x}^{SP}$ for the pair $(X,x) \in D$, defined in Chicco and Mancarella (2008a) as

$$\eta_{CO_2,x}^{SP} \equiv \frac{\mu_{CO_2}^{F}}{\mu_{CO_2}^{X,SP}} \tag{10.26}$$

expression 10.23 becomes

$$\text{PCDER} = 1 - \frac{F_i}{\sum\limits_{(X,x) \in D} \dfrac{X}{\eta_{CO_2,x}^{SP}}} \tag{10.27}$$

For instance, considering a cogeneration plant with useful electricity output W, useful heat output Q, and fuel input F_i, with emission factor $\mu_{CO_2}^{F}$, the emission factors to be used for calculating the CO_2 emission equivalent efficiencies are $\mu_{CO_2}^{W,SP}$ for conventional separate production of electricity and $\mu_{CO_2}^{Q,SP}$ for conventional separate production of heat.

Besides global warming, other aspects can be taken into account to calculate the global emissions. Wang et al. (2010b) consider the acidification potential (AP) as the potential of a substance to build and release H^+ protons, expressed in relative terms with respect to SO_2 as the reference substance, and define the SO_2 equivalents emissions. In the same way, they consider the ozone depletion potential (ODP) as the potential of a substance to deplete the stratospheric ozone, expressed in relative terms with respect to reference substance trichlorofluoromethane (CCl_3F, also identified as R11), and define the R11 equivalent emissions. Correspondingly, let us consider a set **G** of relevant substances and introduce for a substance $g \in$ **G** the emission factors v_g^X (g/kWh) with respect to acidification and o_g^X (g/kWh) with respect to stratospheric ozone depletion. The emission factors related to the reference substances become

$$v_{SO_2eq,g}^X = AP_g \cdot v_g^X \tag{10.28}$$

$$o_{R11eq,g}^X = ODP_g \cdot o_g^X \tag{10.29}$$

In the presence of multiple substances emitted by the same source, the equivalent emissions are obtained by summing up the contributions from the various substances belonging to set **G**:

$$v_{SO_2eq}^X = \sum_{g \in G} v_{SO_2eq,g}^X \tag{10.30}$$

$$o_{R11eq}^X = \sum_{g \in G} o_{R11eq,g}^X \tag{10.31}$$

In this way, it is possible to introduce two global emission indicators written in a form similar to that in Equation 10.23, namely, polygeneration sulfur dioxide emission reductions (PSDER) for the acidification problem

$$PSDER = 1 - \frac{v_{SO_2,eq}^F F}{\displaystyle\sum_{(X,x) \in D} v_{SO_2,eq}^{X,SP} X} \tag{10.32}$$

and polygeneration R11 emission reductions (PR11ER) for the stratospheric ozone depletion problem

$$PR11ER = 1 - \frac{o_{R11,eq}^F F}{\displaystyle\sum_{(X,x) \in D} o_{R11,eq}^{X,SP} X} \tag{10.33}$$

For *local* pollutants, again with reference to the emission factors, it is possible to work out specific *emission balance models*, taking into account the contribution of the pollutant under interest to the local emissions. In this case, the effect of the emissions is limited to a portion of the territory. It is then important to introduce a factor that takes into account that only a fraction of the pollutant emitted by an energy source in separate production can reach the portion of the territory under analysis. For instance, in a simple cogeneration case, given a pollutant g it is possible to consider the factors $\beta_g^{W,SP}$ and $\beta_g^{Q,SP}$ representing the fractions of the pollutants emitted by the separate production sources that are assumed to reach the territorial area of interest. If electricity is produced sufficiently far from that area, so that no amount of pollutant g with specific emissions $\mu_g^{W,SP}$ can reach the area, it may be assumed $\beta_g^{W,SP} = 0$, whereas $\beta_g^{Q,SP} = 1$ could be set up to represent the effect of pollutant g emitted by the boilers located in the area, with specific emissions $\mu_g^{Q,SP}$. The indicator of local emissions reduction (LER) for pollutant g can then be written as

$$
\begin{aligned}
gLER &= 1 - \frac{\mu_g^F \cdot F}{\beta_g^{W,SP}\mu_g^{W,SP} \cdot W + \beta_g^{Q,SP}\mu_g^{Q,SP} \cdot Q} \\
&= 1 - \frac{\mu_g^F}{\beta_g^{W,SP}\mu_g^{W,SP} \cdot \eta_W + \beta_g^{Q,SP}\mu_g^{Q,SP} \cdot \eta_Q}
\end{aligned}
\tag{10.34}
$$

The extension of the above indicator to polygeneration systems yields the polygeneration local emission reduction (PgLER) indicator (related to pollutant g)

$$
PgLER = 1 - \frac{\mu_g^F F}{\sum_{(X,x)\in D} \beta_g^{X,SP}\mu_g^{X,SP} X}
\tag{10.35}
$$

To summarize, in the minimization framework of Equation 10.12, the objective function to be evaluated are

- For equivalent CO_2 emissions:

$$
\mathcal{F}(\mathbf{x}_{sdh}) = 1 - PCDER(\mathbf{x}_{sdh})
\tag{10.36}
$$

- For equivalent sulfur dioxide emissions:

$$
\mathcal{F}(\mathbf{x}_{sdh}) = 1 - PSDER(\mathbf{x}_{sdh})
\tag{10.37}
$$

- For equivalent R11 emissions:

$$
\mathcal{F}(\mathbf{x}_{sdh}) = 1 - PR11ER(\mathbf{x}_{sdh})
\tag{10.38}
$$

- For local emissions of pollutant g:

$$
\mathcal{F}(\mathbf{x}_{sdh}) = 1 - PgLER(\mathbf{x}_{sdh})
\tag{10.39}
$$

Further approaches can be defined under a comprehensive life-cycle assessment framework, incorporating a broad number of impact categories and their hazardous effects on human health, ecosystem quality, and resources, and defining suitable indicators and related objective functions (Carvalho et al., 2011).

10.3.2.3 Economic Performance Indicators

A general indicator of economic performance is built by taking into account the difference between costs and revenues. In the minimization framework of Equation 10.12, with costs $\mathcal{C}(\mathbf{x}_{\mathrm{sdh}})$ and revenues $\mathcal{R}(\mathbf{x}_{\mathrm{sdh}})$, the corresponding formulation has

$$\mathcal{F}(\mathbf{x}_{\mathrm{sdh}}) = \mathcal{C}(\mathbf{x}_{\mathrm{sdh}}) - \mathcal{R}(\mathbf{x}_{\mathrm{sdh}}) \tag{10.40}$$

For a DiT problem, it is possible to consider objective function minimization in each individual time interval (Cardona et al., 2006b; Chicco and Mancarella, 2009b). For instance, taking into account the fuel price ρ_g^{GDS}, the electricity price ρ_i^{EDS} for the electricity bought from the EDS, and the electricity price ρ_o^{EDS} for the electricity sold to the EDS, the objective function components become $\mathcal{C}(\mathbf{x}_{\mathrm{sdh}}) = (\rho_g^{\mathrm{GDS}} F_i + \rho_i^{\mathrm{EDS}} W_i)_{\mathrm{sdh}}$ and $\mathcal{R}(\mathbf{x}_{\mathrm{sdh}}) = (\rho_o^{\mathrm{EDS}}(W_o - W_d))_{\mathrm{sdh}}$. The definition of the time intervals has to be consistent with the time resolutions of the price variations, especially for the electricity markets cleared at each hour (or less, depending on the rules in force in the specific jurisdiction), whereas for gas markets and emission trading the time resolutions are generally longer.

Approaches aimed at simultaneous solution of planning and operation of MG systems are often based on the minimization of the total annual cost (Sakawa et al., 2002; Yokoyama et al., 2002; Oh et al., 2007; Lozano et al., 2009a), maximization of the net present value (Piacentino and Cardona, 2008a), or maximization of the difference between revenues and costs in the daily activity (Aringhieri and Malucelli, 2003). In some cases, the operation problem is specifically formulated as minimization of the variable operational costs (Kong et al., 2005; Weber et al., 2006; Lozano et al., 2009b). Arcuri et al. (2007) propose the maximization of the gross operational margin as the difference between revenues and costs, also introducing an annual tax rate for embedding the operational optimization solutions into a multiyear planning analysis with different scenarios.

Another set of economic indicators can be defined in terms of SS concepts (Piacentino and Cardona, 2008b). For instance, the ratio SS_e between the market (marginal) price of electricity ρ_e (in monetary units per kilowatt-hour) and the variable cost for producing it depends on the market price of fuel ρ_f (in monetary units per standard cubic meter for natural gas, or monetary units per kilogram for diesel):

$$SS_e = \frac{\rho_e}{\rho_f} \frac{\eta_e^{\mathrm{CHP}} \mathrm{LHV}_f}{3600} \tag{10.41}$$

where LHV_f is the fuel lower heating value (in kilojoules per standard cubic meter). According to this definition, operation in TLT mode is convenient for $SS_e > 1$.

Other SS indicators introduced in Piacentino and Cardona (2008a,b) are:

- The thermal total supply spread (TSS_t), obtained from the ratio between the cost of producing 1 kWh of electricity through separate or combined production and the corresponding amount of heat recovered. This indicator can be expressed in the form

$$TSS_t = \frac{\rho_e}{\rho_f} \frac{\eta_e^{CHP}LHV_f}{3600} + \frac{\eta_e^{CHP}\lambda^{CHP}}{\eta_t^{SP}} \tag{10.42}$$

  where λ^{CHP} is the heat-to-power ratio of the CHP unit (Horlock, 1997; Mancarella and Chicco, 2009a).

- The cooling total supply spread (TSS_c), in which profitability of combined cooling production through an absorption chiller with efficiency COP_c is assessed against the electricity saved in separate production through a heat pump with efficiency COP^{SP}. Its expression is

$$TSS_c = \frac{\rho_e}{\rho_f} \frac{\eta_e^{CHP}LHV_f}{3600} \left(1 + \frac{\lambda^{CHP}COP_c}{COP^{SP}}\right) \tag{10.43}$$

The TSS_t and TSS_c indicators represent different margins of convenience existing when heat recovery supplies directly a heating load or is used as input to an absorption chiller to supply the cooling demand, respectively. As the CHP thermal output can also be used for both types of demand, the convenience of operating the CHP corresponds to the condition $\min\{TSS_t, TSS_c\} > 1$, otherwise the indication is to switch off the CHP unit.

The SS ratios are interesting because their values can vary at different time intervals, so that they can be easily introduced in market-price-based frameworks.

Further information of economic relevance can be gathered from calculating the *marginal costs* associated with the change of any operational constraint (Lozano et al., 2009b). The marginal costs are calculated as the dual prices (variations in the objective function resulting by changing a constraint of one unit of a resource) and can be used to identify the operational constraint that can be modified to improve the objective. The dual prices can be calculated by using linear programming solvers, or as Lagrangian multipliers expressing the first-order Kuhn–Tucker optimality conditions (Hemmes et al., 2007; Piacentino and Cardona, 2007). Considering cost minimization, the optimal solution is written as

$$\zeta = H^T \lambda \tag{10.44}$$

where **H** is the system efficiency matrix, and $\boldsymbol{\lambda}$ (Lagrangian multipliers) and $\boldsymbol{\zeta}$ represent the marginal costs of the energy carriers at the output and the input of the MG system, respectively.

10.3.2.4 Environomics Performance Indicators

The environomics approach (Curti et al., 2002a,b; Borchiellini et al., 2002) is based on the construction of a single-objective function, in which environmental costs and benefits referring to CO_2 and other pollutants are included in the formulation of the objective function.

Dealing with environomics, in the minimization framework of 10.12 the performance indicator is built by adding to costs $\mathcal{C}(\mathbf{x}_{sdh})$ and revenues $\mathcal{R}(\mathbf{x}_{sdh})$ a further term $\mathcal{E}(\mathbf{x}_{sdh})$ containing the pollution costs for system manufacture and removal, as well as for preparation and transport of the energy resources, and the pollution costs during operation, so that

$$\mathcal{F}(\mathbf{x}_{sdh}) = \mathcal{C}(\mathbf{x}_{sdh}) + \mathcal{E}(\mathbf{x}_{sdh}) - \mathcal{R}(\mathbf{x}_{sdh}) \tag{10.45}$$

In the operation time frame, the relevant entries are fuel costs, costs related to CO_2 and other pollutants (operational costs, taxes, and penalties), maintenance costs depending on system operation, and benefits from production revenues. The objective function can be expressed in monetary units or, alternatively, in exergy terms (Deng et al, 2008; Roque Díaz et al., 2010).

10.3.3 Multiobjective Formulations

With reference to the formulations introduced in Section 10.3.1.2, some approaches reported in the literature are based on:

1. Transformation of multiple objectives into a *single objective*: This transformation has been formulated for instance with reference to energy efficiency and environmental objectives (Wang et al., 2010b), to economic, environmental, and energy efficiency objectives (Mago and Chamra, 2009; Wang et al., 2010a), or to economic and expected power and heat generation with respect to the corresponding uncertain demand (Chang and Fu, 1998).
2. Handling the multiple objectives *simultaneously*: The multiple objectives considered have been of different types, that is, CO_2 emissions and costs (Burer et al., 2003; Xia et al., 2004; Aki et al., 2006a,b; Pelet et al., 2005; Bernal-Agustìn et al., 2006; Li et al., 2006; Wang et al., 2008), CO_2 emissions, costs, and efficiency (Kavvadias and Maroulis, 2010), energy consumption and local emissions (Boicea et al., 2009), costs, CO_2 emissions, and emissions of other pollutants (Tsay, 2003; Mavrotas et al., 2007; Kavvadias and Maroulis, 2010), also with a stochastic multiobjective formulation with total generation cost, expected power generation deviation, and expected heat generation deviation with respect to the corresponding power and heat demand (Chang and Fu, 1998).

10.4 Constraints

10.4.1 Nature of the Constraints in the Operational Optimization Problem

Equality and inequality constraints have to be included in the overall formulation of the optimization problem. Constraints acting on a single variable or on a function of more variables may be of punctual type (referring to a given time moment) or of integral type (referring to quantities accumulated over time). The details of the constraints encountered in MG operational optimization are provided in the following sections.

For a CiT problem, let us introduce a set of binary variables implementing unit commitment concepts (Arroyo and Conejo, 2000) for equipment $Y \in \mathbf{Y}$, such that $b_h^Y = 1$ if equipment Y is in operation at time interval h, $\bar{b}_h^Y = 1$ if equipment Y is started up at the beginning of time interval h, and $\underline{b}_{-h}^Y = 1$ if equipment Y is shut down at the beginning of time interval h.

10.4.2 Equality Constraints

Equality constraints represent the energy balances referring to each time interval for the various types of energy involved (for instance, fuel, electrical, thermal, and cooling energy, as in the example shown in Figure 10.2). These balances can be seen at both the input and output terminals of the individual plant components $Y \in \mathbf{Y}$, represented in matrix formulation (Chicco and Mancarella, 2009b) as

$$\mathbf{H}^Y \mathbf{v}_i^Y - \mathbf{v}_o^Y = \mathbf{0} \tag{10.46}$$

and at the input and output terminals of the whole plant, expressed as

$$\mathbf{H} \mathbf{v}_i - \mathbf{v}_o = \mathbf{0} \tag{10.47}$$

Another equality constraint represents the uniqueness of the direction in which electricity flows in each time interval. If electricity is bought from the EDS, $W_i > 0$ and $W_o = W_d$, else electricity is sold to the EDS, with $W_i = 0$ and $W_o > W_d$. These equality constraints are associated with the inequality constraints $W_i \geq 0$ and $W_o \geq W_d$ as reported in Section 10.4.3:

$$W_i \cdot \left(W_o - W_d \right) = 0 \tag{10.48}$$

For a CiT problem, the status (on/off) of the units in operation may change depending on the units started up or shut down at the beginning of time interval h. The following equality constraint (Arroyo and Conejo, 2000) expresses the link among the binary variables for unit $Y \in \mathbf{Y}$:

$$\bar{b}_h^Y - \underline{b}_{-h}^Y - b_h^Y + b_{h-1}^Y = 0 \tag{10.49}$$

Further equality constraints refer to the representation of specific aspects, such as the ones indicated in Arroyo and Conejo (2000), for the staircase-like approximation of nonlinear (e.g., exponential) start-up costs or the linear approximation of a nonlinear cost function.

Specific equality constraints acting in a CiT problem depend on *storage* (Geidl and Andersson, 2007b; Hajimiragha et al., 2007), represented by the storage coupling matrix **S** and the vector **z** containing the energy stored for the various energy vectors. Expression 10.47 in the presence of storage is thus extended to obtain

$$\mathbf{H}\mathbf{v}_i - \mathbf{v}_o - \mathbf{S}\frac{d\mathbf{z}}{dt} = 0 \qquad (10.50)$$

where the variation in time of energy stored in a (discrete) time interval h depends on the difference among the energy stored at the two successive time intervals $h-1$ and h, and on the energy losses $\mathbf{z}_h^\ell$ during time interval h:

$$\frac{d\mathbf{z}}{dt} = \mathbf{z}_h - \mathbf{z}_{h-1} + \mathbf{z}_h^\ell \qquad (10.51)$$

Another relevant aspect concerning storage is that the stored energy at the beginning and at the end of the period of analysis (i.e., in the initial condition 0 and at the last time interval H) has to be equal to a predefined value (Rong et al., 2008) or has to be the same (Hajimiragha et al., 2007), namely, in the latter case:

$$\mathbf{z}_0 - \mathbf{z}_H = 0 \qquad (10.52)$$

10.4.3 Inequality Constraints

10.4.3.1 Capacity Constraints

In a DiT problem, capacity constraints are included as lower and upper limits of the energy vectors relevant to each plant component. In general, by introducing the lower and upper limits for equipment $Y \in \mathbf{Y}$ in the vectors $\underline{\mathbf{v}}^Y$ and $\overline{\mathbf{v}}^Y$, respectively, in the matrix formulation the capacity constraints are expressed by

$$\mathbf{H}^Y \mathbf{v}_i^Y - \overline{\mathbf{v}}^Y \leq 0 \qquad (10.53)$$

$$\underline{\mathbf{v}}^Y - \mathbf{H}^Y \mathbf{v}_i^Y \leq 0 \qquad (10.54)$$

For instance, for the CHP the maximum limits $\overline{\mathbf{v}}^{\text{CHP}} = [0, \overline{W}^{\text{CHP}}, \overline{Q}^{\text{CHP}}, 0]^T$ correspond to the rated power for electricity and heat, or the sum of the rated powers if the CHP is composed of a cluster of units. The minimum limits $\underline{\mathbf{v}}^{\text{CHP}} = [0, \underline{W}^{\text{CHP}}, \underline{Q}^{\text{CHP}}, 0]^T$ depend on technical aspects; for instance, for a single CHP unit the lower limit can be set to 50%

of the rated power, and for a cluster of units it can be set to 50% of the rated power of the unit with the smallest size (Chicco and Mancarella, 2009b). In this case, the equipment switch-off condition has to be identified by introducing a dedicated variable (leading to a nonconnected portion of the domain in terms of domain definition through continuous variables), or included as a further operating condition in case of discrete representation of the domain.

With reference to the CCHP scheme shown in Figure 10.2, the limits are set to $\underline{\mathbf{v}}^{AB} = [0,0,0,0]^T$ and $\overline{\mathbf{v}}^{AB} = [0,0,\overline{Q}^{AB},0]^T$ for the AB, $\underline{\mathbf{v}}^{CERG} = [0,0,0,0]^T$ and $\overline{\mathbf{v}}^{CERG} = [0,0,0,\overline{R}^{CERG}]^T$ for the CERG, and $\underline{\mathbf{v}}^{WARG} = [0,0,0,0]^T$ and $\overline{\mathbf{v}}^{WARG} = [0,0,0,\overline{R}^{WARG}]^T$ for the WARG.

In a CiT problem, the definition of the capacity constraints has to take into account the effect of further aspects such as reserve constraints, ramp-rate limits for start-up, shut-down, ramp-up, and ramp-down during operation.

Constraints on maintaining a sufficient amount of *reserves* can be included by reducing the capacity limit to make a predefined amount of reserve $\mathbf{g}_h$ available at time interval h (Kockar et al., 2009). In this case, Equation 10.53 for $Y \in \mathbf{Y}$ is modified as

$$\left(\mathbf{H}^Y \mathbf{v}_i^Y\right)_h + \mathbf{g}_h - \overline{\mathbf{v}}^Y \le 0 \tag{10.55}$$

Following the approach presented in Arroyo and Conejo (2000), the formulation of the capacity constraints shown in Equations 10.53 and 10.54 for equipment $Y \in \mathbf{Y}$ with reference to time interval h can be extended to incorporate the ramp-rate constraints. For a general representation, the limits are indicated here for any energy vector used. In practice, simplified representations can be derived by including only limits referring to a subset of the energy vectors. Considering equipment $Y \in \mathbf{Y}$ at time interval h, let us introduce the vectors $\overline{\mathbf{r}}_h^Y$ and $\underline{\mathbf{r}}_h^Y$ containing the ramp-rate limits during operation for ramp-up and ramp-down, respectively, and the vectors $\overline{\mathbf{s}}_h^Y$ and $\underline{\mathbf{s}}_h^Y$ containing the ramp-rate limits for start-up and shut-down, respectively. In practice, for equipment Y the upper capacity limit at time interval h may be reduced with respect to $\overline{\mathbf{v}}^Y$, taking into account the shut-down ramp-rate if the equipment has to be shut down at next time interval $h + 1$, the start-up ramp-rate if the equipment has to be started up at time interval h, and the ramp-up rate if the equipment is already in operation at time interval h. Likewise, the lower capacity limit at time interval h depends on the ramp-down rate if the equipment is already in operation at time interval $h - 1$, and on the shut-down ramp rate if the equipment is shut down at time interval h.

The upper capacity limits are represented through the following equations:

$$\left(\mathbf{H}^Y \mathbf{v}_i^Y\right)_h - \left(b_h^Y - \underline{b}_{h+1}^Y\right)\overline{\mathbf{v}}^Y - \underline{b}_{h+1}^Y \mathbf{s}^Y \le 0 \tag{10.56}$$

$$\left(\mathbf{H}^Y \mathbf{v}_i^Y\right)_h - \left(\mathbf{H}^Y \mathbf{v}_i^Y\right)_{h-1} - b_{h-1}^Y \overline{\mathbf{r}}^Y - \overline{b}_h^Y \overline{\mathbf{s}}^Y \le 0 \tag{10.57}$$

Likewise, the lower capacity limits are represented as

$$b_h^Y \underline{\mathbf{v}}^Y - \left(\mathbf{H}^Y \mathbf{v}_i^Y \right)_h \leq 0 \tag{10.58}$$

$$\left(\mathbf{H}^Y \mathbf{v}_i^Y \right)_{h-1} - \left(\mathbf{H}^Y \mathbf{v}_i^Y \right)_h - b_h^Y \underline{\mathbf{r}}^Y - \underline{b}_h^Y \underline{\mathbf{s}}^Y \leq 0 \tag{10.59}$$

To prevent start-up and shutdown of equipment *Y* to be superposed, the following constraint is introduced, allowing only values of 0 or 1 for the corresponding binary variables:

$$\overline{b}_h^Y + \underline{b}_h^Y - 1 \leq 0 \tag{10.60}$$

Considering *storage* elements, the upper and lower capacity limits for the various energy vectors are indicated as $\overline{z}$ and $\underline{z}$, respectively, and the corresponding charge and discharge rate limits are indicated as $\overline{z}^+$ and $\overline{z}^-$, respectively (Rong et al., 2008). Including the ramp-rate limits for storage can be avoided if variations in the stored energy are assumed to occur only at the charge and discharge rate limits. In this case, the terms $\overline{z}^+$ and $\overline{z}^-$ are directly embedded in the storage coupling matrix (Geidl and Andersson, 2007b; Hajimiragha et al., 2007). Further limits concerning storage could depend on imposing a minimum level of charge of the storage capacity at the beginning and at the end of the analysis period (Piacentino and Cardona, 2008a).

10.4.3.2 Operational Constraints

For a CiT problem, in addition to the ramp-rate limits included in the upper- and lower-capacity limits, operational constraints may refer to the minimum duration of unit operation in *on* and *off* states. These constraints are relevant to the operation of relatively large thermal units, whose time constants can be significantly long, and for which maintenance reasons could require to limit the on–off or off–on transitions during a given time period.

The implementation of this constraint calls for the introduction of dedicated variables. According to Arroyo and Conejo (2000), let us introduce for equipment $Y \in \mathbf{Y}$ the minimum up time $\overline{\tau}^Y$ and the minimum down time $\underline{\tau}^Y$, together with a counter ς_h^Y indicating the duration for which unit *Y* has been continuously on ($\varsigma_h^Y > 0$) or off ($\varsigma_h^Y < 0$) at the end of time interval *h*. The minimum up-time constraint is then formulated as

$$-\left(\varsigma_{h-1}^Y - \overline{\tau}^Y \right)\left(b_{h-1}^Y - b_h^Y \right) \leq 0 \tag{10.61}$$

and the minimum down-time constraint is expressed by

$$\left(\varsigma_{h-1}^Y + \underline{\tau}^Y \right)\left(b_h^Y - b_{h-1}^Y \right) \leq 0 \tag{10.62}$$

10.4.3.3 Environmental Constraints

Environmental constraints can generally be of integral type, namely, referring to a predefined period of time and as such applied to CiT problems.

Let us consider the vector $\mathbf{e}_h^X$ containing the emissions of an ordered set of GHG or other pollutants emitted during the production of the energy vector X in the time interval $h = 1, \ldots, H$.

The emission vector components can be considered to be subject to a cumulative limit $\overline{\mathbf{e}}^X$, corresponding to the inequality constraint

$$-\overline{\mathbf{e}}^X + \sum_{h=1}^{H} \mathbf{e}_h^X \leq 0 \qquad (10.63)$$

Likewise, the constraint could refer to a minimum acceptable emission reduction with respect to the conventional reference case (Mavrotas et al., 2007).

The inequality constraint formulation can be modified to encompass the presence of a market for emission allowances (Kockar et al., 2009) in which the regulation enables buying or selling the emission allowances indicated as $\vec{\mathbf{a}}_h^X$ (bought) and $\overleftarrow{\mathbf{a}}_h^X$ (sold), thus obtaining

$$-\overline{\mathbf{e}}^X + \sum_{h=1}^{H} \left(\mathbf{e}_h^X + \vec{\mathbf{a}}_h^X - \overleftarrow{\mathbf{a}}_h^X \right) \leq 0 \qquad (10.64)$$

This formulation is also accompanied by constraints on the maximum amount of emissions allowances planned to be bought or sold in each time interval, respectively indicated as $\overline{\vec{\mathbf{a}}}_h^X$ and $\overline{\overleftarrow{\mathbf{a}}}_h^X$, introduced by resorting to the further binary variables $\overleftarrow{b}_h^X$ and $\vec{b}_h^X$ indicating whether the emission allowances are bought or sold, respectively, so that, for $h = 1, \ldots, H$:

$$\vec{\mathbf{a}}_h^X - \vec{b}_h^X \overline{\vec{\mathbf{a}}}_h^X \leq \mathbf{0} \qquad (10.65)$$

$$\overleftarrow{\mathbf{a}}_h^X - \overleftarrow{b}_h^X \overline{\overleftarrow{\mathbf{a}}}_h^X \leq \mathbf{0} \qquad (10.66)$$

with the corresponding exclusivity constraint on the binary variables:

$$\overleftarrow{b}_h^X + \vec{b}_h^X - 1 \leq 0 \qquad (10.67)$$

In addition, the minimum amount of emission allowances bought or sold is null:

$$-\vec{\mathbf{a}}_h^X \leq \mathbf{0} \qquad (10.68)$$

$$-\overleftarrow{\mathbf{a}}_h^X \leq \mathbf{0} \qquad (10.69)$$

If different cumulative limits $\bar{e}^{X_j}$ are set for different territorial areas or jurisdictions $j = 1, \ldots, J$, the formulation of Equation 10.63 is modified to take into account the individual components acting on the specific jurisdiction:

$$-\bar{e}^{X_j} + \sum_{h=1}^{H} e_h^{X_j} \leq 0 \qquad (10.70)$$

Further aspects are discussed in Rong and Lahdelma (2007), addressing multiperiod stochastic optimization for CHP production with CO_2 emission trading through scenario-based analysis.

10.4.3.4 Structural Constraints

With reference to the exemplificative trigeneration scheme of Figure 10.2, a first part of the structural constraints contains the dispatch factor limits within the [0,1] range. In vector form, for the dispatch factors α defined in Section 10.2.2.1, the corresponding inequalities are

$$-\alpha \leq 0 \qquad (10.71)$$

$$\alpha - 1 \leq 0 \qquad (10.72)$$

Another constraint is set up to ensure that the fuel input is positive:

$$-F_i \leq 0 \qquad (10.73)$$

In the representation of the electricity flows, the electricity input W_i from the EDS is restricted to be a positive value, and the possible output to the EDS is given by the difference $(W_o - W_d)$ that has to be positive, as the electricity output W_o from the CCHP system (including the possible electricity input W_i from the EDS) cannot be lower than the electrical load W_d. This leads to the following inequalities:

$$-W_i \leq 0 \qquad (10.74)$$

$$-(W_o - W_d) \leq 0 \qquad (10.75)$$

The equality constraint 10.48 introduced in Section 10.4.2 guarantees that cases with simultaneous input from the EDS and output to the EDS cannot exist in the same time interval.

Furthermore, the electricity input to the CERG must be positive. The corresponding constraint also ensures that there is no possible loop flow of electricity through the CERG input:

$$\left(\alpha_W^{EDS} - 1 \right) W_i \leq 0 \qquad (10.76)$$

In a CiT problem, other inequality constraints may be introduced to represent dedicated formulations such as the staircase-like approximation of nonlinear (e.g., exponential) start-up costs or the linear approximation of a nonlinear cost function (Arroyo and Conejo, 2000).

10.5 Solution Techniques

10.5.1 Classification of the Techniques

The solution techniques depend on the model established for the MG system under analysis. With reference to the classification of the problems indicated in Section 10.3.1, for DiT problems the method is chosen mainly on the basis of the characteristics of convexity, linearity, and presence of integer variables in the problem domain. Nonconvex domains can be handled by defining suitable subproblems on nonoverlapping convex domains covering the entire original domain, or by extending the domain by making it convex (including unfeasible portions of the domain) and verifying *a posteriori* that in the optimal solution no values fall in the extended unfeasible portion of the domain.

Linearization of the efficiency and cost/price curves is generally introduced to avoid the need of using nonlinear solvers, either using piecewise linear curves or surfaces or using piecewise constant approximations. Again, the use of integer variables generated by a gap in the domain of definition of a given variable can be avoided when it is possible to extend the domain of the corresponding variable and check *a posteriori* that the value of the variable in the optimal solution does not fall into the gap.

Complete discretization of the domain of all the variables under analysis (for instance, used within optimization approaches solved with heuristic techniques) allows avoiding the presence of nonconnected domains, as the discrete points in the domain are chosen with the only criterion to be feasible. In the representation of continuous portions of the domain, the discretization step has to be sufficiently reduced to provide acceptable coverage of the domain, taking into account the need for maintaining the computational burden within affordable limits.

For CiT problems, the solutions need to take into account the coupling-in-time of the time intervals. For this purpose, classical solution techniques include multiple time-domain simulations, dynamic programming, multi-interval linear programming, and various heuristics. Constraints of various types can be converted into linear constraints to simplify the computational model; also constraints referring to time-coupling aspects (such as ramp-rate limits, minimum up- and down-time constraints, and so forth) can be converted into linear constraints (Arroyo and Conejo, 2000).

10.5.2 Single-Objective Optimization

Linear programming is adopted when the system model is appropriately linearized and the solution space is convex (Aringhieri and Malucelli, 2003; Kong et al., 2005; Lozano et al., 2009b). Weber et al. (2006) adopt linear programming within a planning and operational approach. A particular case of solution with linear programming is the tricommodity simplex formulated in Rong and Lahdelma (2005), assuming a convex feasible operating region of the trigeneration plant with three energy products in function of the three production levels.

Mixed integer-linear programming (MILP) is one of the most used approaches in the relevant literature, to encompass the presence of regions with linearized characteristics, as well as possible individual points of operation. For instance, Yokoyama et al. (2002) formulate a large-scale MILP problem with binary variables for selection and on/off status of equipment, and continuous variables for capacities and load allocation of equipment. MILP is adopted in Oh et al. (2007) for planning with hourly operational schedules with a branch-and-bound solution technique, in Piacentino and Cardona (2008a) for an integrated optimal planning and operation, in Sakawa et al. (2002), where the coupling among the time periods given by the equipment start-up and shutdown is taken into account and an approximate solution with genetic algorithm is also presented, and in Lozano et al. (2009a).

Mixed-integer nonlinear programming (MINP) is adopted by Curti et al. (2002a) within a comprehensive approach to design, installation, and operation of integrated energy systems, in which the decision variables are defined in a noncontiguous domain and the problem formulated is highly nonlinear; the optimization problem is solved by using a genetic algorithm. Jüdes et al. (2009) handle nonconvex cases through the heuristic branch and cut algorithm, whereas Chicco and Mancarella (2009b) extend the domain of definition of the generated power in order to get continuous variables avoiding the use of integer variables, and resort to a sequential quadratic programming solution.

Lagrangian multipliers are used in Geidl and Andersson (2007a) and Hemmes et al. (2007), with a convex model solved by applying the Karush–Kuhn–Tucker first-order optimality condition to provide the marginal costs of the energy carriers at input and output.

Lagrangian relaxation is used in Rong et al. (2008), to address trigeneration systems with storage. The application of this method calls for defining the feasible operating region for heat and cooling production, in order to compute the corresponding Lagrangian dual solution. The use of Lagrangian relaxation decomposes the overall problem into hourly models (Makkonen and Lahdelma, 2006), then linear programming is adopted to solve a convex model, or MILP to solve a nonconvex model through partitioning of the nonconvex regions into a nonoverlapping set of convex regions.

10.5.3 Multiobjective Optimization

The transformation of a multiobjective problem into a single-objective problem can be obtained by shaping the objective function as indicated in Section 10.3.1.2 and using optimization techniques aimed at solving single-objective functions (Wang et al., 2010b).

From the methodological point of view, further approaches have been implemented, for instance:

- An approach based on goal programming, such as the goal attainment procedure presented in Chang and Fu (1998): This procedure is applicable to convex or nonconvex domains, and is based on defining a set of target goals with corresponding objectives that can be under- or over-achieved. A weight vector controls the degree of under- or over-achievement of each objective, allowing a certain trade-off among the objectives. This approach allows transforming the multiobjective optimization into a scalar optimization problem.

- The approach presented in Tsay (2003), based on the objective function 10.17 and adopting a *minimum least-square error* to drive the solution toward an acceptable optimal value $\tilde{x}$: The stop criterion is set on the basis of satisfaction factors κ_j defined (using the notation reported in Section 10.3.1.2) for the objective functions $j = 1, ..., N$ as

$$\kappa_j = \frac{f_j - f_j(\tilde{x})}{\overset{\vee}{f_j} - \tilde{f}_j} \tag{10.77}$$

The solution process stops if all satisfaction factors are accepted, else the weighting factors in 10.17 are adjusted and the multiobjective calculation continues.

The conceptual core of multiobjective optimization is based on the possibility of calculating the set of compromise solutions located on the Pareto front, going beyond the optimal solutions of the individual objective functions. Generally, for either continuous or discrete combinatorial multiobjective optimization problems, it is practically infeasible to calculate the entire Pareto front. Hence, dedicated algorithms have been set up to provide the *best-known* Pareto front as the computable set of nondominated solutions.

Different techniques have been adopted to construct the best-known Pareto front in the optimization algorithms referring to MG systems operation, namely:

1. The calculation of the *weighted sum* of the individual objectives (Aki et al., 2006b), applicable to a *convex* Pareto front: The weights have no particular meaning, their sum can be restricted to unity, and the nondominated solutions are obtained by considering different weight combinations;
2. The ε-*constrained method* (Aki et al., 2003, 2006a,b; Mavrotas et al., 2007; Wang et al., 2008): This method considers an individual objective as the target to be optimized, and sets for all the other objectives a limit expressed by a threshold ε, then progressively introduces a threshold relaxation and correspondingly upgrades the set of nondominated solutions; this method is also applicable to optimization problems with nonconvex Pareto front; however, it becomes computationally less effective in the presence of many objectives;
3. The *direct construction* through heuristic approaches (Bernal-Agustín et al., 2006; Shukla and Deb, 2007; Boicea et al., 2009; Kavvadias and Maroulis, 2010): The method consists of an iterative process in which at each iteration multiple solutions are generated and the solution set is then reduced to maintain only the nondominated solutions; various approaches using the genetic algorithms have been proposed and effectively used (Deb et al., 2002; Abido, 2003; Konak et al., 2006; Shukla and Deb, 2007).

4. The adoption of *clustering evolutionary algorithms*, with the approach specifically developed by a research group and adopted in various applications (Burer et al., 2003; Xia et al., 2004; Pelet et al., 2005; Li et al., 2006).

After having obtained the points on the best-known Pareto front, personal judgment of the decision-maker is needed to choose the preferred solution. This decision can be assisted by using suitable numerical techniques to rank the compromise solutions, such as those based on multiattribute decision-making concepts (Li, 2009; Wang et al., 2009).

10.6 Example of Application

10.6.1 Trigeneration System Demand and Configuration

The optimization example presented here refers to a DiT problem for a trigeneration system exploited in a tertiary sector application to cover electricity, heat, and cooling needs. The trigeneration system configuration is the one indicated in Figure 10.2 (Section 10.2). The electricity, heat, and cooling demand is summarized by considering a typical day of an intermediate season, partitioned into hourly time periods (Chicco and Mancarella, 2008c, 2009b; Kavvadias and Maroulis, 2010). Electricity prices are taken from a real case in Italy (single national prices in the Italian electricity market on May 16, 2007). The trigeneration system is seen as a price-taker, and the same price $\rho_i^{EDS} = \rho_o^{EDS}$ is considered for electricity input and output. The gas price is assumed constant at $\rho_g^{GDS} = 24$ €/MWh during the day, as it may be set in relatively long-term contractual provisions (e.g., annual contracts). Figure 10.3 shows the corresponding hourly energy demands and prices. At each time period, the user's energy demand array $\mathbf{v}_d = [W_d, Q_d, R_d]^T$ is assigned as an input data.

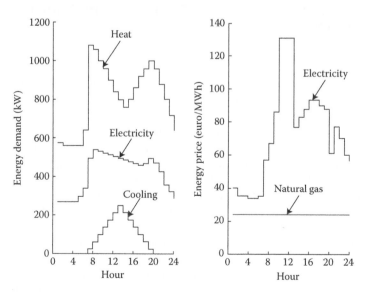

FIGURE 10.3 Energy demands and prices.

The trigeneration system equipment has the following characteristics:

- A CHP unit with an electrical capacity of 700 kW$_e$, a thermal capacity of 1050 kW$_t$, electrical efficiency $\eta_W^{CHP} = 0.3$, and thermal efficiency $\eta_Q^{CHP} = 0.45$.
- An AB with a capacity of 1500 kW$_t$ and thermal efficiency $\eta_Q^{AB} = 0.85$.
- A CERG with a capacity of 600 kW$_c$ and COP$^{CERG} = 3$.
- A WARG with a capacity of 600 kW$_c$ and COP$^{WARG} = 0.9$.

The rating of the trigeneration equipment components is not necessarily optimized from the design point of view, as this example shown could be seen as an operational optimization embedded into the external loop of a planning problem aimed at sizing the trigeneration system components.

The decision variables are the dispatch factors indicated in Equation 10.8, as well as the fuel input F_i and the electricity input W_i, and are included in the array $\mathbf{x} = [\alpha_W^{EDS} \alpha_F^{GDS} \alpha_W^y \alpha_Q^{CHP} \alpha_Q^{AB} \alpha_Q^y F_i W_i]^T$.

To simplify the descriptions, constant efficiency and COP values are assumed for the trigeneration equipment. This leads to using the system efficiency matrix $\mathbf{H}$ reported in Section 10.2.2.2. The presence of products among the decision variables leads to the formulation of a nonlinear optimization problem, in which the nonlinearity comes from the decision variables, and not from the equipment characteristics. The numerical solution is obtained with nonlinear optimization techniques (Chicco and Mancarella, 2008c, 2009b). The maximum capacity constraint of the equipment is enforced during the calculations, and the lower limit is set to zero to avoid a nonconnected domain. The solutions obtained will be checked *a posteriori* for feasibility regarding the lower limit.

10.6.2 Energy Costs and Energy Efficiency Optimizations for a Single Time Period

Single-objective optimizations are carried out by using the minimum energy cost and the maximum TPES indicator as individual objectives. In the latter case, energy efficiency of the trigeneration system is compared with separate production of conventional electricity, heat and cooling generation serving the same user's demand. Separate production efficiencies are set to $\eta_e^{SP} = 0.4$ for electricity production, $\eta_t^{SP} = 0.85$ for heat production, and COP$^{SP} = 3$ for the reference electric chiller.

The details of the solution of single-objective optimization for a single hour (hour 12:00) are shown in this section. The average power (or hourly energy) demand values are $W_d = 504$ kW$_e$ for electricity, $Q_d = 840$ kW$_t$ for heat, and $R_d = 212.5$ kW$_c$ for cooling. The initial values of the dispatch factors are set up to obtain a feasible initial solution.

With minimum energy cost optimization, the solution obtained (Figure 10.4) maintains the CHP at its maximum output to benefit from selling electricity to the EDS. This solution depends on the relatively high electricity price at which the local trigeneration system can sell electricity to the EDS, thus making it convenient to use more fuel from the GDS to make the CHP operate at its maximum capacity. The electrical output from the CHP goes to the electrical load and the excess is sent to the EDS, whereas the thermal output from the CHP supplies to the thermal load and the WARG. The AB

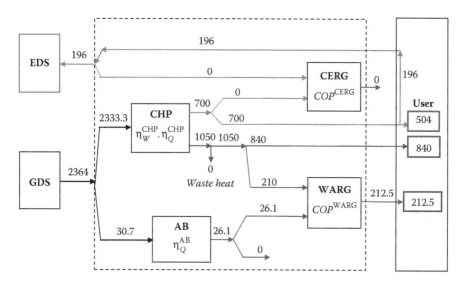

FIGURE 10.4 Optimal solution for hour 12:00 with minimum energy costs (numerical values in kW).

supplies additional heat to the WARG to serve the cooling load. In this solution, the CERG is not supplied.

Conversely, with maximum TPES optimization, the solution (Figure 10.5) leads to limit the fuel input, and the CHP is exploited to ensure thermal demand coverage (corresponding to the TLT mode). The cooling demand is covered by the CERG, supplied by the CHP with a further small contribution from the EDS. The WARG remains inactive.

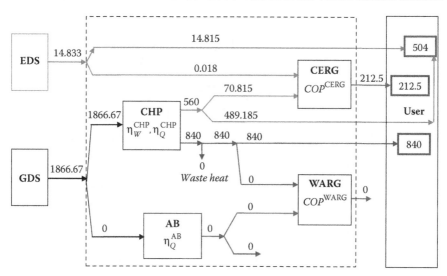

FIGURE 10.5 Optimal solution for hour 12:00 with maximum trigeneration primary energy saving indicator (numerical values in kW).

In this case, providing electricity from the EDS to supply the cooling load through the CERG proves to be most energy-efficient than taking fuel from the GDS.

The reported solutions are clearly different for the two optimization criteria, even though both solutions are optimal and supply the same user's energy demand of electricity, heat, and cooling. These solutions depend on the numerical values of the input data, and cannot be used to draw general conclusions for trigeneration system applications. In any case, these solutions point out a remarkable difference among the results from different types of optimization, in fact corresponding to conflicting solutions. As such, there is room for applying multiobjective optimization and seeking the compromise solutions located on the best-known Pareto front, as shown in Section 10.6.4.

10.6.3 Energy Costs and Energy Efficiency Optimizations for a Typical Day

The optimizations carried out in the previous section can be repeated for different time periods, leading to a set of operational solutions whose characteristics may change according to the evolution in time of the energy loads and prices. In the stepwise representations of the following figures, the solutions shown for a certain hour correspond to the step indicated from that hour to the successive one. For instance, the solutions for hour 12:00 correspond to the step from hour 12:00 to hour 13:00.

Results of hourly calculations are shown in Figure 10.6 (energy costs) and in Figure 10.7 (TPES indicator), in which the values obtained from the optimizations are compared with the ones referring to classical TLT and ELT operation strategies.

The electricity, heat, and cooling energy flows are reported in Figures 10.8 through 10.10, respectively. Furthermore, Figures 10.11 and 10.12 show the evolution in time of

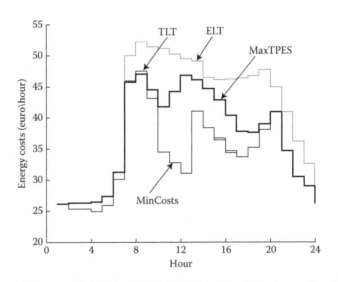

FIGURE 10.6 Energy costs for the different hourly solutions.

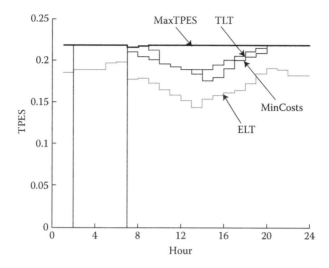

FIGURE 10.7 Values of the trigeneration primary energy saving indicator for the different hourly solutions.

the dispatch factors in the optimal cases of energy cost minimization and TPES maximization, respectively.

The above results strictly depend on the evolution in time of the different energy loads and prices. In this context, it can be seen that the maximum TPES solution provides clear indications on the use of the CHP to supply the thermal user's demand (with backup needed from the AB only at hours 07:00 and 08:00) and on the use of the CERG to supply the cooling user's demand (the WARG is always inactive). The CHP is always switched on and provides electricity and heat at loading levels higher than 50% of its capacity (even though the minimum switch-on operational limit is set up to zero), and during most of the day there is an amount of electricity sent to the EDS.

In the minimum energy cost solution, during the night the CHP is switched off from 02:00 to 07:00, losing in this case any advantage in terms of energy efficiency (TPES = 0). During the day, the minimum cost is generally obtained by maintaining the CHP in operation at its maximum output. Correspondingly, the user's cooling demand is mainly covered by the WARG, with a few exceptions with CERG operation at hours 07:00 and 08:00 (at which the CHP operates at its capacity limits) and at hour 13:00 (at which the WARG output reaches its capacity limit), and an additional condition at hour 09:00 (at which the user's cooling demand is partially supplied from CERG and partially from WARG).

As far as the classical TLT and ELT operation strategies, in the typical day analyzed the ELT strategy is almost never convenient. In the central hours of the day, the minimum energy cost solution is relatively similar to the solution obtained by applying the TLT strategy, being at the same time substantially different from the one that provides the maximum TPES.

The overall results indicate that from hour 20:00 to 1:00 the results obtained by using the minimum energy costs or maximum TPES provide similar information. In the other

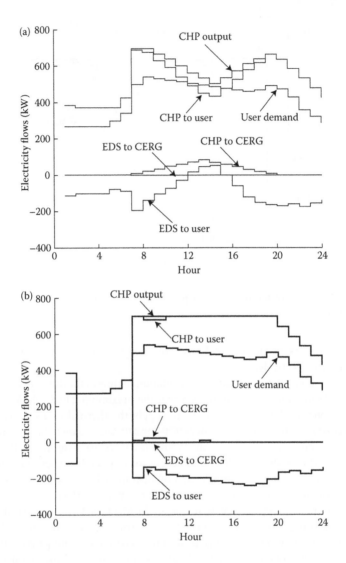

FIGURE 10.8   Electricity flows in the optimal solutions. (a) Maximum TPES solutions. (b) Minimum enery costs solutions.

hours, the information is evidently conflicting. During the hours at which energy cost minimization and TPES maximization are clearly conflicting, it is possible to gain further insights by performing multiobjective optimization, as shown in the following section.

10.6.4 Multiobjective Optimization

The determination of some compromise solutions belonging to the best-known Pareto front for hour 12:00 is shown here as an example. The calculations have been carried out

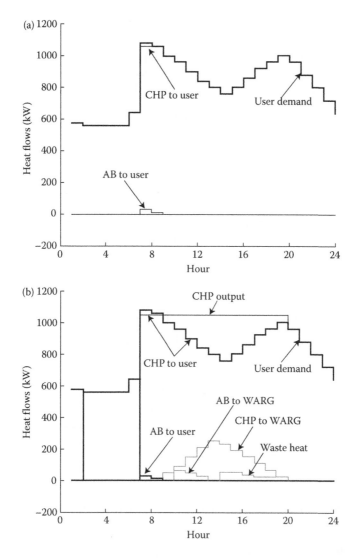

FIGURE 10.9 Heat flows in the optimal solutions. (a) Maximum TPES solutions. (b) Minimum energy costs solutions.

by using the NSGA-II algorithm (Deb et al., 2002), with suitable adaptations to handle the presence of the equality and inequality constraints.

The example presented here has been constructed by using a population of 200 chromosomes, evolving within an iterative process stopping when the best-known Pareto front does not exhibit significant changes for 10 successive generations. To discuss the characteristics of possible compromise solutions, Figure 10.13 shows two (nondominated) compromise solutions belonging to the best-known Pareto front. Table 10.1 reports the objective functions and the dispatch factors obtained for these nondominated solutions, as well as the

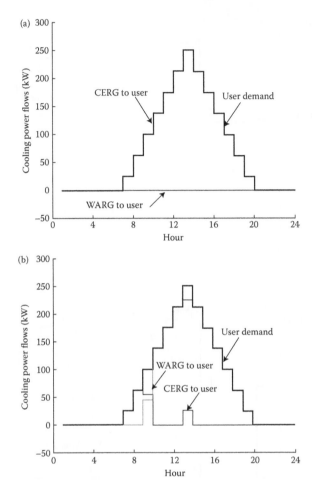

FIGURE 10.10 Cooling power flows in the optimal solutions. (a) Maximum TPES solutions. (b) Minimum energy costs solutions.

results of the individual optimizations illustrated in Section 10.6.2. The solutions are ordered in ascending value of energy costs, clearly resulting in corresponding descending values of 1 – TPES (minimized) or ascending values of TPES (maximized) because of the nondominant characteristics of the solution points located on the best-known Pareto front.

The solution obtained for compromise solution no. 1 is shown in Figure 10.14. The two compromise solutions share the characteristics of selling electricity to the EDS (but to a lower extent with respect to the minimum energy cost solution) as well as the presence of a nonnegligible amount of waste heat (8.59 kW$_t$ in compromise solution no. 1, and 29.28 kW$_t$ in compromise solution no. 2), and the use to a different extent of both CERG and WARG to supply the user's cooling demand. No lower operational limits have been imposed on CERG and WARG, so that compromise solution no. 2 (Figure 10.15) indicated a limited WARG contribution to serve the cooling load.

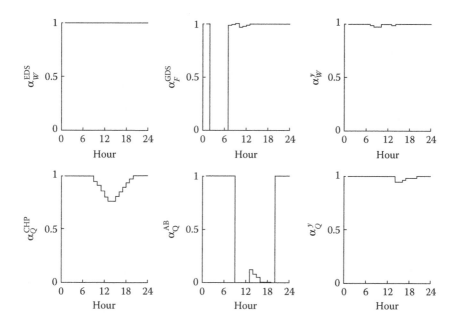

FIGURE 10.11 Dispatch factors for the minimum energy cost solutions.

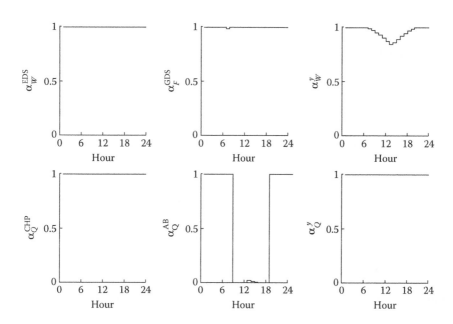

FIGURE 10.12 Dispatch factors for the maximum trigeneration primary energy saving solutions.

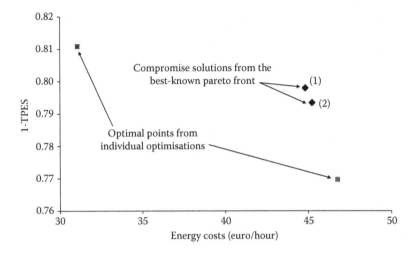

FIGURE 10.13 Pareto front obtained from NSGA-II versus single optimal points from individual optimizations.

The results shown in this section confirm the variety of optimal and compromise solutions that can be obtained for the trigeneration scheme used in this case study application. These solutions may be activated by establishing suitable control schemes using the dispatch factors as control variables. The effectiveness of application of the optimal solutions found clearly depends on the estimated loads and on performance characteristics of the equipment.

10.7 Conclusions

In local MG plants, the interactions among the local equipment give some possibility to adjust the energy flows of various energy vectors inside the plant, leading to increasing operational flexibility. In the numerical model of the MG system, the local interactions are represented through dispatch factors acting as decision variables. Given the energy loads, in an MG system multiple solutions with different energy flows can be found, correspond-

TABLE 10.1 Optimal Solutions from Individual Optimizations and Multiobjective Optimization (NSGA-II) for Hour 12:00

Solution	Dispatch Factors						Energy Cost (€/h)	1–TPES	TPES
	α_F^{GDS}	α_W^y	α_Q^{CHP}	α_Q^{AB}	α_Q^y	α_W^{EDS}			
Minimum energy cost	0.9870	1	0.8	0	1	1	31.08	0.8109	0.1891
Compromise no. 1	0.9724	0.9091	0.9683	0.2537	0.9901	1	44.82	0.7979	0.2021
Compromise no. 2	0.9999	0.8823	0.9920	0.2565	0.9666	1	45.25	0.7935	0.2065
Maximum TPES	1	0.8735	1	0	1	0.9988	46.74	0.7697	0.2303

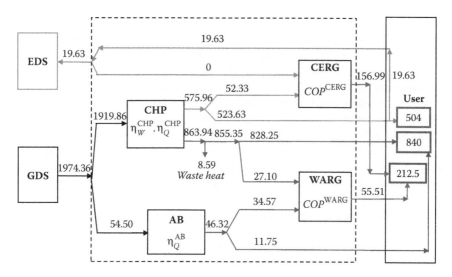

FIGURE 10.14 Compromise solution no. 1 for hour 12:00 from NSGA-II (numerical values in kW).

ing to different energy inputs. Among these solutions, it is possible to set up suitable optimizations to identify the most convenient solutions according to specified objective functions with appropriate constraints. The different and often conflicting nature of the objectives paves the way to the formulation of multiobjective optimization problems.

This chapter has provided an integrated view of objective functions, constraints, and solution methods used to address optimal MG system operation. Examples have been shown with reference to trigeneration systems, to highlight some details of the formulation and solution of optimization problems. More generally, local MG systems

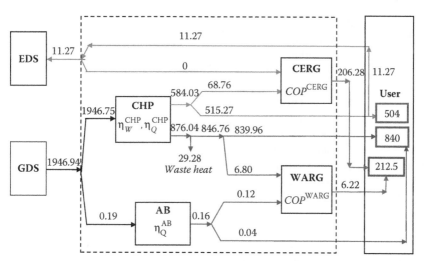

FIGURE 10.15 Compromise solution no. 2 for hour 12:00 from NSGA-II (numerical values in kW).

may interact with other local sources (e.g., from renewable energy) and are interconnected among them through the energy networks, thus integrating MG with the distributed energy resources encompassing distributed generation, distributed storage, and demand response, forming the wider distributed MG paradigm (Tsikalakis and Hatziargyriou, 2007; Yusta et al., 2007; Chicco and Mancarella, 2009a).

For a given plant structure, the analysis and optimization procedures can be run offline to identify the most convenient operational strategies. Then, the optimization results have to be applied to actual MG system operation, activating the operational strategies through suitable control systems. For this purpose, specific control structures and tools have to be provided through exploitation of information and communication technology concepts and applications. The integration of information and communication technology plays a crucial role in the optimal coordination and management of MG systems interacting through various energy distribution networks and microgrids (Marnay et al., 2008; Tsikalakis and Hatziargyriou, 2008), which opens the way to an innovative multienergy smart grid framework.

Further developments for MG optimization are being enabled by enhanced availability of real-time data for the different energy vectors. In this respect, challenging aspects are the formulation and solution of integrated objective functions, including technical, economic, environmental, and social aspects in a comprehensive optimization framework, also taking into account data uncertainty (for instance, through stochastic optimization). The numerical approaches to solve the optimization problems using these integrated objective functions have to be deeply refined, dealing with computational complexity issues for making these solutions more easily tractable.

Nomenclature

Acronyms

AB	Auxiliary boiler
AHP	Absorption heat pump
AP	Acidification potential
CCHP	Combined cooling, heat, and power
CERG	Compression electrical refrigerator group
CHP	Combined heat and power
CiT	Coupled-in-time
COP	Coefficient of performance
DiT	Decoupled-in-time
DMG	Distributed multigeneration
EDC	Engine-driven chiller
EDHP	Engine-driven heat pump
EDS	Electrical distribution system
EHP	Electric heat pump
ELT	Electrical load tracking
GAHP	Gas absorption heat pump
GARG	Gas absorption refrigerator group
GDS	Gas distribution system

GHG	Greenhouse gas
GWP	Global warming potential
LER	Local emission reduction
LHV	Lower heating value
MILP	Mixed-integer linear programming
MINP	Mixed-integer nonlinear programming
ODP	Ozone depletion potential
PCO2ER/	
PCDER	Polygeneration CO_2/carbon dioxide emission reduction
PR11ER	Polygeneration R11 emission reduction
PSDER	Polygeneration sulphur dioxide emission reduction
PPES	Polygeneration primary energy saving
SS	Spark spread
TLT	Thermal load tracking
TPES	Trigeneration primary energy saving
TSS	Total supply spread
WARG	Water absorption/adsorption refrigerator group

Symbols
Variables

b	binary variable
c	(subscript) cooling
d	day type index
e	(subscript) electrical
f	objective function; (subscript) fuel
g	generic GHG
h	time interval index
i	(subscript) input
j	jurisdiction index
m	mass
n	number of days
o	(subscript) output
s	season index
t	time; (subscript) thermal
u	(superscript) useful
w	weighting factor
x	type of energy
y	(superscript) cogeneration
l	losses
D	number of day types
F	fuel
H	number of time intervals
J	number of jurisdictions
M	number of solution points in multiobjective optimization
N	number of objective functions in multiobjective optimization
Q	heat

R	cooling
S	number of seasons
W	electricity
X	generic energy vector
Y	generic multigeneration equipment
α	dispatch factor
β	fraction of pollutant
η	efficiency
κ	satisfaction factor
λ	heat to power ratio
μ	emission factor for global warming and local pollution
ν	emission factor for the acidification problem
o	emission factor for the stratospheric ozone depletion problem
ρ	price
τ	time duration
ς	time duration counter

Sets and global functions:

D	set with energy vector and type of energy
G	set of pollutants
X	set of energy vectors
Y	set of multigeneration equipment or components
$\mathcal{C}$	costs
$\mathcal{E}$	pollution costs
$\mathcal{F}$	function of the decision variables
$\mathcal{R}$	revenues

Vectors and matrices:

$\mathbf{a}$	emission allowances
$\mathbf{e}$	emissions
$\mathbf{f}$	objective functions
$\mathbf{g}$	reserves
$\mathbf{h}$	equality constraints function
$\mathbf{r}$	reserve; ramp-rate limits during operation
$\mathbf{s}$	start-up/shutdown ramp-rate limits
$\mathbf{v}$	energy vector
$\mathbf{x}$	decision variables
$\mathbf{z}$	energy stored
$\mathbf{H}$	efficiency matrix
$\mathbf{S}$	storage coupling matrix
$\boldsymbol{\alpha}$	dispatch factors
$\boldsymbol{\beta}$	decision variables subvector
$\boldsymbol{\lambda}$	Lagrangian multipliers
$\boldsymbol{\psi}$	inequality constraints function
$\boldsymbol{\zeta}$	marginal values

References

Abido MA, 2003. Environmental/economic power dispatch using multiobjective evolutionary algorithms. *IEEE Transactions on Power Systems* 18(4): 1529–1537.

Aki H, Oyama T, Tsuji K, 2003. Analysis of energy pricing in urban energy service systems considering a multiobjective problem of environmental and economic impact. *IEEE Transactions on Power Systems* 18(4): 1275–1282.

Aki H, Oyama T, Tsuji K, 2006a. Analysis of energy service systems in urban areas and their CO_2 mitigations and economic impacts. *Applied Energy* 83(10): 1076–1088.

Aki H, Yamamoto S, Ishikawa Y, Kondoh J, Maeda T, Yamaguchi H, Murata A, Ishii I, 2006b. Operational strategies of networked fuel cells in residential homes. *IEEE Transactions on Power Systems* 21(3): 1405–1414.

Arcuri P, Florio G, Fragiacomo P, 2007. A mixed integer programming model for optimal design of trigeneration in a hospital complex. *Energy* 32: 1430–1447.

Aringhieri R, Malucelli F, 2003. Optimal operations management and network planning of a district heating system with a combined heat and power plant. *Annals of Operations Research* 120: 173–199.

Arroyo JM, Conejo AJ, 2000. Optimal response of a thermal unit to an electricity spot market. *IEEE Transactions on Power Systems* 15(3): 1098–1104.

Badami M, Portoraro A, 2009. Performance analysis of an innovative small-scale trigeneration plant with liquid desiccant cooling system. *Energy & Buildings* 41(11): 1195–1204.

Beihong Z, Weiding L, 2006. An optimal sizing method for cogeneration plants. *Energy and Buildings* 38: 189–195.

Bernal-Agustín JL, Dufo-López R, Rivas-Ascaso DM, 2006. Design of isolated hybrid systems minimizing costs and pollutant emissions. *Renewable Energy* 31:2227–2244.

Boicea AV, Chicco G, Mancarella P, 2009. Optimal operation of a microturbine cluster with partial-load efficiency and emission characterization. *Proceedings of IEEE Power Tech 2009*, Bucharest, Romania, June 28–July 2, 2009, paper 109.

Borbely AM, Kreider JF, ed., 2001. *Distributed Generation—The Power Paradigm of the New Millennium*. CRC Press, Boca Raton, FL.

Borchiellini R, Massardo AF, Santarelli M, 2002. Carbon tax vs CO_2 sequestration effects on environomic analysis of existing power plants. *Energy Conversion and Management* 43: 1425–1443.

Burer M, Tanaka K, Favrat D, Yamada K, 2003. Multi-criteria optimization of a district cogeneration plant integrating a solid-oxide fuel-cell/gas turbine combined cycle, heat pumps and chiller. *Energy* 28(6): 497–518.

Calì M, Borchiellini R, 2004. District heating networks calculation and optimization, in Frangopoulos CA, ed., *Exergy, Energy System Analysis, and Optimization. Encyclopedia of Life Support Systems (EOLSS)*. Eolss Publishers, Oxford, UK.

Canova A, Chicco G, Genon G, Mancarella P, 2008. Emission characterization and evaluation of natural gas-fueled cogeneration microturbines and internal combustion engines. *Energy Conversion and Management* 49(10): 2900–2909.

Capasso A, Grattieri W, Lamedica R, Prudenzi A, 1994. A bottom-up approach to residential load modeling. *IEEE Transactions on Power Systems* 9(2): 957–964.

Cardona E, Piacentino A, 2003. A methodology for sizing a trigeneration plant in Mediterranean areas. *Applied Thermal Engineering* 23: 1665–1680.

Cardona E, Piacentino A, 2005. Cogeneration: A regulatory framework toward growth. *Energy Policy* 33: 2100–2111.

Cardona E, Piacentino A, Cardona F, 2006a. Matching economical, energetic and environmental benefits: An analysis for hybrid CHCP-heat pump systems, *Energy Conversion and Management* 47(20): 3530–3542.

Cardona E, Sannino P, Piacentino A, Cardona F, 2006b. Energy saving in airports by trigeneration. Part II: Short and long term planning for the Malpensa 2000 CHCP plant. *Applied Thermal Engineering* 26(14–15): 1437–1447.

Cârdu M, Baica M, 2002. Regarding the greenhouse gas emissions of thermopower plants. *Energy Conversion and Management* 43: 2135–2144.

Carpaneto E, Chicco G, 2008. Probabilistic characterisation of the aggregated residential load patterns. *IET Generation, Transmission and Distribution* 2(3): 373–382.

Carrión M, Philpott AB, Conejo AJ, Arroyo JM, 2007. A stochastic programming approach to electric energy procurement for large consumers. *IEEE Transactions on Power Systems* 22(2): 744–754.

Carvalho M, Serra LM, Lozano MA, 2011. Geographic evaluation of trigeneration systems in the tertiary sector. Effect of climatic and electricity supply conditions. *Energy* 36: 1931–1939.

Chang CS, Fu W, 1998. Stochastic multiobjective generation dispatch of combined heat and power systems. *IEE Proceedings—Generation Transmission Distribution* 145(5): 583–591.

Chicco G, Mancarella P, 2006. From cogeneration to trigeneration: Profitable alternatives in a competitive market. *IEEE Transactions on Energy Conversion* 21: 265–272.

Chicco G, Mancarella P, 2007a. Trigeneration primary energy saving evaluation for energy planning and policy development. *Energy Policy* 35(12): 6132–6144.

Chicco G, Mancarella P, 2007b. Enhanced energy saving performance in composite trigeneration systems. *Proceedings IEEE PowerTech 2007*, Lausanne, Switzerland, July 1–5, 2007, paper 526.

Chicco G, Mancarella P, 2008a. A unified model for energy and environmental performance assessment of natural gas-fueled poly-generation systems. *Energy Conversion and Management* 49(8): 2069–2077.

Chicco G, Mancarella P, 2008b. Assessment of the greenhouse gas emissions from cogeneration and trigeneration systems. Part I: Models and indicators. *Energy* 33(3): 410–417.

Chicco G, Mancarella P, 2008c. Optimal operational strategies for multi-generation systems. *Proceedings Power Systems Computational Conference*, Glasgow, Scotland, July 14–18, 2008, paper 528.

Chicco G, Mancarella P, 2008d. Evaluation of multi-generation alternatives: An approach based on load transformations. *Proceedings IEEE Power and Energy Society General Meeting* 2008, Pittsburgh, PA, July 20–24, 2008, paper 08GM1283.

Chicco G, Mancarella P, 2009a. Distributed multi-generation: A comprehensive view. *Renewable and Sustainable Energy Reviews* 13(3): 535–551.

Chicco G, Mancarella P, 2009b. Matrix modelling of small-scale trigeneration systems and application to operational optimization. *Energy* 34(3): 261–273.

Costa A, Paris J, Towers M, Browne T, 2007. Economics of trigeneration in a kraft pulp mill for enhanced energy efficiency and reduced GHG emissions. *Energy* 32(4): 474–481.

Curti V, von Spakovsky MR, Favrat D, 2002a. An environomic approach for the modeling and optimization of a district heating network based on centralized and decentralized heat pumps, cogeneration and/or gas furnace. Part I—Methodology. *International Journal of Thermal Sciences* 39(7): 721–730.

Curti V, von Spakovsky MR, Favrat D, 2002b. An environomic approach for the modeling and optimization of a district heating network based on centralized and decentralized heat pumps, cogeneration and/or gas furnace. Part II—Application. *International Journal of Thermal Sciences* 39(7): 731–741.

Danny Harvey LD, 2006. *A Handbook on Low-energy Buildings and District Energy Systems: Fundamentals, Techniques, and Examples.* James and James, UK.

Deb K, Pratap A, Agarwal S, Meyarivan T, 2002. A fast and elitist multiobjective genetic algorithm: NSGA-II. *IEEE Transactions on Evolutionary Computation* 6(2): 182–197.

Deng SJ, Jiang W, 2005. Levy process-driven mean-reverting electricity price model: The marginal distribution analysis. *Decision Support Systems* 40: 483–494.

Deng J, Wang R, Wu J, Han G, Wu D, Li S, 2008. Exergy cost analysis of a micro-trigeneration system based on the structural theory of thermoeconomics. *Energy* 33: 1417–1426.

Deng J, Wang RZ, Han GY, 2011. A review of thermally activated cooling technologies for combined cooling, heating and power systems. *Progress in Energy and Combustion Science* 37(2): 172–203.

Fumo N, Mago PJ, Chamra LM, 2009. Emission operational strategy for combined cooling, heating, and power systems. *Applied Energy* 86(11): 2344–2350.

Geidl M, 2007. Integrated modeling and optimization of multi-carrier energy systems, PhD. dissertation no. 17141, Swiss Federal Institute of Technology (ETH) Zürich, Switzerland.

Geidl M, Andersson G, 2007a. Optimal power flow of multiple energy carriers. *IEEE Transactions on Power Systems* 22(1): 145–155.

Geidl M, Andersson G, 2007b. Optimal coupling of energy infrastructures. *Proceedings of IEEE PowerTech*, Lausanne, Switzerland, 2007, pp. 1398–1403.

Geidl M, Koeppel G, Favre-Perrod P, Klöckl B, Andersson G, Fröhlich K, 2007. Energy hubs for the future. *IEEE Power and Energy Magazine* 5(1): 25–30.

Grekas DN, Frangopoulos CA, 2007. Automatic synthesis of mathematical models using graph theory for optimisation of thermal energy systems. *Energy Conversion and Management* 48(11): 2818–2826.

Hajimiragha A, Cañizares C, Fowler M, Geidl M, Andersson G, 2007. Optimal energy flow of integrated energy systems with hydrogen economy considerations. *Proceedings of 2007 iREP Symposium—Bulk Power System Dynamics and Control: VII, Revitalizing Operational Reliability*, August 19–24, 2007, Charleston, SC.

Hemmes K, Zachariah-Wolff JL, Geidl M, Andersson G, 2007. Towards multi-source multi-product energy systems, *International Journal of Hydrogen Energy* 32(10–11): 1332–1338.

Herman R, Kritzinger JJ, 1993. The statistical description of grouped domestic electrical load currents. *Electric Power Systems Research* 27: 43–48.

Hernandez-Santoyo J, Sanchez-Cifuentes A, 2003. Trigeneration: An alternative for energy savings. *Applied Energy* 76(1–3): 219–227.

Heteu PMT, Bolle L, 2002. Economie d'énergie en trigénération. *International Journal of Thermal Sciences* 41: 1151–1159.

Jüdes M, Vigerske S, Tsatsaronis G, 2009. Optimization of the design and partial-load operation of power plants using mixed-integer nonlinear programming: Chapter 9, in Kallrath J, Pardalos PM, Rebennack S, Scheidt M, eds, *Optimization in the Energy Industry*. Springer-Verlag, Berlin Heidelberg.

Kavvadias KC, Maroulis ZB, 2010. Multi-objective optimization of a trigeneration plant. *Energy Policy* 38(2): 945–954.

Kavvadias KC, Tosios AP, Maroulis ZB, 2010. Design of a combined heating, cooling and power system: Sizing, operation strategy selection and parametric analysis. *Energy Conversion and Management* 51(4): 833–845.

Kockar I, Conejo AJ, McDonald JR, 2009. Influence of the emissions trading scheme on generation scheduling. *Electrical Power and Energy Systems* 31: 465–473.

Koeppel G, Korpås M, 2008. Improving the network infeed accuracy of non-dispatchable generators with energy storage devices. *Electric Power Systems Research* 78: 2024–2036.

Konak A, Coir DW, Smith AE, 2006. Multi-objective optimization using genetic algorithms: A tutorial. *Reliability Engineering and System Safety* 91: 992–1007.

Kong XQ, Wang RZ, Huang XH, 2005. Energy optimization model for a CCHP system with available gas turbines. *Applied Thermal Engineering* 25(2–3): 377–391.

Krause T, Andersson G, Fröhlich K, Vaccaro A, 2011. Multiple-energy carriers: Modeling of production, delivery, and consumption. *Proceedings of the IEEE* 99(1): 15–27.

Kreider JF, ed., 2001. *Handbook of Heating, Ventilation and Air Conditioning*. CRC Press, Boca Raton, FL.

Kuhn V, Klemes J, Bulatov I, 2008. MicroCHP: Overview of selected technologies, products and field test results. *Applied Thermal Engineering* 28: 2039–2048.

Horlock JH, 1997. *Cogeneration-Combined Heat and Power (CHP)*. Krieger, Malabar, FL.

Lazzarin R, Noro M, 2006. District heating and gas engine heat pump: Economic analysis based on a case study. *Applied Thermal Engineering* 26: 193–199.

Li H, Maréchal F, Burer M, Favrat D, 2006. Multi-objective optimization of an advanced combined cycle power plant including CO_2 separation options. *Energy* 31(15): 3117–3134.

Li X, 2009. Study of multi-objective optimization and multi-attribute decision-making for economic and environmental power dispatch. *Electric Power Systems Research* 79(5): 780–795.

Liu XH, Geng KC, Lin BR, Jiang Y, 2004. Combined cogeneration and liquid-desiccant system applied in a demonstration building. *Energy and Buildings* 36: 945–953.

Lozano MA, Ramos JC, Carvalho M, Serra LM, 2009a. Structure optimization of energy supply systems in tertiary sector buildings. *Energy and Buildings* 41(10): 1063–1075.

Lozano MA, Carvalho M, Serra LM, 2009b. Operational strategy and marginal costs in simple trigeneration systems. *Energy* 34(11): 2001–2008.

Mago PJ, Chamra LM, 2009. Analysis and optimization of CCHP systems based on energy, economical, and environmental considerations. *Energy and Buildings* 41: 1099–1106.

Mago PJ, Chamra LM, Ramsay J, 2010. Micro-combined cooling, heating and power systems hybrid electric-thermal load following operation. *Applied Thermal Engineering* 30(8–9): 800–806.

Maidment GG, Prosser G, 2000. The use of CHP and absorption cooling in cold storage. *Applied Thermal Engineering* 20: 1059–1073.

Makkonen S, Lahdelma R, Asell A, Jokinen A, 2003. Multi-criteria decision support in the liberalized energy market. *Journal of Multi-Criteria Decision Analysis* 12: 27–42.

Makkonen S, Lahdelma R, 2006. Non-convex power plant modelling in energy optimisation. *European Journal of Operational Research* 171: 1113–1126.

Mancarella P, 2006. From cogeneration to trigeneration: Energy planning and evaluation in a competitive market framework. PhD. dissertation, Politecnico di Torino, Torino, Italy.

Mancarella P, 2009. Cogeneration systems with electric heat pumps: Energy-shifting properties and equivalent plant modelling. *Energy Conversion and Management* 50: 1991–1999.

Mancarella P, Chicco G, 2008. Assessment of the greenhouse gas emissions from cogeneration and trigeneration systems. Part II: Analysis techniques and application cases. *Energy* 33(3): 418–430.

Mancarella P, Chicco G, 2009a. *Distributed Multi-generation Systems: Energy Models and Analyses*. Nova Science Publishers, New York.

Mancarella P, Chicco G, 2009b. Global and local emission impact assessment of distributed cogeneration systems with partial-load models. *Applied Energy* 86(10): 2096–2106.

Mancarella P, Chicco G, 2010. Distributed cogeneration: Modeling of environmental benefits and impact, Chapter 1, in Gaonkar, DN, ed., *Distributed Generation*, pp. 1–26. InTech, Vukovar, Croatia.

Marnay C, Venkataramanan G, Stadler M, Siddiqui AS, Firestone R, Chandran B, 2008. Optimal technology selection and operation of commercial-building microgrids. *IEEE Transactions on Power Systems* 23(3): 975–982.

Martens A, 1998. The energetic feasibility of CHP compared to the separate production of heat and power. *Applied Thermal Engineering* 18: 935–946.

Mavrotas G, Georgopoulou E, Mirasgedis S, Sarafidis Y, Lalas D, Hontou V, Gakis N, 2007. An integrated approach for the selection of best available techniques (BAT) for the industries in the greater Athens area using multi-objective combinatorial optimization. *Energy Economics* 29: 953–973.

Meunier F, 2002. Co- and tri-generation contribution to climate change control. *Applied Thermal Engineering* 22: 703–718.

Minciuc E, Le Corre O, Athanasovici V, Tazerout M, 2003. Fuel savings and CO_2 emissions for tri-generation systems. *Applied Thermal Engineering* 23: 1333–1346.

Oh SD, Lee HJ, Jung JY, Kwak HY, 2007. Optimal planning and economic evaluation of cogeneration system. *Energy* 32(5): 760–771.

Onovwiona HI, Ugursal VI, 2006. Residential cogeneration systems: Review of the current technology. *Renewable and Sustainable Energy Reviews* 10(5): 389–431.

Pelet X, Favrat D, Leyland G, 2005. Multiobjective optimisation of integrated energy systems for remote communities considering economics and CO_2 emissions. *International Journal of Thermal Sciences* 44(12): 1180–1189.

Piacentino A, Cardona F, 2007. On thermoeconomics of energy systems at variable load conditions: Integrated optimization of plant design and operation. *Energy Conversion and Management* 48(8): 2341–2355.

Piacentino A, Cardona F, 2008a. EABOT—Energetic analysis as a basis for robust optimization of trigeneration systems by linear programming. *Energy Conversion and Management* 49(11): 3006–3016.

Piacentino A, Cardona F, 2008b. An original multi-objective criterion for the design of small-scale polygeneration systems based on realistic operating conditions. *Applied Thermal Engineering* 28(17–18): 2391–2404.

Porteiro J, Míguez JL, Murillo S, López LM, 2004. Feasibility of a new domestic CHP trigeneration with heat pump: II. Availability analysis. *Applied Thermal Engineering* 24: 1421–1429.

Rong A, Lahdelma R, 2005. An efficient linear programming model and optimization algorithm for trigeneration. *Applied Energy* 82(1): 40–63.

Rong A, Lahdelma R, 2007. CO_2 emissions trading planning in combined heat and power production via multi-period stochastic optimization. *European Journal of Operational Research* 176(3): 1874–1895.

Rong A, Lahdelma R, Luh PB, 2008. Lagrangian relaxation based algorithm for trigeneration planning with storages. *European Journal of Operational Research* 188(1): 240–257.

Roque Díaz P, Benito YR, Parise JAR, 2010. Thermoeconomic assessment of a multi-engine, multi-heat-pump CCHP (combined cooling, heating and power generation) system—A case study. *Energy* 35(9): 3540–3550.

Sakawa M, Kato K, Ushiro S, 2002. Operational planning of district heating and cooling plants through genetic algorithms for mixed 0–1 linear programming. *European Journal of Operational Research* 137: 677–687.

Salgado F, Pedrero P, 2008. Short-term operation planning on cogeneration systems: a survey. *Electric Power Systems Research* 78(5): 835–848.

Shivakumar K, Narasimhan S, 2002. A robust and efficient NLP formulation using graph theoretic principles for synthesis of heat exchanger networks. *Computers and Chemical Engineering* 26(11): 1517–1532.

Shukla PK, Deb K, 2007. On finding multiple Pareto-optimal solutions using classical and evolutionary generating methods. *European Journal of Operational Research* 181: 1630–1652.

Tozer R, James RW, 1998. Heat powered refrigeration cycles. *Applied Thermal Engineering* 18: 731–743.

Tsay MT, 2003. Applying the multi-objective approach for operation strategy of cogeneration systems under environmental constraints. *Electrical Power and Energy Systems* 25: 219–226.

Tsikalakis AG, Hatziargyriou ND, 2007. Environmental benefits of distributed generation with and without emissions trading. *Energy Policy* 35(6):3395–3409.

Tsikalakis AG, Hatziargyriou ND, 2008. Centralized control for optimizing microgrids operation. *IEEE Transactions on Energy Conversion* 23(1), 241–248.

Uche J, Serra L, Sanz A, 2004. Integration of desalination with cold-heat-power production in the agro-food industry. *Desalination* 166: 379–391.

Valero A, Lozano M, 1997. An introduction to thermoeconomics: Chapter 8 in Boehm RF, ed., *Developments in the Design of Thermal Systems*. Cambridge University Press, Cambridge, UK.

Voorspools KR, D'haeseleer WD, 2003. The impact of the implementation of cogeneration in a given energetic context. *IEEE Transactions on Energy Conversion* 18: 135–141.

Wang C, Larsson M, Ryman C, Grip CE, Wikström JO, Johnsson A, Engdahl J, 2008. A model on CO_2 emission reduction in integrated steelmaking by optimization methods. *International Journal of Energy Research* 32(12): 1092–1106.

Wang JJ, Jing YY, Zhang CF, Zhao JH, 2009. Review on multi-criteria decision analysis aid in sustainable energy decision-making. *Renewable and Sustainable Energy Reviews* 13: 2263–2278.

Wang JJ, Jing YY, Zhang CF, 2010a. Optimization of capacity and operation for CCHP system by genetic algorithm. *Applied Energy* 87(4): 1325–1335.

Wang J, Zhai Z, Jing Y, Zhang C, 2010b. Optimization design of BCHP system to maximize to save energy and reduce environmental impact. *Energy* 35(8): 3388–3398.

Wang Y, Lior N, 2007. Fuel allocation in a combined steam-injected gas turbine and thermal seawater desalination system. *Desalination* 214: 306–326.

Weber C, Maréchal F, Favrat D, Kraines S, 2006. Optimization of an SOFC-based decentralized polygeneration system for providing energy services in an office-building in Tōkyō. *Applied Thermal Engineering* 26(13): 1409–1419.

Wu DW, Wang A, 2006. Combined cooling, heating and power: A review. *Progress in Energy and Combustion Science* 32: 459–495.

Yokoyama R, Hasegawa Y, Ito K, 2002. A MILP decomposition approach to large scale optimization in structural design of energy supply buildings. *Energy Conversion and Management* 43: 771–790.

Yusta JM, Khodr HM, Urdaneta AJ, 2007. Optimal pricing of default customers in electrical distribution systems: Effect behavior performance of demand response models. *Electric Power Systems Research* 77(5–6): 548–558.

Xia H, Koyama M, Leyland G, Kraines S, 2004. A modularized framework for solving an economic-environmental power generation mix problem. *International Journal of Energy Research* 28(9): 769–784.

Ziher D, Poredos RZ, 2006. Economics of a trigeneration system in a hospital. *Applied Thermal Engineering* 26:680–687.

Index